NATURAL VEGETATION OF OREGON AND WASHINGTON

The Oregon State University Press is proud to announce the publication of a reprint edition of the classic volume on regional ecology, *Natural Vegetation of Oregon and Washington* by Jerry F. Franklin and C.T. Dyrness, with the addition of a bibliographic supplement. This supplement identifies the major advances in research and our understanding of the vegetation of the Pacific Northwest since *Natural Vegetation of Oregon and Washington* was first published in 1973 by the USDA Forest Service, and contains more than 500 citations, bringing the bibliography up to date through 1987.

Natural Vegetation of Oregon and Washington has long been recognized as a model for ecological writing. The vegetational zones of the region and their environmental relations are described and examined in detail, including the composition and succession of each. In addition, the volume contains information on unusual habitats, physiography, geology, and soils, and is illustrated with over 200 photographs. Appendices define soil types, list scientific and common names of plants, and provide a general subject index.

Natural Vegetation of Oregon and Washington by Jerry F. Franklin and C.T. Dyrness. ISBN 0-87071-356-6. 464 pages, 8 x 10½ inches, paperback. $22.95.

Oregon State University Press
101 Waldo Hall
Corvallis, OR 97331

Cal

NATURAL VEGETATION OF OREGON AND WASHINGTON

Jerry F. Franklin and C.T. Dyrness
*The classic volume on regional ecology,
with more than 500 new bibliographic entries*

NATURAL VEGETATION OF OREGON AND WASHINGTON

NATURAL VEGETATION OF OREGON AND WASHINGTON

Jerry F. Franklin and C.T. Dyrness

Oregon State University Press

Printed in the United States of America

PREFACE

Natural Vegetation of Oregon and Washington had its genesis in a hastily prepared introduction to the plant communities of the Pacific Northwest for participants in the XI International Botanical Congress in 1969. When supplies of this first document were exhausted within a year, it was decided not simply to reprint it, but to prepare a revised and expanded volume including a great deal of additional information and numerous additional photographs. The resulting volume, *Natural Vegetation of Oregon and Washington*, was published by the USDA Forest Service Pacific Northwest Forest and Range Experiment Station in 1973.

Major advances have occurred in our understanding of the vegetation of the region since that time, and another major revision of the book is needed. Significant new information is now available in such fields as autecology and population biology, vegetation description and classification, successional processes, and ecosystem functioning. In addition, a comprehensive description of the vegetational series now exists, allowing a much more consistent book organization. New chapters are needed on disturbance ecology and ecosystem processes.

Unfortunately, however, major revision is not possible at this time. In the face of ongoing demand for the book—which has been out of print for several years—the Oregon State University Press has decided to reprint *Natural Vegetation of Oregon and Washington* with the addition of a bibliographic supplement which brings the literature citations up to date and helps identify some of the major advances in our understanding of the vegetation of the Pacific Northwest.

Jerry F. Franklin

Acknowledgments

Preparation of *Natural Vegetation of Oregon and Washington* involved contributions from many individuals which the authors gratefully acknowledge. Many of the deficiencies in the original work were indicated by Jack Major (University of California at Davis) whose review was a major guide in revision. R. Daubenmire (Washington State University), R. Fonda (Western Washington State College), D. Thornburgh (Humboldt State College), W. Moir, D.R.M. Scott (University of Washington), and D. Zobel (Oregon State University) also made numerous valuable comments and criticisms. Portions of the revised manuscript were reviewed by G. Douglas (University of Alberta), R. Emmingham (Oregon State University), R. Fonda, J. Henderson (Utah State University), A.R. Kruckeberg (University of Washington), G. Simonson (Oregon State University), and R. Waring (Oregon State University). R. Carkin (Pacific Northwest Forest and Range Experiment Station) and G.M. Hawk (Coniferous Forest Biome, US/IBP) were of major assistance in checking scientific names and preparing tables. Most of the photographic aerial obliques were taken by Wally Guy, photographer for the Station, who also prepared final prints for the majority of photographs. Glenda Faxon of the Station staff typed most of the manuscript and kept track of many other details during its assembly. G. Hansen's editorial staff in the Experiment Station deserve special credit for their extensive work in producing this book, particularly B.J. Bell, J.C. Etheridge, M.I. Hoyt, and D.E. Thompson.

In addition, we are grateful to the following people for providing many of the citations in the bibliographic supplement: Jim Agee, Roger del Moral, Angie Evenden, Richard Fonda, Bob Frenkel, Sarah Greene, David Hibbs, Art Kruckeberg, Richard Mack, Joe Means, Len Volland, and Don Zobel.

CONTENTS

CHAPTER I. INTRODUCTION

The Pacific Northwest is among the more diverse regions of North America in environment and vegetation. Oregon and Washington, the heart of this region, encompass wet coastal and dry interior mountain ranges, miles of coastline, interior valleys and basins, and high desert plateau (fig. 1). Moisture, temperature, and substrate vary greatly. Natural vegetation types range from dense coastal forests of towering conifers through woodland and savanna to shrub steppe.

The ecology and plant geography of the region have been studied by scientists for over half a century. Major contributors have included W. S. Cooper, R. Daubenmire, H. P. Hansen, L. A. Isaac, V. J. Krajina, D. B. Lawrence, T. T. Munger, M. E. Peck, C. V. Piper, E. H. Reid, and G. B. Rigg. Unfortunately, most of the knowledge which has been gathered is fragmented—dispersed through journals, books, theses, and unpublished files of data.

We present here a generalized account of the major vegetation types within the States of Oregon and Washington, an integration of the scattered information into a regional account. Published articles, theses, and personal data files are the source materials. The unevenness of coverage is unfortunate but unavoidable; some plant formations have been studied in great detail, and other communities or locales have received cursory or no study.

The purpose is threefold: (1) to outline major phytogeographic units and suggest how they fit together and relate to environmental factors; (2) to direct the interested reader to sources of detailed information on the environment and vegetation of the Pacific Northwest, since such information cannot be provided in an account of this size; and (3) to illustrate the major plant communities with photographs.

We hope this outline will enable the scientist or student new to the Pacific Northwest to better understand what he is seeing and how his various observations are related. Perhaps it will also provide some new insights to readers more familiar with the region.

Format, Definitions, and Nomenclature

We have followed the same general format used in our earlier publication (Franklin and Dyrness 1969). The general geologic, edaphic, and climatic features of the region are considered in Chapter II. The broad vegetational patterns and their significance are outlined in Chapter III; readers interested in this subject might also review Daubenmire (1969a). The forest vegetation is considered in Chapters IV, VI, and VII, and the grassland and sagebrush steppe in Chapters VIII and IX. Vegetation found in the valleys of western Oregon is discussed in Chapter V. We conclude with chapters on timberline vegetation (X) and unusual habitats and localities (XI).

Appended to this report are indexes for individual taxa, subjects, and community types which have been recognized and named. We have tried to standardize community names whenever possible using scientific names, slashes (/) between taxa of different life forms or layers and hyphens (-) between taxa of the same life form or layers.

Ecology is fraught with specialized and ambiguous terminology. We have followed Daubenmire's (1968a) definitions in most cases, especially when dealing with synecological terminology, such as climax, association, etc. In some cases, this was not possible due to uncertainty on our part as to the exact meaning of the original authors. The reader unfamiliar with such terms might particularly consider pages 27 to 32, 229 to 237, and 259 to 262 of Daubenmire (1968a) for orientation. The glossaries of Carpenter (1956), Hanson (1962), and Habeck and Hartley (1968) are also helpful.

Plant nomenclature in this paper generally follows these sources: trees, Little (1953); other vascular plants, Hitchcock et al. (1955, 1959, 1961, 1964, 1969) or Peck (1961) for taxa not covered in the former; mosses, Lawton (1965, 1971); and lichens, Howard (1950).

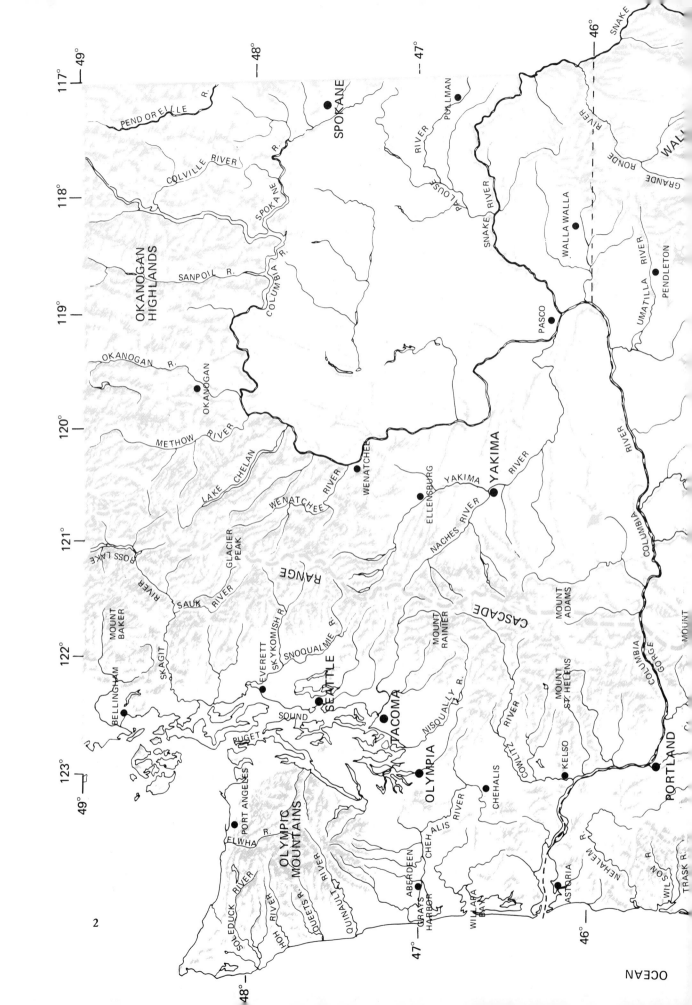

49°
117° ⌐ 49°
118° ⊢ 48°
PEND OREILLE   R.
COLVILLE   RIVER
SPOKANE   R.
SPOKANE
119° ⊢ 47°
OKANOGAN   HIGHLANDS
SANPOIL   R.
COLUMBIA   R.
PALOUSE   RIVER
PULLMAN
SNAKE   RIVER
SNAKE
GRANDE   RONDE   RIVER
WALLA
120° ⊢
OKANOGAN   R.
OKANOGAN
METHOW   RIVER
WALLA WALLA
PASCO
UMATILLA   RIVER
PENDLETON
46°
121° ⊢
LAKE   CHELAN
WENATCHEE   RIVER
WENATCHEE
ELLENSBURG
YAKIMA   RIVER
YAKIMA
RIVER
GLACIER   PEAK
RANGE
NACHES   RIVER
COLUMBIA   RIVER
ROSS LAKE
RIVER
SAUK   RIVER
MOUNT   ADAMS
122° ⊢
MOUNT   BAKER
SKAGIT
SKYKOMISH R.
SNOQUALMIE   R.
EVERETT
CASCADE
MOUNT   RAINIER
MOUNT   ST. HELENS
MOUNT
COLUMBIA   GORGE
SEATTLE
PUGET   SOUND
TACOMA
NISQUALLY   R.
COWLITZ   RIVER
KELSO
123° ⌐ 49°
OLYMPIA
CHEHALIS
PORTLAND
PORT ANGELES
ELWHA   R.
OLYMPIC   MOUNTAINS
SOLEDUCK   RIVER
HOH   RIVER
QUEETS R.
QUINAULT   RIVER
CHEHALIS   RIVER
CHEHALIS
ABERDEEN
GRAYS   HARBOR
WILLAPA   BAY
WILLSON
NEHALEM   RIVER
ASTORIA
NOS   R.
TRASK R.
46°
47°
48°
2
OCEAN

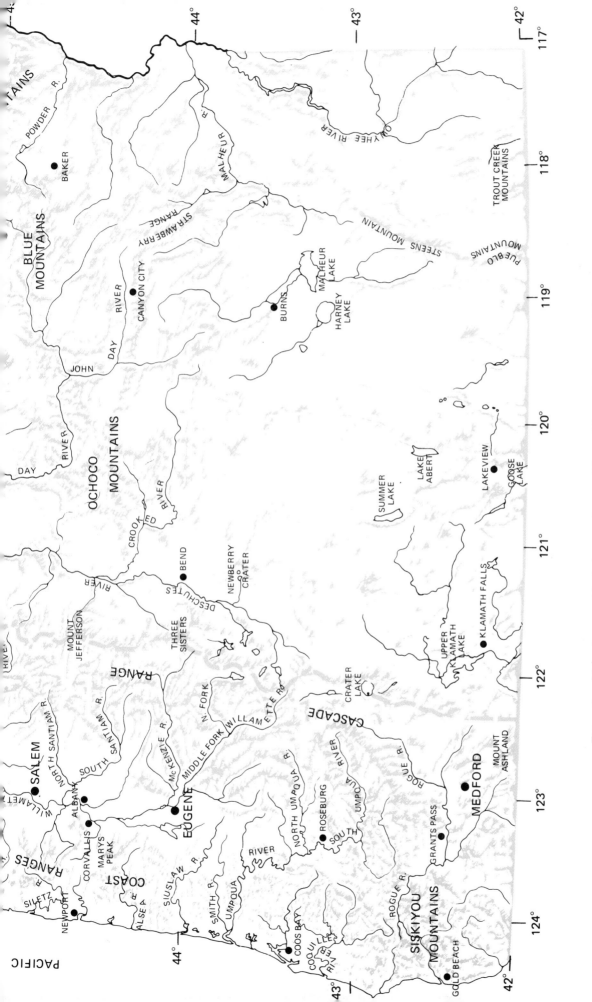

Figure 1. — Major topographic features and some cities and towns in Oregon and Washington.

3

Whenever possible, nomenclature from older ecological studies cited has been updated. In an appendix are listed the complete scientific names for taxa mentioned along with common names locally applied to many of the plants.

Paleobotany, Paleoecology, and Floristic Evolution

The evolution of the flora and plant formations of Oregon and Washington is a fascinating subject and contributes significantly to understanding the present vegetational mosaic. Unfortunately, it is not within the scope of this paper. We recommend readers interested in these subjects consult the following: Wolfe (1969), Chaney (1938, 1948) and Axelrod (1958) on paleobotany, Hansen (1947) and Heusser (1960) on postglacial vegetation changes and development, and Daubenmire (1947), Mason (1947), and Detling (1968) on evolution of plant formations and floras. Some interesting floristic comparisons of this region with other parts of the world are found in Schofield (1969) and Graham (1972).

CHAPTER II. ENVIRONMENTAL SETTING

Physiography, Geology, and Soils

Since geology, physiography, and at least some aspects of soils are interrelated, we will consider these three environmental features together. In Oregon and Washington, all three present highly varied and complex patterns. Landforms vary from level river valleys and lava plains to precipitous mountain slopes. Elevations range from sea level to over 4,450 meters. The geologic complexity is bewildering in many areas, with formations dating from the Paleozoic era (over 400 million years old) to Recent. Vulcanism has dominated the shaping of much of the landscape, but sedimentary and metamorphic rocks also abound.

Since climate and vegetation are added to landform and geology as factors in soil formation, the tremendous variety of soils is not surprising. For example, zonal great soil groups range from Camborthids (Desert soils) in arid southeastern Oregon to Cryorthods (Podzol soils) in the cool, humid climate of the northern Cascade Range. The most striking soil changes within short distances are found on eastern slopes of the Cascade Range. Here Haplorthods and Haploxerolls (Podzol and Chestnut soils) are separated by only a few kilometers because of abrupt changes in precipitation and concomitant changes from forest to grass-shrub vegetation.

The great relief in extensive mountainous areas within Oregon and Washington perpetuates many soils in a state of profile immaturity. Soils on steep slopes are constantly influenced by gravitational instability expressed as soil creep or landslides, often severely limiting profile development. Consequently, many mountain soils are regosolic or lithosolic, lacking genetic horizons except for a thin A. In these areas, effects of the parent rock on soil properties are major and those of climate and vegetation are minimal. Areas of these immature soils are typically characterized by extensive rock outcroppings.

Volcanic activity along the crest of the Cascade Range during Pleistocene and Recent times has extensively influenced regional soils. Large tracts at higher elevations in the Cascades are mantled with deposits of pumice and volcanic ash which, because of their youth, generally exhibit little genetic development. In many areas, such as the southern Cascade Range of Washington, there have been several depositions of volcanic ejecta; soil profiles often have three or four buried horizon sequences.

Pumice and volcanic ash soils also occur well beyond the Cascade Range. Distance from their source and orientation of these deposits are largely functions of wind direction and velocity during eruptions. Recently, it has become apparent that many, if not most, of the soils of the Pacific Northwest have had some influence from aerially deposited volcanic ejecta. Amounts of incorporated ash and pumice are often small, however, and detectable only through detailed soil mineralogic or micromorphological investigations.

For descriptive purposes, the two-State area has been separated into 15 physiographic provinces (fig. 2). The divisions used are largely those outlined by Baldwin (1964), Fenneman (1931), and Easterbrook and Rahm (1970). Naturally, in many instances, boundaries separating provinces are arbitrary and gradual transitions exist. However, the provinces are broad stratifications of relatively homogeneous areas and reduce complexity to more manageable proportions.

Geologic information for this section is from several primary sources: for Oregon, Baldwin (1964) and "Geologic Map of Western Oregon West of the 121st Meridian" (Wells

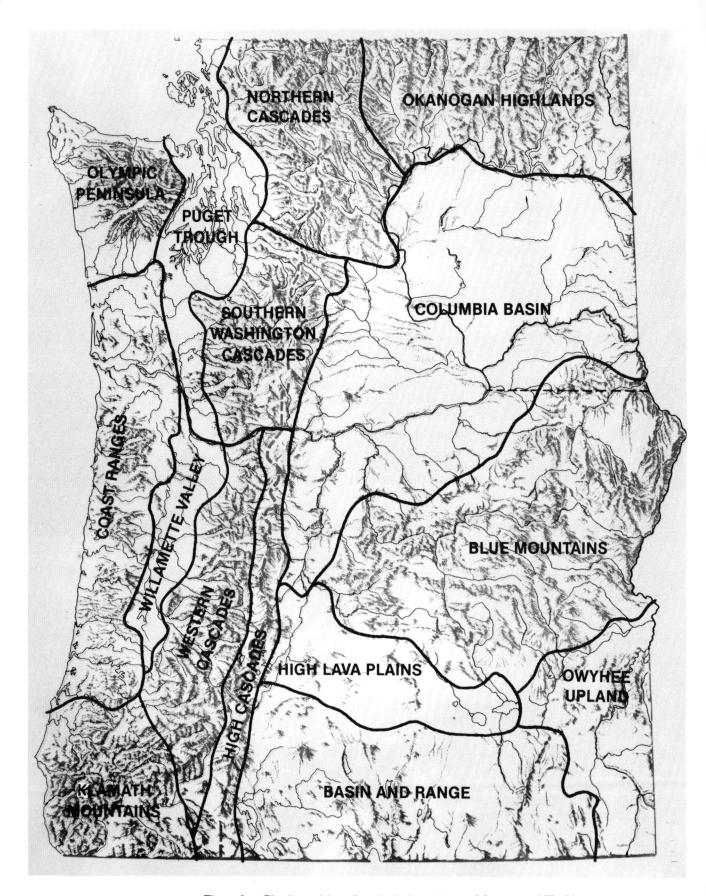

Figure 2. — Physiographic and geological provinces of Oregon and Washington.

6

and Peck 1961); for Washington, Campbell (1953) and "Geologic Map of Washington" (Huntting et al. 1961). Other pertinent references include Williams (1942), Danner (1955), Foster (1960), Snavely and Wagner (1963), Fiske et al. (1963), Peck et al. (1964), Mackin and Cary (1965), Wise (1970), Dott (1971), and Beaulieu (1971).

Great soil group names in the text follow the National Cooperative Soil Survey Classification of 1967, with 1938 classification names shown in parentheses. Names and distribution of great groups defined in the 1967 Soil Classification System were largely obtained from Appendix IV of the "Columbia—North Pacific Region Comprehensive Framework Study" (Pacific Northwest River Basins Commission 1969) and from the soils sheet in the National Atlas (Soil Conservation Service 1969). Information pertaining to the 1938 Soil Classification System is based largely on "Soils of the Western United States" (Western Land Grant Universities et al. 1964). Other references consulted include Knox (1962), Youngberg (1963), and numerous soil survey reports for localized areas.

Table 1 shows the relative importance of various great soil groups within the 15 physiographic provinces.

Table 1. — Principal great soil groups within the 15 physiographic provinces of Oregon and Washington

| Province | 1938 Classification System[1] | | 1967 Classification System[2] | |
	Widespread great soil groups[3]	Less abundant great soil groups	Widespread great groups[3]	Less abundant great groups
Olympic Peninsula	Sols Bruns Acides Reddish Brown Lateritic Lithosol	Podzol Brown Podzolic Alpine Turf Alpine Meadow Humic Gley Alluvial Regosol Rockland[4]	Xerochrepts Dystrochrepts Haplumbrepts	Haplorthods Cryorthods Rockland[4] Haplohumults Haploxerolls Haplaquepts Udifluvents
Coast Ranges	Reddish Brown Lateritic Sols Bruns Acides Regosol Lithosol	Noncalcic Brown Prairie Grumusol Humic Gley Alluvial	Haplumbrepts Haplohumults	Dystrochrepts Xerumbrepts Haplorthods Haplaquepts
Klamath Mountains	Reddish Brown Lateritic	Sols Bruns Acides Noncalcic Brown Western Brown Forest Podzol Prairie Grumusol Humic Gley Alluvial Lithosol Rockland[4]	Haplohumults Haploxerults	Haplumbrepts Haploxeralfs Xerochrepts Dystrochrepts Hapludalfs Haploxerolls Chromoxererts
Willamette Valley	Prairie Planosol Alluvial	Gray-Brown Podzolic Chernozem Reddish Brown Lateritic Grumusol Humic Gley	Argixerolls Haplohumults	Haploxerolls Xerumbrepts Haplaquolls Albaqualfs
Puget Trough	Brown Podzolic Regosol Alluvial Reddish Brown Lateritic	Gray Wooded Prairie Sols Bruns Acides Humic Gley Lithosol	Haplorthods Haplohumults	Xerumbrepts Argixerolls Haplaquepts Xeropsamments Haploxerolls Vitrandepts

7

Table 1. — Principal great soil groups within the 15 physiographic provinces of Oregon and Washington (continued)

Province	1938 Classification System[1]		1967 Classification System[2]	
	Widespread great soil groups[3]	Less abundant great soil groups	Widespread great groups[3]	Less abundant great groups
Northern Cascades	Podzol Brown Podzolic Lithosol	Western Brown Forest Gray Wooded Chestnut Alpine Turf Alpine Meadow Regosol Rockland[4]	Cryorthods Haplorthods Xerochrepts	Rockland[4] Haploxerolls Haplumbrepts Fragiorthods Haploxeralfs Haplaquepts Xeropsamments
Southern Washington Cascades	Brown Podzolic Podzol Lithosol Regosol	Reddish Brown Lateritic Gray-Brown Podzolic Sols Bruns Acides Western Brown Forest Chestnut Humic Gley Alluvial Rockland[4]	Vitrandepts Dystrandepts Haplorthods	Haplohumults Hapludalfs Rockland[4] Cryorthods Haploxeralfs Haploxerolls Xeropsamments
Western Cascades	Brown Podzolic Regosol Reddish Brown Lateritic	Podzol Sols Bruns Acides Noncalcic Brown Prairie Humic Gley Alluvial Lithosol Rockland[4]	Haplumbrepts Xerumbrepts	Haploxerults Argixerolls Haploxeralfs Cryorthods Haplohumults Haplorthods Cryumbrepts Rockland[4]
High Cascades	Regosol Brown Podzolic	Podzol Western Brown Forest Lithosol Rockland[4]	Vitrandepts Cryorthods	Haplorthods Cryumbrepts Rockland[4] Haplumbrepts
Okanogan Highlands	Brown Podzolic Gray Wooded Lithosol	Western Brown Forest Chernozem Brown Humic Gley Alluvial Regosol Rockland[4]	Xerochrepts Cryorthods	Vitrandepts Haploxerolls Haploxeralfs Haplorthods Xeropsamments Haplaquolls Rockland[4] Durixerolls
Blue Mountains	Western Brown Forest Regosol Lithosol	Brown Podzolic Reddish Brown Lateritic Chernozem Chestnut Prairie Alpine Turf Alpine Meadow Humic Gley Alluvial Rockland[4]	Argixerolls Vitrandepts	Fragiorthods Haploxerolls Palexerolls Durixerolls Haplargids Xerorthents (shallow) Rockland[4]
Columbia Basin	Brown Chestnut Chernozem Prairie Sierozem Regosol	Planosol Humic Gley Solonetz Solonchak Alluvial Lithosol Rockland[4]	Haploxerolls Argixerolls Camborthids Torripsamments	Argiudolls Haplargids Xerorthents Xerorthents (shallow) Rockland[4] Durixerolls Haplaquolls Albaqualfs Haploxeralfs

Table 1. — Principal great soil groups within the 15 physiographic provinces of Oregon and Washington (continued)

Province	1938 Classification System[1]		1967 Classification System[2]	
	Widespread great soil groups[3]	Less abundant great soil groups	Widespread great groups[3]	Less abundant great groups
High Lava Plains	Brown Chestnut Lithosol Regosol (pumice)	Solonetz Solonchak Humic Gley Alluvial Regosol Rockland[4]	Haplargids Camborthids Vitrandepts	Durargids Haploxerolls Natrargids Haplaquolls Haplaquepts Haplorthents Durorthids Rockland[4]
Basin and Range	Brown Chestnut Lithosol Regosol (pumice) Western Brown Forest	Chernozem Prairie Reddish Brown Lateritic Sierozem Desert Humic Gley Solonetz Solonchak Alluvial Regosol Rockland[4]	Haplargids Durargids Vitrandepts	Argixerolls Camborthids Durorthids Haplaquolls Haplaquepts Haploxerolls Haplorthents Rockland[4]
Owyhee Upland	Brown Chestnut Lithosol	Sierozem Desert Solonetz Humic Gley Alluvial Regosol Rockland[4]	Haplargids Haploxerolls	Argixerolls Durargids Haplaquolls Natrargids Camborthids Rockland[4]

[1] Based largely on "Soils of the Western United States" (Western Land Grant Universities et al. 1964).
[2] Based largely on information in "Columbia — North Pacific Region Comprehensive Framework Study," Appendix IV, "Land and Mineral Resources" (Pacific Northwest River Basins Commission 1969).
[3] Listed in approximate order of importance.
[4] A miscellaneous land type in which rock outcrops or rock rubble dominate the landscape.

Olympic Peninsula Province

The Olympic Peninsula Province is made up of a central core of the rugged Olympic Mountains surrounded by almost level lowlands. On the east, the lowland strip is 3 to 16 kilometers wide and is part of the Puget Trough. The lowland strips are very narrow on the north, but 16 to 32 kilometers wide on the west and 48 kilometers wide along the south side of the peninsula. Most ridges in the Olympic Mountains are 1,200 to 1,500 meters in elevation with some higher peaks attaining elevations of 2,100 to 2,420 meters. Glaciation has strongly influenced landforms. All main river valleys are broad and U-shaped, and all major peaks are ringed with cirques, many containing active glaciers. The extremely high precipitation (perhaps as high as 6,350 millimeters per year in the interior) has caused rapid downcutting by streams, resulting in many precipitous mountain slopes (fig. 3).

Geologically, the mountainous portion of the Olympic Peninsula Province is made up of two volcanic belts encircling a large interior area containing sedimentary rocks. The volcanic belts bound the peninsula on the north and east sides, and as far west as Lake Quinault on the south. The outer belt, by far the thickest, is comprised of basalt flows and

Figure 3.—The rugged Olympic Mountains viewed from Hurricane Ridge (south of Port Angeles, Washington); the glacially carved valley contains the Elwha River *(photo courtesy Olympic National Park).*

breccias of Eocene age. Between the two volcanic belts lies a generally thin band of argillite and graywacke, also Eocene. The inner volcanic belt is very thin and discontinuous and consists of altered basalt, "pillow" lava, and flow breccia deposited late in the Mesozoic era or perhaps during the Paleocene epoch. The rugged interior of the peninsula is almost exclusively comprised of sedimentary rocks deposited late in the Mesozoic or very early in the Tertiary period. These rocks are largely graywacke, with some interbedded slate, argillite, and volcanic rocks.

The less mountainous area along the north edge of the peninsula is a complex of Oligocene and Miocene sandstones, some interbedded with siltstone and conglomerates. In addition, glacial drift occurs in fairly large deposits near Sequim and Port Angeles and west of Ozette Lake. The broad, level areas along the western and southern margins of the peninsula have been interpreted as marine terraces or glacial outwash fans.

Until recently, almost nothing was known about the soils in the mountainous interior of the Olympic Peninsula. However, in 1969 Fonda and Bliss and in 1970 Kuramoto and Bliss described a variety of soils developed under both coniferous forest and subalpine meadow vegetation and derived from the sedimentary parent materials of the interior. Forested soils were reported to be Brown Podzolics[1] (Spodosols) and Lithosols (Entisols).

---

[1] Great soil groups mentioned in the text are described briefly in the Appendix.

10

The Spodosols (probably Haplorthods) have a very dark grayish-brown silt loam surface horizon underlain by a dark yellowish-brown sandy clay loam B horizon. Soils developed under alpine meadow vegetation were classed within the Spodosol, Mollisol, and Inceptisol orders. The Spodosols were reported to have thin, gray sandy loam B2 horizons. The Inceptisols had very dark grayish-brown sandy loam subsurface horizons.

The deeper, well-developed soils derived from basalt are generally classified as Haplohumults (Reddish Brown Lateritic soils). Most often these soils are characterized by a reddish-brown silt loam or silty clay loam surface horizon underlain by a silty clay loam or silty clay subsoil showing the effects of clay accumulation. The sandstone parent materials situated along the northern edge of the peninsula give rise to Haplumbrepts (Western Brown Forest soils). These are moderately deep soils with thick, dark-colored surface horizons. Surface textures are silt loam or silty clay loam, and the subsoil is either silty clay loam or silty clay textured.

A large variety of soils have formed in glacial till and outwash, depending on such factors as particle size and degree of compaction in parent materials. The majority of upland soils derived from glacial till have been classed as Xerochrepts (Regosols). Such soils generally have a loam-textured surface horizon overlying a gravelly sandy loam substratum. Soils developed in till or glacial outwash on terraces are in most areas either Dystrochrepts (Sols Bruns Acides) or Haplorthods (Brown Podzolic soils). Textures range from gravelly silt loam to clay loam or silty clay loam. The Haplorthods often have a gravelly cemented layer at a depth of approximately 1 meter.

Alluvial soils occupying terraces along west-flowing rivers such as the Quinault, Queets, Hoh, and Soleduck are classed as Udifluvents. These are deep silt loam to very fine sandy loam soils which are moist throughout the year.

Coast Ranges Province

The Coast Ranges Province extends from the middle fork of the Coquille River in Oregon northward into southwestern Washington where it includes the area known as the Willapa Hills. The entire southern section of the province is topographically mature—i.e., steep mountain slopes with ridges that are often extremely sharp (fig. 4). Excepting the area drained by the Wilson and Trask Rivers, the proportion of steep slopes decreases in the northern section of the Coast Ranges. Mountain passes are generally located on the eastern border of the range due to faster rates of headward erosion by the numerous westward-flowing streams. Elevations of main ridge summits in the province range from about 450 to 750 meters. Scattered peaks, often capped with intrusive igneous rocks, rise well above surrounding ridges. Marys Peak, 1,249 meters high, is the highest peak in the Coast Ranges.

Geology south of the Salmon and Yamhill Rivers differs substantially from that to the north. Geologic history of the southern Coast Ranges began during early Eocene times with deposition of "pillow" basalts near the present town of Alsea. Later in the Eocene, the vast sedimentary beds of the Tyee formation, which make up by far the largest portion of this section of the Coast Ranges, were deposited under marine conditions. The Tyee formation, largely composed of rhythmically bedded, tuffaceous, and micaceous sandstone, occurs throughout the southern Coast Ranges and is virtually the only rock present in the central portion. Also during the Eocene, other smaller marine sedimentary formations were laid down, mostly to the south and along the coast. Scattered igneous intrusions, largely gabbro, occurred during the Oligocene and cap many of the most prominent peaks (e.g., Marys Peak, Prairie Peak, and Grass Mountain). During the Miocene, localized depositions of both sedimentary and volcanic rocks occurred which are now exposed near Newport and Coos Bay. Most of the spectacular coastal headlands are made up of Miocene basalt (fig. 5). The Pliocene epoch saw no new depositions, the principal activity at this

Figure 4.—The Oregon Coast Ranges west of Eugene, Oregon; note accordant ridge crests and extensive stream dissection. The highest peak in the background is Roman Nose Mountain (elevation, 870 m.).

Figure 5.—Typical coastal headland showing the basalt flows which, because they are more resistant to erosion than adjacent sedimentary formations, are responsible for these prominent features along the Oregon coast (Cascade Head, Oregon).

time apparently being the rapid erosion of the tremendously thick beds of sediments. Pleistocene deposits, generally sandy in nature, were laid down along the coast during a period of rising sea level. This general rise of sea level, following the melting of glacial ice, also drowned the mouths of coastal rivers.

As in the southern section, all rock formations in the northern Coast Ranges are Tertiary. Eocene formations are widespread and include both volcanic and sedimentary rocks. Eocene siltstone and sandstone are found along and to the south of the Yamhill River near Vernonia, Oregon, and in the Willapa Hills of southwestern Washington. Eocene volcanic rocks, largely basalt with some tuffs and breccias, occupy extensive areas northeast of Tillamook and in the Willapa Hills. Oligocene sedimentary formations, including siltstone, shale, and sandstone, are found near Vernonia, along the Nehalem River, and, to a limited extent, in the Willapa Hills. During the Miocene epoch, extensive basalt flows occurred in the most northerly section of the Oregon Coast Ranges and in the Willapa Hills. Near the Columbia River in Oregon, these flows are classified with the extremely widespread Columbia River Basalt. The Plio-Pleistocene was largely a period of erosion, with streams excavating their valleys as the ranges were slowly uplifted.

Soils developing in the very extensive deposits of sandstone exhibit a wide range of characteristics despite the fact most are classified as Haplumbrepts (Western Brown Forest soils). On steep, smooth mountain slopes they tend to be shallow, stony loam textured, and brown or yellowish-brown in color. Deeper soils derived from sandstone colluvium occupy uneven, benchy slopes that generally exhibit some degree of continuing instability. On broad ridgetops, soils from sandstone parent materials tend to be deep, with a B horizon showing some clay accumulation and a thick surface horizon of high organic matter content. Sandstone soils which show maximum profile development and are low in bases are classed as Haplohumults (Reddish Brown Lateritic soils). These soils have a much more reddish color, a silty clay loam-textured A horizon, and a silty clay B horizon.

Soils developed from siltstone or shale parent materials resemble those derived from sandstone in some respects, but generally they are noticeably finer textured. Typically, they have a silt loam surface horizon and a silty clay or clay-textured B horizon. Those with thick, dark-colored A horizons are generally classed as Haplumbrepts, whereas soils with light-colored surface horizons have been classified as Dystrochrepts (Sols Bruns Acides).

Over most of the Coast Ranges Province, soils derived from basalt are Haplumbrepts. Deep, well-developed profiles are reddish-brown in color and are relatively stone free. Surface textures are generally clay loam and the subsoil, a silty clay loam. Most often, however, basaltic soils tend to be fairly shallow and stony. Some Haplohumults on basalt parent materials are found in the southern portion of the province.

Sand dune areas are common along the Oregon coast. Soils on old, stabilized dunes are most often Haplorthods (Podzols and Brown Podzolic soils). These soils range from excessively drained to poorly drained and are characteristically loamy sand to fine sand textured. Soils derived from alluvium along major streams are most commonly Haplaquepts (Low-Humic Gley soils) and Haplumbrepts. The Haplaquepts are dominantly very poorly drained silt loams over silty clay loam soils formed in silty alluvium.

Klamath Mountains Province

The Klamath Mountains Province encompasses a complex of ranges in southwestern Oregon and northern California. The northernmost portion of range in Oregon is also commonly identified as the Siskiyou Mountains. This region is logically set apart from the remainder of southern Oregon by the boundary separating its pre-Tertiary rocks from Tertiary rock formations outside the area. The pre-Tertiary rocks of this province probably include the oldest in Oregon.

The Klamath Mountains Province is largely a region of rugged, deeply dissected terrain (fig. 6). Mountain crests, comprised of steeply folded and faulted pre-Tertiary strata, vary in elevation from 600 meters near the coast to approximately 1,200 meters in the east. Ridge accordance suggests an ancient and now greatly dissected peneplain. Many peaks rise above this summit peneplain. The highest of these monadnocks in Oregon is 2,280-meter Mount Ashland which rises 1,060 meters above the general level of its surroundings.

The geologic history of the Klamath Mountains began during the Paleozoic era with deposition of volcanic tuffs and sedimentary rocks which were subsequently metamorphosed, largely into schists. A period of erosion and folding followed until late in the Triassic period when a large series of volcanic and sedimentary rocks were deposited near Medford and Grants Pass. These rocks have all undergone extensive metamorphism into various types of schists, gneisses, marbles, and other metavolcanic or metasedimentary rocks. These rock types outcrop east of Gold Beach and at other scattered locations throughout the province. During the Jurassic period, sandstones, siltstones, and shales were laid down along the coast and in a belt extending from the southwestern corner of Oregon across the province in a generally northeasterly direction. Most of these deposits have undergone very little alteration. These rock strata were intruded with ultramafic rocks such as peridotite and dunite during late Jurassic or very early Cretaceous times. The intrusions have largely been altered to serpentine which now appears in elongated, stringerlike outcrops, generally associated with fault zones (fig. 7). Other rocks which were intruded at approximately the same time include a variety of granitics—diorite, quartz diorite, granodiorite, and granite. The largest areas of these rocks are found north and south of Grants Pass, east of Oregon Caves, south of Ashland, southwest of Tiller, and between Grants Pass and Gold Beach in the Pearsoll Peak area. The early Cretaceous period saw additional depositions of sediments—rocks which now appear as grayish-green arkosic sandstone and siltstone.

Figure 6.—The Klamath Mountains in Oregon showing characteristic, deep dissection and knifelike ridges.

Figure 7.—Unstable terrain underlain by serpentine and peridotite in the Klamath Mountains of southwestern Oregon.

Rock strata within the province were greatly modified by folding and deformation during the middle Cretaceous period. Apparently, the Klamath Mountains were truncated and underwent peneplanation during the Miocene and Pliocene epochs. Subsequent erosion and stream dissection have given rise to the mature topography which characterizes the area today.

Soils of the Klamath Mountains Province fall into two main groupings—those in the western portion and those in the east. Soils in the eastern half of the province generally reflect the effects of drier conditions, especially during the summer months, whereas soils to the west tend to remain moist for a considerably longer period of the year.

Upland soils in the western half of the province are, for the most part, Haplohumults (Reddish Brown Lateritic soils). Parent materials for these soils include both sedimentary and basic igneous rocks. They are moderately deep (1 to 2 meters to bedrock), reddish-brown soils possessing a silt loam or silty clay loam A horizon underlain by a silty clay B horizon. Scattered upland areas of peridotite or serpentine bedrock most often have reddish-colored soils which are classed as Hapludalfs (Gray-Brown Podzolic soils) or Xerochrepts (Regosols). Such soils are invariably unproductive, having very shallow and stony profiles.

The western portion of the province contains a wide variety of valley-bottom soils. The most important well-drained soils derived from alluvium on terrace landforms are Dystrandepts and Haplumbrepts. The Dystrandepts (Ando soils) have a thick, very dark-colored silt loam surface horizon underlain by a silty clay loam subsoil. Haplumbrepts (Western Brown Forest soils) on terraces tend to be deep, well drained, and silt loam or loam textured. The most common poorly drained streamside soils are Haplaquepts (Low-Humic Gley soils). These wet soils tend to have silt loam surfaces and silty clay loam subsoils.

In the eastern half of the province, the principal upland soils are Haploxerults (Reddish Brown Lateritic soils) which are continuously dry for a long period of the year. Generally, these are reddish-brown soils derived from sedimentary parent materials having bedrock

14

within 1 meter of the surface. Texture tends to be loam in the surface and clay loam in the subsoil. Haploxeralfs (Noncalcic Brown soils) are less widely distributed and are generally derived from metamorphic rocks. These are relatively shallow, gravelly clay loam soils which are high in bases. Granitic parent materials usually give rise to Xerochrepts (Regosols) of low fertility. A typical granitic soil has a coarse sandy loam A horizon underlain by a loamy very fine sand substratum.

Soils on flood plains and alluvial fans in the eastern section of the Klamath Mountains Province are principally well-drained Haploxeralfs and Haploxerolls (Prairie soils). Chromoxererts (Grumusols) occur on alluvial fans and bottomlands in areas surrounding Roseburg and Medford. These are clay-textured soils which undergo considerable expansion and contraction with wetting and drying.

Willamette Valley Province

The Willamette valley is a broad structural depression oriented north-south and situated in Oregon between the Coast Ranges on the west and the Cascade Range on the east. The valley is approximately 200 kilometers long, extending from the Columbia River to Cottage Grove where the two mountain ranges converge. Valley width generally ranges from 30 to 50 kilometers. Topographically, the valley is characterized by broad alluvial flats separated by groups of low hills (e.g., Portland, Chehalem, Eola, Salem, and Coburg Hills) (fig. 8).

Figure 8.—The Willamette valley in Oregon is characterized by broad, almost level, alluvial terrain interrupted by low basalt hills.

15

The valley floor has a very gentle, north-facing slope; elevation increases from 50 meters at Salem to only 129 meters at Eugene, 130 kilometers to the south. As a result, the Willamette River is a sluggish stream with many meanders, especially from Oregon City southward.

The Willamette valley is bordered on the west by a variety of sedimentary and volcanic rocks of Eocene age. They include submarine pillow basalts, conglomerates, and tuffaceous sandstones and siltstones which are actually eastward extensions of Coast Ranges formations. In the southern portion of the valley, these Eocene rock formations probably extend under valley fill materials to the western margin of the Cascade Range. Marine sedimentary rocks of Oligocene and Miocene age outcrop along the eastern margin of the valley. Columbia River basalt (Miocene) is found as far south as the Salem area and caps the Portland, Salem, and Eola Hills. Similar early Miocene basalt flows occur in the Eugene area where, for example, they are found in the Coburg Hills. A westward extension of the Cascade Andesites of Plio-Pleistocene age outcrops near Gresham and caps many of the hills near Oregon City.

The floor of the northern Willamette valley is underlain by thick, nonmarine sedimentary deposits of Plio-Pleistocene age. These deposits are present but not as thick in the southern part of the valley. Following the Illinoian glaciation late in the Pleistocene epoch, the entire valley as far south as Eugene was drowned by water and partially filled with silt to a depth of about 30 meters. Later, near the close of Wisconsin glaciation (10,000 to 15,000 years ago), the valley was again flooded because of an ice dam on the Columbia River. Evidences of this flooding include ice-carried erratics as far south as Harrisburg and a thin covering of silt to a maximum elevation of 122 meters. Recent alluvial deposits occur along the Willamette River in areas where it has cut into Pleistocene lakebeds.

Soils on the valley floor, derived from silty alluvial and lacustrine deposits, were formed under dominantly grassland vegetation. Soil morphology largely reflects effects of landform position and soil drainage. Well-drained soils situated on the Willamette River flood plain are deep, moderately dark-colored and range from sandy loam to silty clay loam in texture, with the fine-textured soils occupying the highest positions. These soils generally lack subsurface horizons of clay accumulation. Soils occupying terrace positions generally exhibit a greater degree of profile development, typically having silt loam surface horizons underlain by silty clay loam subsurface horizons. Closed depressions on terraces contain very poorly drained Albaqualfs (Planosols). These soils are characterized by a bleached, light-colored surface horizon, usually silt loam in texture, and an abrupt transition to a clay-textured subsoil.

Soils derived from igneous and sedimentary rocks situated along the edges of the valley and on low hills are similar to those found in the adjacent Coast Ranges and Western Cascades Provinces. A typical, well-drained soil derived from basalt colluvium possesses a reddish-brown silty clay loam surface and a clay-textured subsurface horizon, with depth to bedrock averaging 1.5 meters.

Principal soils on the valley floor are Argixerolls (Prairie soils) with lesser amounts of Haploxerolls (Prairie soils), Haplaquolls (Humic Gley soils), and Albaqualfs. On adjacent uplands and rolling hills, soils are largely Haplohumults (Reddish Brown Lateritic soils) with some Xerumbrepts (Regosols) and Haplumbrepts (Western Brown Forest soils).

Puget Trough Province

The Puget Trough Province extends the entire length of Washington from the Canadian border to Oregon where the Willamette valley is its physiographic and geologic continuation. The northern half of the province includes Puget Sound, and the southern half is largely the Cowlitz River valley and upper basin of the Chehalis River. Relief is moderate, and elevations of the trough floor seldom exceed 160 meters.

The northern, or Puget Sound basin, portion of the province is a depressed, glaciated area which is now partially submerged. The geology and topography resulted almost entirely from a lobe of the cordilleran icecap which pushed into the area from the north during the Pleistocene epoch (Folsom 1970, Mark and Ojamaa 1972). There were apparently several glacial epochs, the Vashon glaciation being most recent. The terminal moraine of the Vashon glacier is located approximately 16 to 24 kilometers south of Olympia. Inside the moraine is a large area, sloping gently toward Puget Sound and containing many lakes and poorly drained depressions underlain by glacial drift. Glacial deposits range from very porous gravels and sands to a hard till in which substantial clay and silt are mixed with coarser particles. For approximately 50 kilometers south of the terminal moraine (as far south as Toledo), the area is largely covered by outwash sands and gravels which were sluiced southward by the melting Vashon glacier.

Tertiary rock formations are exposed in the southern portion of the province (south of Toledo). Topographic characteristics, however, are similar to those further north. The majority of the area is made up of Eocene basalt flows and flow breccia. Smaller areas of Miocene and Pliocene nonmarine sedimentary rocks are found south of Toledo and east of La Center. Immediately north of the Columbia River is a substantial area of Pleistocene lacustrine deposits, similar to those found in the Willamette valley.

The majority of soils in the Puget Sound basin (northern) portion of the province are formed in glacial materials under the influence of coniferous forest vegetation. Haplorthods (Brown Podzolic soils) are most common. These soils generally have at least moderately thick forest floor layers with some development of an H layer. A thin, weakly developed A2 horizon is typical beneath the humus layer. This is underlain by an iron- and humus-enriched B horizon which is usually reddish-brown in color. Soil texture is commonly gravelly sandy loam, and profile depth averages about 1 meter. Underlying materials are either loose gravels and sands or hard, cemented till. Associated soils in the northern section of the province include Xerumbrepts (Regosols), Haplaquepts (Low-Humic Gley soils), Xeropsamments (Regosols), and Haploxerolls (Prairie and Chestnut soils).

In the southern portion of the province, Haplohumults (Reddish Brown Lateritic soils) are the most common under forest vegetation. These soils, which may be derived from either basic igneous or sedimentary parent materials, have well-aggregated silt loam to clay loam surface horizons underlain by B horizons generally showing evidence of clay accumulation. Soils developed under grassland vegetation are also common in this portion of the province. Often occupying terrace landforms, these soils are classed as Argixerolls (Prairie soils). Typically, these are deep, silt loam-textured soils lacking subsurface clay accumulation. The southeastern section of the province also contains Vitrandepts—regosolic soils containing considerable quantities of pumice particles.

Northern Cascades Province

The Northern Cascades Province extends north from Snoqualmie Pass to the Canadian border. Unlike the southern Cascades, these mountains are to a large extent comprised of ancient sedimentary rocks, most of which are folded and at least partially metamorphosed. Intrusions of large granitic batholiths are also common.

The province is a topographically mature area of great relief. Valleys are uniformly very deep and steep sided (fig. 9). An outstanding feature of the main eastward- and westward-flowing streams is their low gradient to within 6 or 7 kilometers of the main divide. On this basis, it may be concluded that the northern Cascades are as deeply dissected as is possible with their present elevation and that their relief within the current erosion cycle is at a maximum. Another striking topographic feature is the approximately uniform elevation of the main ridgetops. Near the middle of the range, this level varies from 1,800 to 2,600 meters. Towering above this relatively even crest are two dormant volcanoes—Mount Baker and Glacier Peak. In addition to the volcanoes, there are several granitic peaks of exceptional height.

Many ridges and peaks have glacial features. There are literally hundreds of cirques; some peaks, ringed by cirques, have been eroded to matterhorns. In addition, main east-west valleys probably owe their very low gradients to glaciation. Today, this portion of the Cascades contains more active glaciers than any other area within the continental United States (fig. 10) (Post et al. 1971).

The geologic history of the area began late in the Paleozoic era with deposition of clastic marine sediments in a constantly falling geosyncline. These were metamorphosed during a period of compression and folding when the sea withdrew during the Jurassic period. Products of this metamorphism include argillite, slate, phyllite, schist, greenschist, and greenstone—rock types widely distributed throughout the Northern Cascades Province. Gneisses, which occupy a large portion of the central section of the province, are also products of this same period of metamorphism.

Figure 9.—The Northern Cascades Province of Washington is characterized by unusually deep dissection and maximum relief; the steep-sided, U-shaped, main valleys are the result of glaciation.

During early Cretaceous times, another geosynclinal trough was formed and the area reinvaded by the sea. Resulting rocks include mainly graywacke, siltstone, slate, phyllite, and argillite. These rocks, perhaps not as widespread as the older deposits, are located (1) north of Mount Baker, (2) between the Skykomish and Stillaguamish Rivers, (3) north of the town of North Bend, and (4) in the northeastern portion of the province (fig. 11). Sometime during the late Cretaceous period the sea withdrew. Sediment deposition continued on the west, however, where there was a broad plain with meandering rivers. Here, continental sedimentary formations were laid down during late Cretaceous and early Paleocene times.

The Cascade Range was gradually uplifted during the Pliocene epoch. However, prior to this time large quantities of granitic rocks intruded the preexisting strata. Large masses of these rocks, including granite, granodiorite, and quartz diorite of Tertiary age, outcrop near the crest of the range in both the southern and northern portions of the province. In addition, older, Mesozoic granite rocks occupy large areas to the east.

The volcanic peaks of Mount Baker and Glacier Peak were built up during the Pleistocene

Figure 10.—Extensive glaciers and snowfields are significant alpine features in the Northern Cascades Province; view of Mount Shuksan (2,781 m.) with Mount Baker (3,285 m.) in the background.

epoch, chiefly of andesite flows. At the same time, glacial till was deposited in virtually every major valley. These deposits are highly variable and may range from fine to coarse texture. Generally, the fine material is glaciolacustrine in origin.

Extreme variability of parent materials combines with effects of extensive glaciation to produce a soil pattern in the Northern Cascades Province bewildering in complexity. Residual rock is either frequently covered by or intimately mixed with glacial materials. In addition, the rapid pace of geologic erosion on steep slopes severely restricts soil formation over large areas where Rocklands and extremely shallow stony soils predominate. Thus, diverse parent materials, large amounts of glaciation, and precipitous slopes are largely responsible for the complex soil pattern.

Knowledge of the soils of the Northern Cascades Province is limited. However, in 1970, the U.S. Forest Service issued a report of a soil survey of the Mount Baker National Forest (Snyder and Wade 1970). Soils of the National Forest were classed into four main

Figure 11.—An area of Cretaceous sedimentary rock (Swauk sandstone) northwest of Wenatchee, Washington; the steeply tilted beds of sandstone are clearly visible in this area of highly erodible soils.

groupings: (1) deep glacial soils, (2) deep glacial lake-deposited soils (lacustrine), (3) deep residual soils, and (4) shallow residual soils. The deep glacial soils were typically well-drained gravelly loamy sands to loams situated in valleys and on side slopes to an elevation of 1,375 meters. Subsoils were generally weakly or moderately compact till. Deep glacial soils were generally classed as either Cryorthods (Podzols) or Fragiorthods (Podzols with fragipans). Soils developed on glaciolacustrine deposits occurred in valley bottom and toeslope positions and tended to be finer textured than till-derived soils, ranging from loams to silty clay loams. These soils were classed as Haplumbrepts (Western Brown Forest soils) or Cryorthods. Deep residual soils, derived from sedimentary, schist, and granitic rocks, were found on steep midslope and toeslope landforms. These were classed as Cryorthods and were generally gravelly sandy loam to silt loam in texture. Shallow residual soils were situated on ridgetops and steep side slopes unaffected by glacial deposits and were developed on sedimentary and metamorphic rock types. These soils exhibited very little profile development and were gravelly sandy loam to loam textured.

Other soils found in the Northern Cascades Province west of the main divide include Xerochrepts and Haplorthods. Xerochrepts (Regosols) encompass those widely distributed, poorly developed soils which are usually stony loams less than 1 meter in depth. Common parent materials are glacial till and metamorphic rocks such as argillite. Haplorthods (Brown Podzolic soils) are also generally loam textured and may be derived from a variety of rocks, ranging from sedimentary to granitic. Principal soils derived from alluvium along main west-flowing streams are Xeropsamments (Regosols), Haplaquepts (Low-Humic Gley soils), and Haploxerolls (Prairie soils).

Soils east of the Cascade crest reflect the drier conditions under which they were formed. Probably most abundant are Haploxerolls (Chestnut and Brown soils) formed on a variety of parent materials but generally influenced to some extent by volcanic ash and, in some areas, loess. Textures range from stone-free silt loams to very cobbly loams. Other soils present in the eastern portion of the province include Xerochrepts (Regosols) and Haploxeralfs (Noncalcic Brown soils).

20

Southern Washington Cascades Province

The Southern Washington Cascades Province extends south from Snoqualmie Pass to the Columbia River. Unlike the Northern Cascades Province, andesite and basalt flows dominate with only minor amounts of igneous intrusive, sedimentary, and metamorphic rocks.

The area is characterized by generally accordant ridge crests separated by steep, deeply dissected valleys. The general ridge elevation is approximately 2,000 meters in the northern section and 1,200 meters in the southern. An extensive area around Mount Adams is composed mainly of recent lava flows; it comprises a gently sloping plateau at 900- to 1,500-meter elevation, differing markedly from the rest of the province. Three dormant volcanoes dominate the surrounding landscape: Mount Rainier (4,392 m.), Mount Adams (3,801 m.), and Mount St. Helens (2,948 m.) (fig. 12). Probably more geologic investigations have been carried out in the vicinity of Mount Rainier than on any other dormant volcano. See, for example, the recent reports by Crandell (1969a, 1969b, 1971) on the geology of Mount Rainier National Park.

Figure 12.—Mount St. Helens in the Southern Washington Cascades Province; the Kalama River valley in the foreground contains an extensive mudflow deposit which can be traced in this photograph to its origin on the slopes of the mountain. This photograph was taken prior to the eruption of Mount St. Helens in May 1980.

At least 90 percent of the Southern Washington Cascades Province is made up of andesite and basalt flows with their associated breccias and tuffs. These lava extrusions have been classified into four rather generalized age classes: (1) Eocene to lower Oligocene, (2) upper Oligocene to lower Miocene, (3) middle Miocene, and (4) Pleistocene to Recent. The Eocene-Oligocene volcanic rocks are described as bedded andesitic breccias with interbedded andesite and basalt which have been considerably altered by faulting and folding. These rocks are widely distributed from Mount Rainier southward, largely west of the Cascade Range crest. The Oligocene-Miocene volcanic rocks are mainly andesite flows and flow breccias which, to a large extent, retain a fresh look and are in a horizontal position. These rocks outcrop in a large area north of Mount Rainier and at scattered locations to the south. The Miocene deposits are Columbia River basalt which extends into the province from the east. Pleistocene to Recent deposits include the andesite flows and pyroclastics which comprise the slopes of Mount Rainier, Mount Adams, and Mount St. Helens. Pleistocene to Recent vesicular basalt lavas are especially widespread in the vicinity of Mount Adams.

Tertiary granitic rocks are found only at scattered locations. The largest outcroppings are north and east of Mount Rainier. Another area of granodiorite is southwest of Randle. Small deposits of sedimentary rocks are located southwest of Mount Rainier and in the northeastern corner of the province. These include Eocene and Miocene sandstone, siltstone, and shale.

Areas adjacent to the three volcanic peaks are generally mantled with pumice deposits of variable age, origin, and thickness. Deposits near Mount St. Helens sometimes exceed 300 centimeters. The most recent pumice deposit in the vicinity of Mount Rainier is thought to have been laid down 100 to 150 years ago during the volcano's last eruption.

Pleistocene glacial activity was widespread in the Southern Cascades Province, although most were small alpine glaciers. At its maximum extension, however, the Vashon glacier pushed into the flanks of the Cascades, impounding long lakes in several mountain valleys. Today deep glaciolacustrine deposits mark the limits of these lakes.

Fewer soils in the Southern Washington Cascades Province are developed in glacial materials and, because of generally less rugged topography, there are smaller areas of Rockland and stony skeletal soils than in the Northern Cascades Province. The most widespread soils are derived from a combination of parent materials consisting of both pumice and basalt and andesite. In some instances, the surface layers consist of a series of depositions of unmixed eolian volcanic ash and pumice overlying residual soil materials. Soils in areas of deep pumice deposits are most often Cryorthods (Podzols) or Haplorthods (Brown Podzolics). Cryorthods on pumice typically have gray A2 horizons over a reddish gravelly coarse sand B2ir horizon. Probably the most common soils developed on mixed pumice and basic igneous parent materials are Inceptisols, specifically Dystrandepts and Vitrandepts (Regosols). These are poorly developed soils having only weakly differentiated horizons. Textures of such soils range from gravelly sandy loam to silt loam.

Well-developed Haplohumults (Reddish Brown Lateritic soils) are common in the western portion of the province. These soils are derived from basalt and andesite and are generally characterized by a reddish-brown loam to clay loam surface horizon underlain by a B horizon of clay loam or silty clay loam texture. Glacial or glaciofluvial materials often give rise to moderately well drained to imperfectly drained Haplorthods (Brown Podzolic soils) or Hapludalfs (Gray-Brown Podzolic soils). Such soils are generally gravelly sandy loam in texture, and B horizons are iron stained and often contain iron concretions.

Soils developed in alluvial deposits along the major west-flowing streams are typically Xeropsamments (Regosols) and Haploxerolls (Prairie soils). These are coarse-textured soils, commonly ranging from loamy sand to sandy loam.

East of the Cascade crest, soils are largely Haploxeralfs (Noncalcic Brown soils) and Haploxerolls (Chestnut soils). These soils, derived from andesite, sandstone, or glacial till, are also influenced to a considerable extent by volcanic ash or loess or both. Soil textures are most often silt loams and loams.

Western Cascades Province

The Cascade Range in Oregon is divisible into two distinct physiographic provinces. The High Cascades Province, on the east, includes all major peaks of the range (e.g., Mount Hood, Mount Jefferson, Three Sisters) and originated during late Pliocene and Pleistocene epochs. The Western Cascades Province consists of older volcanic flows and pyroclastics laid down during the Oligocene and Miocene.

The relief of the Western Cascades Province is generally rugged in the eastern portions (fig. 13), but slopes are more gentle in the west. Over much of the area there is a striking accordance of main ridge crests at an average elevation of about 1,500 meters. Elevations higher than 1,500 meters are uncommon, and only a few peaks exceed 1,800 meters.

During the Oligocene and Miocene epochs, numerous volcanic eruptions and effusions

Figure 13.—The Western Cascades of Oregon are deeply dissected, with the higher ridges generally composed of thick flows of andesite; this view shows Twin Buttes in the foreground and the Three Sisters, located in the High Cascades Province, in the background.

produced deposits of basalts, andesites, and pyroclastic rocks, frequently in a complex pattern. Pyroclastic rocks in this area include tuffs, breccias, and agglomerates. Besides these extrusive igneous rocks, a small amount of granite outcrops in several places, notably along the McKenzie River. Subsequent alteration by alpine glaciation occurred during the Pleistocene epoch. Glaciation is evidenced by widely spaced deposits of glacial drift and the characteristic U-shape of the major valley drainages.

Pyroclastics are abundant in the central portion of the Western Cascades Province. Between the McKenzie and South Umpqua Rivers, approximately three-fourths of the total area is made up of pyroclastic rocks. To the north, breccias are almost entirely absent except in the Collawash River and certain sections of the Santiam River drainages. South of the Umpqua basin, pyroclastics remain common to the Rogue River which marks their southernmost boundary.

Basalt and andesite are the most common bedrock materials in the Western Cascades Province. A large proportion of the province is made up of andesite from the North Fork of the Willamette River northward to the Clackamas River. Also, south of the South Umpqua River, andesite is again the most common rock. Basalt has a more scattered occurrence and is generally found along the western margin of the Western Cascades Province.

Glacial deposits are widely distributed throughout the Western Cascades Province, with the majority concentrated in the valleys of major streams. Tracing the exact extent of mountain glaciation in this area is very difficult because many morainal features have been obliterated by subsequent stream dissection.

Soils in a large portion of the Western Cascades Province can be placed into two major groups—soils developed from pyroclastic parent materials (largely tuffs and breccias) and those derived from basic igneous rocks (mainly basalt and andesite). These parent material groupings have generally produced contrasting soil types. Since tuffs and breccias are readily weatherable, soils from these materials tend to be deep and fine textured, especially on gentler slopes. Pyroclastic soils are frequently imperfectly drained, and mass soil movements (e.g., slumps and earthflows) are common. Well-developed soils from tuffs and breccias typically possess moderately thick, dark-brown clay loam A horizons and olive-brown to reddish-brown silty clay or silty clay loam B horizons. On steep slopes, poorly developed gravelly clay loam soils are most common. Well-developed pyroclastic soils are classified as Haploxerults (Reddish Brown Lateritic soils); and those less well developed, containing abundant stone fragments, are generally classed as Haplumbrepts (Western Brown Forest soils) or Xerumbrepts (Regosols).

Soils derived from basalt and andesite are generally well drained and tend to be stonier and coarser textured than those from pyroclastic parent materials. These soils are more stable and not as subject to mass erosion. On steep slopes, soils from basalt and andesite are relatively poorly developed, consisting of dark-brown gravelly loam or sandy loam surface horizons. Especially at higher elevations, surface layers often contain noticeable amounts of aerially deposited volcanic ash and pumice. On gentle slopes, and especially in the southern and western portions of the province, soils developed on basic igneous rocks are often deep and well developed. These soils are reddish-brown in color, with a loam or clay loam surface horizon, and a clay loam or clay subsoil. Well-developed soils are generally classed as Argixerolls (Prairie and Western Brown Forest soils) in the southern portion of the province, and Haplohumults (Reddish Brown Lateritic soils) in the western portion. Less well-developed soils derived from basalt and andesite are usually Haplumbrepts; and immature soils with little profile development are Xerumbrepts (Regosols).

Glacial till soils are most abundant in the northern portion of the province. These soils are generally stony and/or gravelly loams or sandy loams and are classified as Cryorthods (Podzols), Haplorthods (Podzols and Brown Podzolic soils), or Cryumbrepts (Tundra soils).

High Cascades Province

The High Cascades Province is essentially an area of rolling terrain interrupted at intervals by glaciated channels, some quite deep, carrying westward-flowing streams. The area is dotted with volcanic peaks and cones rising to 50 to 1,600 meters above the surrounding area (fig. 14). The major peaks are, from north to south, Mount Hood (3,427 m.), Mount Jefferson (3,199 m.), and the Three Sisters (3,062-3,157 m.). The general elevation of the sometimes broad, gently sloping portion is approximately 1,500 to 1,800 meters.

The High Cascades Province is geologically young; some flows of lava (scoriaceous basalt) are only several hundred years old. The most extensive depositions were extruded from volcanic vents during the late Pliocene and Pleistocene epochs. These flows are of gray olivine basalts and olivine-bearing andesites with subordinate amounts of dense porphyritic pyroxene andesites. Scattered over the area are younger flows comprised of andesites and basalts which are dated as upper Pleistocene and Recent epochs. Most major

Figure 14.—The High Cascades Province of Oregon is essentially a gently sloping, high plateau area containing scattered volcanic peaks composed of andesite and smaller "cinder cones"; peaks on the skyline are Mount Jefferson (3,199 m.) on the left and Three-Fingered Jack (2,392 m.) on the right; the cratered cinder cones in the foreground comprise Sand Mountain.

peaks in the area are made up of olivine-bearing andesite and originated during the upper Pleistocene epoch. The smaller cones, commonly called cinder cones, are generally comprised of gray to red basaltic and andesitic pyroclastic rocks.

Bedrock in the High Cascades Province is frequently obscured by a mantle of pumice and ash from several volcanic eruptions. The most extensive deposition of these materials resulted from the explosive culminating eruption of Mount Mazama, which occurred about 6,600 years ago.

Glacial deposits are also locally abundant, especially adjacent to some of the higher peaks. For example, the flanks of Mount Hood are typically mantled with deep deposits of glacial till.

The High Cascades Province is an area dominated by immature soils developed in volcanic ejecta and soils showing more profile development which are derived from glacially deposited materials. Soils on glacial till are most abundant in the northern section of the province and are generally classified as Cryorthods, Haplorthods, or Cryumbrepts. The Cryorthods (Podzols) are deep and well drained with gray, stony or gravelly sandy loam surface horizons underlain by reddish-brown, stony or gravelly loam subsoils. Haplorthods (Brown Podzolic soils) are similar but show less development of a gray A2 horizon. Typical soil textures are a gravelly loam surface over a very stony loam subsoil. Cryumbrepts (Regosols) tend to be much shallower to bedrock (1 meter or less) and are generally very stony sandy loam in texture.

In the central and southern portions of the province, extensive areas are mantled with deposits of volcanic ejecta, such as pumice, cinders, and ash. Soils in these materials generally exhibit little profile development and are classed as Vitrandepts (Regosols). Typically, a thin, dark-colored A1 horizon of sandy loam or loamy sand texture is underlain by a transitional AC horizon which grades into the unaltered coarse sand or gravelly sand parent material.

Okanogan Highlands Province

The Okanogan Highlands Province is characterized by moderate slopes and broad, rounded summits. In this respect, it differs markedly from the rugged Northern Cascades Province on the west. Excepting the main river valleys, much of the province lies above 1,200 meters. There is a scattering of peaks which attain elevations of over 2,400 meters. The province is made up of several upland areas separated by a series of broad, north-south river valleys. These south-flowing rivers are, from west to east, the Okanogan, Sanpoil, Columbia, Colville, and Pend Oreille (Clark Fork) Rivers.

Virtually the entire province was repeatedly covered by glacial ice during the Pleistocene epoch. As a result, deposits of glacial drift are found throughout the area. Although many of these deposits occur only intermittently in stream valleys, they are widespread in the eastern portion of the province, especially north of Spokane. In some of the main valleys, glaciolacustrine sediments form a series of terraces on valley walls.

The Okanogan Highlands Province is, in many respects, geologically similar to the Northern Cascades Province. There is an almost bewildering variety of rock types, ranging in age from Precambrian to late Tertiary. The Precambrian rocks, consisting largely of phyllite, are restricted to the eastern portion of the province. Lying above these deposits are rocks of the Cambrian, Ordovician, Devonian, and Mississippian periods of the Paleozoic era. Rock types represented in these formations include quartzite, graywacke, slate, argillite, phyllite, greenstone, and some limestone. The most abundant rock types in the province are granitics, which were deposited during the Mesozoic era. These rocks, which include granite, quartz monzonite, quartz diorite, and granodiorite, occupy most of the area in the western section of the province and are interspersed with the Paleozoic rocks to the east. Tertiary depositions are largely confined to areas adjacent to main river valleys. Bedded

Tertiary sediments are found near the Okanogan, Sanpoil, Columbia, and Pend Oreille Rivers. Later in the Tertiary period, these same areas were influenced by eruptions of andesite and basalt lavas. Columbia River Basalt (Miocene) extends across the Columbia River into the province in an area south of Okanogan.

In the Okanogan Highlands Province, the soil pattern is closely tied to elevation. In the mountainous areas away from the major river valleys, soils derived from granitic parent materials are Xerochrepts (Regosols) and Cryorthods (Podzols). The Xerochrepts, having only very weak horizon development, are most often cold, acid, stony or gravelly sandy loams with a total depth to bedrock of 1 meter or less. The Cryorthods tend to be silt loam textured and have substantially more profile development, generally consisting of a thin, light-colored A2 horizon underlain by a B2 horizon of high iron content. High elevational soils from glacial materials often have considerable amounts of volcanic ash incorporated into the surface horizon and are therefore classified as Vitrandepts (Regosols in pumice or ash). Because of the high ash content, surface layers are silt loam textured; subsoils are generally gravelly loam.

At lower elevations, along the margins of river valleys and the southern boundary of the province, soils reflect the drier climate and transitional forest-grassland vegetation. The most abundant parent material in these locations is glacial till, and Haploxerolls (Chernozem soils) are the predominant soils. These soils typically have a dark, moderately thick A1 horizon underlain by a B horizon which shows little increase in clay content and is distinguished largely by changes in color and structure. Soil textures are commonly sandy loam to loam. Other soils in these foothill positions include Vitrandepts and Haploxeralfs (Noncalcic Brown soils).

Soils at lowest elevations, occupying terraces and flood plains along major rivers, are most often coarse textured and well drained to excessively drained. Parent materials generally consist of glacial outwash sands and gravels. The most important soil great groups present in these locations are Haploxerolls, Vitrandepts, Haplorthods (Podzols and Brown Podzolic soils), and Xeropsamments (Regosols).

Blue Mountains Province

The Blue Mountains Province is made up of several ranges of mountains separated by faulted valleys and synclinal basins. The mountainous areas include the Ochoco, Blue, and Wallowa Mountains, as well as the Strawberry, Greenhorn, and Elkhorn Ranges. Relief within the various mountain ranges is highly variable. Moderate slopes are common within the Blue and Ochoco Mountains, whereas the heavily glaciated Wallowa Mountains exhibit the greatest relief (fig. 15). Maximum elevations range from about 2,100 meters in the Ochoco Mountains to 2,900 meters at Eagle Cap in the Wallowa Mountains. Valley elevations are about 750 meters in the vicinity of the Ochocos and 900 meters in the broad basin between the Blue and Wallowa Mountains (near La Grande and Baker, Oregon). Spectacular Hells Canyon comprises the eastern boundary of the province. This canyon, occupied by the Snake River, ranges up to 1,660 meters in depth and is 24 kilometers wide at its broadest point.

Geologically, the Blue Mountains Province may be conveniently separated into eastern and western units, with the dividing line a short distance east of John Day, Oregon. The western Blue Mountains contain outcrops of some of the oldest rocks in Oregon. These are Paleozoic formations (Mississippian and Pennsylvanian) of limestone, mudstone, and sandstone which outcrop along tributaries of the upper Crooked River. Ultramafic intrusions, some altered to serpentine, occur in the Strawberry Range. Triassic and Jurassic formations are located near the communities of Suplee and Izee and consist of a wide range of rocks such as conglomerate, sandstone, siltstone, shale, and limestone. The Clarno (Eocene) and John Day (Oligocene) are two formations widely known because of their

Figure 15.—The Wallowa Mountains of Oregon are more deeply dissected and have steeper slopes than other sections of the Blue Mountains Province; snow-capped peaks in the background include Eagle Cap, the highest in the range.

abundant vertebrate fossils. Located along the lower John Day River, these formations are composed largely of breccia and varicolored tuffs. Columbia River basalt, a thick formation extruded in many sheets during the Miocene epoch, occupies large areas within the western Blue Mountains. Late Miocene and Pliocene formations are also present and consist of bedded tuffs and silts.

The eastern portion of the Blue Mountains Province spans a large part of the geologic time scale. Paleozoic formations of the Permian period are widespread near Baker and Sumpter and consist of schists, limestone, slate, argillite, tuff, and chert. As in the western section, these formations often have ultramafic and mafic intrusions ranging from peridotite to gabbro. Triassic sedimentary formations (sandstone, siltstone, shale, and limestone) are common but are not continuous because of erosion and subsequent burial by Tertiary rocks. Triassic limestone and argillaceous beds are especially prominent in the Wallowa Mountains where they outcrop on many ridge crests. Granitic stocks, perhaps extensions of the great Idaho Batholith, are found in the Wallowa Mountains, near Baker and Sumpter and along the John Day River. Columbia River basalt is widespread along the north slope of the Blue Mountains, forms the mass of the range between Pendleton and La Grande, and is also found to the south. Thus, it is inferred that uplift of the Blue Mountains occurred after the deposition of these lavas during the Miocene epoch. Alluvial deposits of sand and gravel, dating from the Pliocene and Pleistocene, cover the floors of many basins. Also during the Pleistocene, glaciation was widespread in both the Blue and Wallowa Mountains, as shown by numerous cirques, glacial lakes, and moraines.

Following deposition of the most recent lava flows, much of the area within the central and northern portions of the Blue Mountains was covered by a layer of aerially deposited volcanic ash and fine pumice. Subsequent erosion has largely removed the ash from south-facing slopes; however, other locations are typically mantled by the material. In addition, many upland areas, especially in the eastern portion of the province, were mantled by loess deposits, probably during the late Pleistocene.

Soils of the Blue Mountains Province may most conveniently be grouped into two main units—those at moderate to high elevations which were formed under dominantly forest vegetation, and soils at generally lower elevations which were formed under grassland or shrub-grassland vegetation. Forested soils developed in volcanic ash are almost completely restricted to broad ridgetops and north-facing slopes where the ash mantle thickness varies from approximately ½ to 1 meter. These Vitrandepts (Regosols) are characteristically dark brown in color and fine sandy loam to silt loam in texture. Fragiorthods (Brown Podzolic soils with fragipans) also occur on mountainside slopes under forest vegetation. Since most often these soils are derived from loess, they tend to be deep and silt loam in texture. Soils of the forest-grassland transition at moderate to high elevations are generally within the Mollisol order, with the principal great groups being Argixerolls (Prairie soils), Haploxerolls (Chestnut and Chernozem), and Palexerolls (Prairie soils). These soils are formed in loess and basic igneous rock materials, with depth to bedrock averaging less than 1 meter. Surface horizons are generally comprised of dark-brown silt loam, and subsoil textures range from silty clay loam to clay.

The most widespread soils supporting grassland and shrub-grassland vegetation are Argixerolls (Prairie soils and Chernozems). In the western portion of the province, these soils are commonly derived from ancient lake-deposited sediments, with profiles consisting of a clay loam surface horizon over a clay-textured subsoil. In the eastern section of the province, the Argixerolls are developed from loess and basic igneous rocks. In these locations, soils tend to have silt loam surface horizons and clay loam subsoils. Other grassland soils of more limited extent include Haploxerolls (Brown soils) and Haplargids (Sierozems).

Soils formed in alluvium along major streams are largely Haploxerolls (Chestnut and Brown soils). Soil texture ranges from silt loam to silty clay loam, and soil drainage varies from well drained to poorly drained. In some bottomland locations, notably in the eastern portion of the province, soils having a hardpan at depths of less than 1 meter (Durixerolls) are common.

Columbia Basin Province

The Columbia Basin is the largest single province; it occupies an extensive area south of the Columbia River between the Cascade Range and Blue Mountains in Oregon and roughly two-thirds of the area east of the Cascades in Washington. Topography varies from very gently undulating to moderately hilly. Steep slopes are of limited occurrence and restricted to isolated basaltic buttes or canyons cut by some of the major rivers; for example, the Deschutes River in Oregon (fig. 16). Over most of the area, elevations range from 300 to 600 meters, although they are less than 150 meters adjacent to the Columbia River.

Although early Tertiary rocks are found at scattered locations in Oregon, the important geologic events in the Columbia Basin Province began during the Miocene epoch with the vast outpouring of lavas making up the Columbia River Basalt formation. This huge basalt layer covers over 500,000 square kilometers in Washington, Oregon, and Idaho and underlies virtually the entire province. The Columbia River Basalt formation, ranging in total thickness from 600 to over 1,500 meters, is made up of numerous individual flows about 8 to 30 meters thick. Bottom portions of individual flows are dense, dark-gray basalt, but near upper margins the basalt becomes scoriaceous. In some areas, deformation of the Columbia River basalt produced ridges and hills during the Pleistocene epoch.

In the central portion of eastern Washington's Columbia Basin Province is a unique geologic feature—the Channeled Scablands. This is a gigantic series of dry, deeply cut channels in Columbia River basalt (fig. 17) which form an extensive and complex drainage

Figure 16.—Canyons cut in basalt by the Deschutes and Crooked Rivers in the southern portion of the Columbia Basin Province; west of Madras, Oregon.

Figure 17.—General view of Channeled Scablands in the Columbia basin of central Washington; this area is characterized by numerous dry channels cut in Columbia River basalt and generally shallow, stony soils (from "Landforms of Washington," Easterbrook and Rahm, by permission).

network. Many of the deeply entrenched drainageways diverge upstream only to converge again further downstream. Perhaps the best known feature in the Channeled Scablands is Grand Coulee with its spectacular Dry Falls. Although the origin of these puzzling features is still debated, Bretz (1959) probably offered the most satisfactory theory. He suggests that floodwaters, pouring from glacial Lake Missoula (western Montana) as a result of dam failure during the Pleistocene epoch, were responsible for cutting the channels.

Plio-Pleistocene deposits cover the Columbia River basalt over extensive areas. The most widespread deposit is the Palouse loess which mantles an elliptical area 160 kilometers long in southeastern Washington. This material, deposited during the Pleistocene epoch, is made up of massive, tan-colored silt which may be over 45 meters thick. The Palouse area is characterized by smoothly rolling hills (fig. 18) and soils of high fertility which are generally used for wheat and pea production. Similar buff-colored, structureless silt deposits are found in Oregon near Moro and in Grass Valley. In addition, a large area near Boardman, Oregon, is underlain by unconsolidated sand apparently of Pleistocene age. Similar deposits, probably glaciolacustrine in origin, are located west of Walla Walla and near Toppenish, Washington.

Although virtually all soils in the Columbia Basin Province have been formed under grassland or shrub-grassland vegetation, a wide variety of soils is present. Most of the broad soil differences correlate with annual precipitation. In general, precipitation is heaviest along the margins of the basin and gradually decreases toward the central portion.

Figure 18.—Rolling Palouse Hills, composed of deep loess deposits, near Pullman, Washington, Columbia Basin Province (*photo courtesy H. W. Smith*).

As a result, four distinct soil regions, forming a roughly concentric circular pattern, may be identified within the province.

The first soil region is located along all Columbia Basin Province boundaries with the exception of the west but is best expressed in the Palouse Hills near the Washington-Idaho border. The climate is subhumid, with annual precipitation ranging from 400 to 600 millimeters. In this area Argixerolls predominate; under the old classification system, these soils were classed as Prairie soils and Chernozems. Argixerolls derived from Palouse loess showing maximum profile development (Prairie soils) possess a thick, dark-colored A1 horizon of silt loam texture overlying a clay loam or silty clay loam B2 horizon having well-defined prismatic structure. Typically, calcium carbonate has been leached to levels well below the base of the solum. Less well-developed soils on loess (Chernozems) generally have shallower profiles, less clay accumulation, and subangular blocky structure in the B2 horizon and a zone of calcium carbonate in the B3, usually within 1 meter of the soil surface. Other soils encountered in this region include Haploxerolls (Chestnut and Brown soils), Albaqualfs (Planosols), Haploxeralfs (Noncalcic Brown soils), and Xerorthents (Regosols).

The second soil region is adjacent but generally at lower elevations and more arid, receiving 230 to 400 millimeters of annual precipitation. The principal soils are Haploxerolls (Chestnut soils) derived from loess. Typically, these soils have moderately thick, brown silt loam A1 horizons over light-brown silt loam B horizons with incipient prismatic structure. A zone of calcium carbonate accumulation commonly is present in the B3 horizon. Other soils present in smaller amounts include Argixerolls (Prairie soils), Durixerolls (Prairie soils with hardpan), Xerorthents (shallow; Lithosols), and Rockland.

The third soil region roughly encircles the central portion of the basin and also is semiarid with 230 to 400 millimeters of annual precipitation. Parent materials are principally loess and sandier windblown materials with lesser amounts of basalt. Poorly developed Haploxerolls (Brown soils) are the most common soils in the region. They possess a moderately thick, dark grayish-brown, loam-textured A horizon which is low in organic matter content. B horizons show little clay accumulation and a B3ca is generally present. Xerorthents (shallow; Lithosols) are also common in this region since it encompasses a large portion of the Channeled Scablands. Other soils of some importance include Camborthids (Sierozems), Argixerolls (Prairie soils), Haplargids (Sierozem and Desert soils), Haplaquolls (Humic Gley soils), Durixerolls (Prairie soils with hardpan), and Rockland.

The fourth soil region includes desertic soils of the lower bowllike center of the Columbia Basin Province. This is an area of arid climatic conditions receiving 100 to 230 millimeters of precipitation annually. Here, the dominant soils are Camborthids (Sierozems). These soils have thin, light-colored A horizons over B horizons which may be darker than the A and usually contain larger amounts of clay. A carbonate-enriched horizon, that may be cemented, occurs in the lower part of the B horizon. Other soils present include Haploxerolls (Chestnut and Brown soils), Xerorthents (shallow; Lithosols), Haplargids (Sierozem and Desert soils), Haplaquolls (Humic Gley soils), Torripsamments (Regosols), and Rockland.

High Lava Plains Province

The High Lava Plains Province of central Oregon is characterized by young lava flows of moderate relief interrupted by scattered cinder cones and lava buttes. As a result of porous bedrock and scanty rainfall, many streams are seasonal. Undrained basins containing playa lakes, some dry and others with fluctuating levels, are common. Several basins, now dry, contained extensive lakes during the Pleistocene epoch (e.g., Fort Rock valley and Christmas and Fossil Lakes). Most of the province has a base elevation of about 1,200 meters above sea level.

Geologic formations in the High Lava Plains Province consist largely of Pliocene and Pleistocene lavas, tuffs, and alluvium. In many areas, Quaternary valley fill deposits overlie the older volcanic flows. These are comprised of alluvium and lake deposits plus eolian sediments, all of which were derived from the volcanic rocks of the uplands.

Evidences of extensive volcanic activity during Pleistocene and Recent times are abundant, especially in the western portion of the province. The largest volcanic peak is the Paulina Peak shield volcano which contains Newberry Crater. Pumice, resulting from an eruption of this volcano about 4,000 years ago, mantles an extensive area to the north and east of the crater. Deposits of Mount Mazama pumice are also widespread in the same general area, as well as to the south in the Basin and Range Province. Broad areas of Pleistocene lava flows are a notable feature in the vicinity of Bend. In addition, several outstanding examples of Recent lava flows are situated south of Bend at Lava Butte (fig. 19) and east of Fort Rock.

Soils of the High Lava Plains Province are similar to those occurring in the Basin and Range and Owyhee Upland Provinces to the south and east. Uplands in the central

Figure 19.—Lava Butte south of Bend, Oregon—an outstanding example of a Recent volcanic cone and associated lava flows located in the western portion of the High Lava Plains Province.

33

and eastern portions, supporting shrub-grassland vegetation, are largely mantled with Camborthids, Haplargids, and Durargids. The Camborthids (Sierozem soils), limited to the central portion of the province, are formed in pumice and water-laid materials which have been deposited in layers ½ to 1 meter thick over basalt bedrock. These soils commonly have a sandy loam surface horizon and a loamy subsoil lacking accumulations of calcium carbonate. The Haplargids (Sierozem and Brown soils), the most widely distributed upland soil, are derived from basalt or tuff and most commonly have very stony loam textures. Depth to bedrock ranges from approximately ½ to 2 meters. The Durargids (Sierozem and Brown soils with duripans) are similar to the Haplargids but have silica-cemented hardpans at depths of about one-half meter.

In several locations, notably in the vicinity of Malheur and Harney Lakes, soils are formed from lacustrine deposits in old lake basins. Here, Natrargids (Solonetz soils) comprise one of the most important soil great groups. These are deep, silty soils possessing a subsurface horizon of clay and sodium accumulation. Other soils found in ancient dry lakebeds include Haplaquepts (Low-Humic Gley soils), Durorthids (Regosols with hardpan), Camborthids, and Durargids.

Soils derived from silty alluvium also occur at scattered locations throughout the province. These are generally poorly drained soils of silt loam to silty clay loam texture which occupy flood plain landforms. The most important great groups are Haplaquolls (Humic Gley soils), Haploxerolls (Chestnut and Brown soils), and Haplorthents (Alluvial soils).

In the western portion of the province, adjacent to the High Cascades Province, regosolic soils developed on pumice support open coniferous forest vegetation. Pumice soils are most extensive east of Newberry Crater where the younger Newberry pumice frequently overlies preexisting pumice deposits from Mount Mazama. These unusual soils (Vitrandepts) are briefly described in the following section covering the Basin and Range Province.

Basin and Range
and
Owyhee Upland Provinces

Southeastern Oregon has been divided into two physiographic provinces: Basin and Range and Owyhee Upland. Although topography differs, the two provinces are similar geologically.

The Basin and Range Province is characterized by fault-block mountains enclosing basins with internal drainage (fig. 20). The Owyhee Upland Province exhibits considerably less faulting and, in general, may be described as a north-facing basin which is drained by the Owyhee River. Elevations in both provinces range from about 1,200 meters to 2,930 meters atop Steens Mountain. Except for slopes of the fault-block mountains, much of the area is rolling with low relief. Since annual precipitation in the area averages only 180 to 300 millimeters, most streams are intermittent, and numerous undrained basins contain shallow, saline lakes.

Excepting small amounts of Paleozoic and Mesozoic formations which outcrop in the Pueblo and Trout Creek Mountains, virtually all rocks date from Miocene to Recent epochs. The western Basin and Range Province is made up largely of Miocene to Recent flows of basalt, pyroclastics, and alluvial sediments. Further east, two rock assemblages are prominent: (1) Miocene flows of rhyolite, dacite, and andesite near Abert Rim and Paisley; and (2) altered basalt and andesite flows and tuffs overlain by tuffaceous sedimentary rocks in an area just east of Lakeview. Principal fault-blocks in the area (Winter Ridge, Abert Rim, and Steens Mountain) are capped with Miocene flows of basalt. At the base of the Steens Mountain fault scarp are tuffaceous sedimentary rocks as well as flows

Figure 20.—Hart Mountain, a typical fault-block mountain in the Basin and Range Province of southeastern Oregon; note the internally drained depressions containing shallow, saline lakes.

of rhyolites, andesites, and dacites. Steens Mountain is also of interest because of evidences of extensive glaciation—glacial carved channels and cirque basins at the head of virtually every drainage (fig. 21).

To the east, near the Owyhee River, are Miocene and Pliocene beds of tuffaceous sedimentary rocks capped by flows of rhyolite and basalt. In addition, thick beds of quartzose sandstone, siltstone, and conglomerate outcrop near the mouth of the Owyhee River. The most recent volcanic activity in the area occurred during the Pleistocene epoch and resulted in basalt flows of limited extent at Diamond (fig. 22) and Cow Lakes Craters.

Figure 21.—Kiger Gorge, a glacially carved valley penetrating Steens Mountain, eastern Basin and Range Province, Oregon.

Soils of these provinces may conveniently be divided into two main groups—those in the west which developed under forest vegetation and those in the east associated with grassland-shrub vegetation. A tree-covered high plateau area in the northwestern corner of the Basin and Range Province is mantled with extensive deposits of Mount Mazama pumice. Although most was aerially deposited originally, the pumice has been reworked by water in some areas or was deposited in glowing avalanches which swept down slopes near the volcano during the eruption. Soils derived from pumice have immature, regosolic profiles consisting of a moderately thick surface layer with some organic matter accumula-

36

tion overlying nearly unweathered, yellow- and buff-colored pumice gravel and sand. These Vitrandepts (Regosols) are slightly acid and, due to the high porosity of the pumice particles, have water-holding and cation exchange capacities far greater than generally expected in such coarse material.

Upland soils to the east which support grassland-shrub vegetation are dominantly Haplargids and Durargids. The Haplargids (mostly Sierozem and Brown soils) are derived from basalt and generally have a very stony loam surface horizon underlain by either a clay or stony loam subsoil. The Durargids (Sierozem and Brown soils with hardpans) are also developed on basaltic parent materials and are characterized by a very stony loam surface horizon over a clay subsoil. A silica-cemented hardpan is present at depths of 2 to 5 decimeters. Rhyolite and dacite parent materials south and west of Paisley give rise to Argixerolls (Prairie soils) typically having stony loam A horizons and stony silty clay loam subsoil.

Scattered throughout the area are a number of ancient dry lakebeds with deep silty lacustrine deposits. A wide variety of soils is found in these areas. Principal well-drained soil groups are Camborthids (Sierozems and Desert soils), Durargids, and Durorthids (Regosols with hardpan). These soils are generally silt loam textured and, almost without exception, include a silica-cemented hardpan within 1 meter of the surface. Poorly drained soils on lake-deposited sediments are most often Haplaquolls (Humic Gley soils) and Haplaquepts (Low-Humic Gley soils).

Figure 22.—Diamond Craters, source of Pleistocene basalt flows in the Owyhee Upland Province of southeastern Oregon.

Flood plain soils from recent alluvium include Haplaquolls, Haploxerolls (Chestnut and Brown soils), and Haplorthents (Alluvial soils). These soils, having a variety of textures, are mostly poorly drained and have seasonally high water tables. Well-drained soils derived from alluvium are situated on alluvial fans and terraces and are generally classed as Durargids.

Climate

The varied climates of Oregon and Washington result from complex interplay between maritime and continental airmasses and the mountain ranges, particularly the Cascade Range that divides the States into eastern and western parts. Climatic data for representative stations in both areas are provided in table 2; this table, the isohyetal map (fig. 23), and the isoline maps for January mean minimum and July mean maximum temperatures (figs. 24 and 25) can be consulted during the following discussion.

Table 2. — Climatic data from some representative weather stations in Oregon and Washington

Area and Station	Eleva-tion	Lati-tude	Longi-tude	Temperature					Precipitation		
				Average annual	Average January	Average January minimum	Average July	Average July maximum	Average annual	June through August	Average annual snowfall
	Meters			- - - - - - - - - Degrees C. - - - - - - - - -					Millimeters		Centi-meters
West of Cascade Range:											
Quinault, Wash.[1]	72	47°28′	123°51′	10.6	3.8	1.2	17.3	23.8	3,371	244	30
Otis, Oreg.[1]	49	45°02′	123°56′	10.3	5.3	2.00	15.3	20.1	2,496	163	--
Bellingham, Wash.[2]	34	48°47′	122°29′	9.5	2.7	--1.4	16.1	23.3	853	102	25
Seattle, Wash.[2]	34	47°39′	122°18′	11.6	4.5	1.1	18.7	24.1	888	79	29
Portland, Oreg.[2]	9	45°32′	122°40′	12.6	4.6	1.4	20.3	25.8	1,076	70	31
Medford, Oreg.[2]	400	42°22′	122°52′	11.4	1.9	−1.2	22.2	31.8	502	36	19
East of Cascade Range:											
Spokane, Wash.	718	47°37′	117°31′	8.8	−3.7	--7.8	23.9	28.7	437	58	147
Yakima, Wash.	323	46°34′	120°32′	9.9	--2.5	--8.9	21.7	31.3	200	29	64
Pendleton, Oreg.	455	45°41′	118°51′	11.6	.1	−4.3	23.1	31.2	314	42	47
Klamath Falls, Oreg.	1,249	45°12′	121°47′	9.1	−1.4	−6.4	20.4	29.6	357	40	104

[1] Coastal station.
[2] Station in lee of Coast Ranges.

Source: U. S. Weather Bureau (1965a, 1965b).

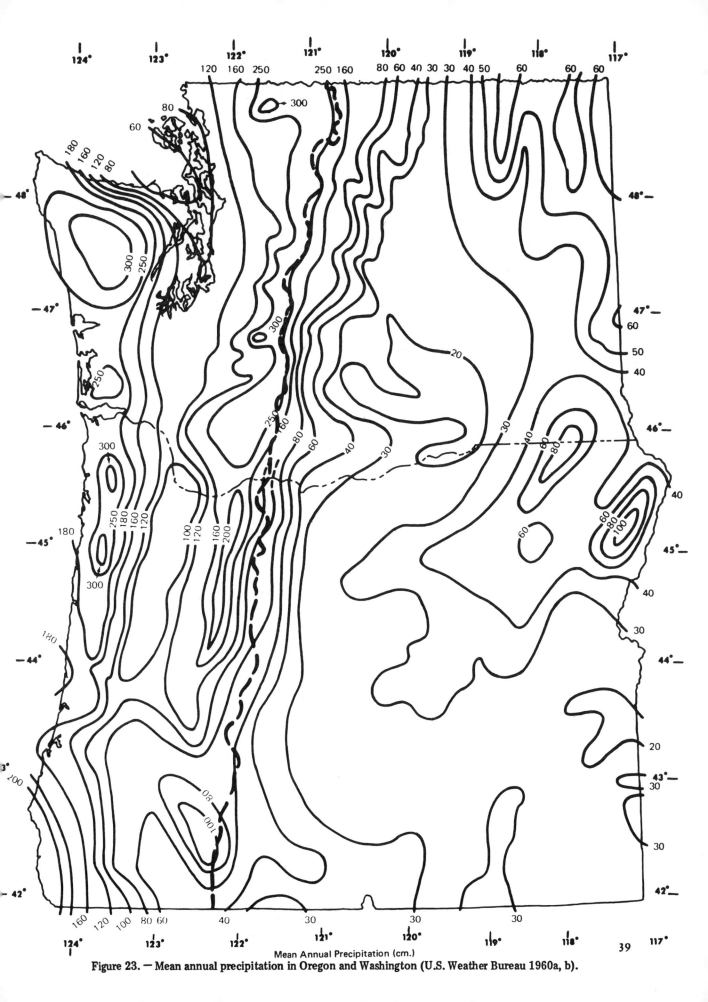

Mean Annual Precipitation (cm.)

Figure 23. — Mean annual precipitation in Oregon and Washington (U.S. Weather Bureau 1960a, b).

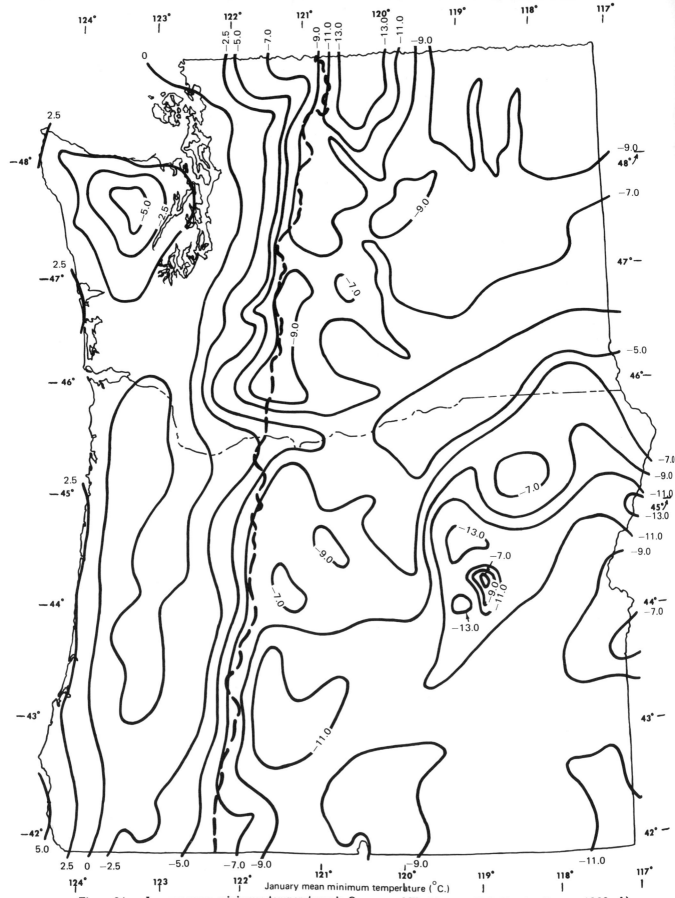

Figure 24. — January mean minimum temperatures in Oregon and Washington (U.S. Weather Bureau 1960a, b).

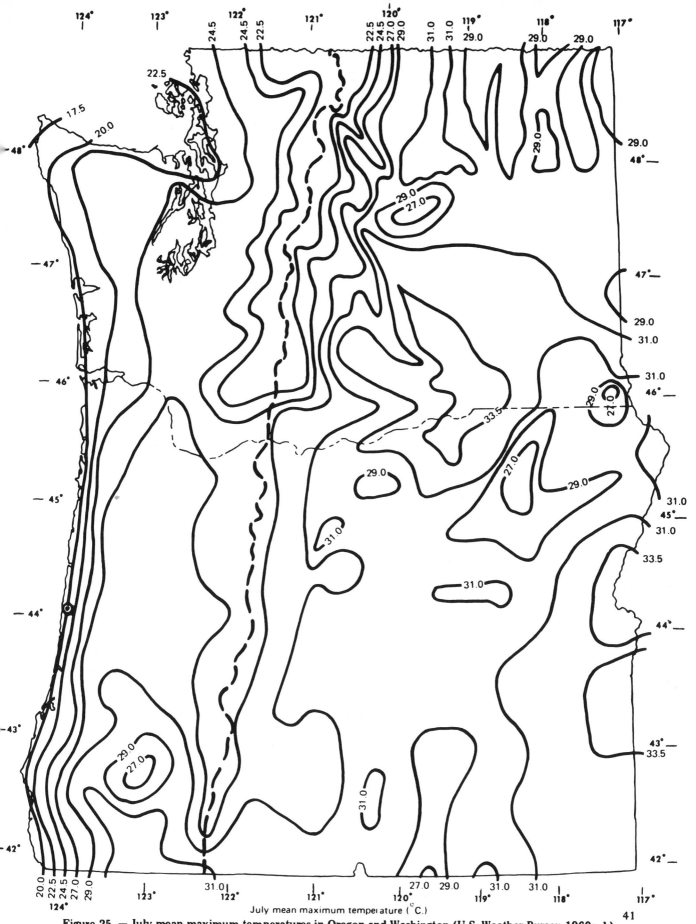

Figure 25. — July mean maximum temperatures in Oregon and Washington (U.S. Weather Bureau 1960a, b).

41

Western Oregon and Washington have a maritime climate, characterized by:

1. mild temperatures with prolonged cloudy periods, muted extremes, and narrow diurnal fluctuations (6° to 10°C.);
2. wet, mild winters, cool, relatively dry summers; and a long, frost-free season; and
3. heavy precipitation (typically 1,700 to 3,000 millimeters or more on the coast and 800 to 1,200 millimeters in the Puget-Willamette trough), 75 to 85 percent of which occurs between October 1 and March 31, mostly as rain.

Most precipitation is cyclonic, the result of low-pressure systems that approach from the Pacific Ocean on the dominant westerlies. During summers, storm tracks are shifted to the north, and high-pressure systems bring fair, dry weather for extended periods.

There are some important variations in the climate of the western lowlands as a result of the coastal mountains and of latitude. Coastal mountains are responsible for the drier and less muted climate of the Willamette valley, Puget trough, and interior valleys of southwestern Oregon. The maritime airmasses are blocked from these areas to varying degrees, and precipitation declines markedly in resultant rain shadows (table 2). At the same time, there is a general latitudinal increase in precipitation from south to north. Consequently, the interior valleys of southwestern Oregon typically have hot, dry climates (see Medford in table 2).

Eastern Oregon and Washington combine features of both maritime and continental climates. Temperatures are milder than those in the Great Plains since the Rocky Mountains buffer the full brunt of the continental airmasses. Still, temperatures fluctuate more widely than west of the Cascades, diurnal fluctuations of 10° to 16° C. being typical. Winters are colder, summers are hotter, and frost-free seasons are shorter. Precipitation is still primarily cyclonic in origin but is considerably less than to the west since the area lies in the rain shadow of the Cascade Range; annual precipitation is typically 250 to 500 millimeters. Precipitation is not quite as seasonal, only 55 to 75 percent of it occurring between October 1 and March 31, but summers (June through August) are very dry (30 to 70 millimeters). A high proportion of the annual precipitation falls as snow, which is relatively uncommon in the coastal areas.

Mountain masses have profound effects on the climatic regime. As mentioned, the Cascade Range is an extremely important barrier to the movement of maritime and continental airmasses. Within the Cascade Range, elevation has a primary effect on local climate. Precipitation and snowfall increase and temperatures decrease rapidly with increasing elevation on both western and eastern slopes of the range (fig. 26). Similar phenomena occur more locally with smaller mountain masses. For example, precipitation is very high on the western slopes of the coastal ranges (Olympic Mountains, Coast Ranges, and Siskiyou Mountains), and rain shadows occur to the east. In eastern Oregon and Washington, increases in precipitation and decreases in temperature are associated with mountain masses such as the Blue Mountains and Okanogan Highlands.

Some details of climate associated with individual vegetation types will be included in later chapters. The reader should keep in mind (1) the basically mild, summer-dry regional climate and (2) the blocking effects of mountain masses on westerly winds of maritime and northeasterly winds of continental airmasses.

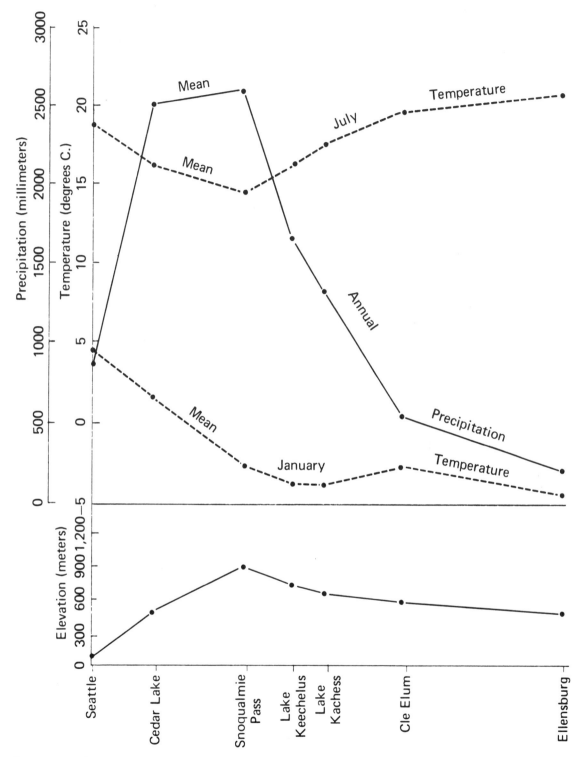

Figure 26. — Climatic cross section of the Cascade Range in the vicinity of Snoqualmie Pass, Washington (47°25′ N. lat.); distance from Seattle to Ellensburg approximately 152 km. (data from U.S. Weather Bureau 1956).

CHAPTER III. MAJOR VEGETATIONAL AREAS

Vegetation—natural plant communities—can be organized in numerous ways. Unfortunately, no single system is completely satisfactory either in providing a logical "cell" for all community types or a structure suitable for all users of a volume of this type. It is much the same problem that exists in plant taxonomy, i.e., structuring a linear system when a multidimensional classification is really necessary.

We begin our classifications by recognizing four major groupings (fig. 27): (1) forests, (2) grasslands and shrub-grass communities (hereafter referred to as steppe and shrub-steppe, respectively), (3) interior valleys of western Oregon (Chapter V)[1] and (4) timberline (subalpine parklands) and alpine regions (Chapter X).[2] The two broad physiognomic divisions of forest and steppe can be further divided geographically. Distinctive forest regions are found in western Washington and northwestern Oregon (Chapter IV), interior southwestern Oregon (Chapter VI), and in eastern Washington and Oregon (Chapter VII). Steppe and shrub-steppe are separable into those found in the Columbia Basin Province (primarily in eastern Washington) (Chapter VIII) and in central and southeastern Oregon (Chapter IX).

---

[1] Here and throughout this book, western Washington or Oregon refers to the region west of the crest of the Cascade Range and eastern Washington or Oregon to the area east of the crest.

[2] Communities found on unique, specialized habitats or in geographic anomalies form a fifth group; they are considered in Chapter XI and will not be discussed further here.

Legend

FORESTED REGIONS
    Picea sitchensis Zone

    Tsuga heterophylla Zone

        Puget Sound area

    Mixed Conifer and Mixed Evergreen Zones

    Pinus ponderosa Zone (broad sense)

        Pumice region

    Abies grandis and *Pseudotsuga menziesii* Zones

    Subalpine forests (including *Abies amabilis,*
        A. lasiocarpa, A. magnifica shastensis,
        and *Tsuga mertensiana* Zones)

INTERIOR VALLEYS OF WESTERN OREGON
    Willamette valley

    Umpqua and Rogue valleys

STEPPE REGIONS
    STEPPE (without *Artemisia tridentata*)

    SHRUB-STEPPE (with *Artemisia tridentata*)

    DESERT SHRUB

    Juniperus occidentalis Zone

TIMBERLINE AND ALPINE REGIONS

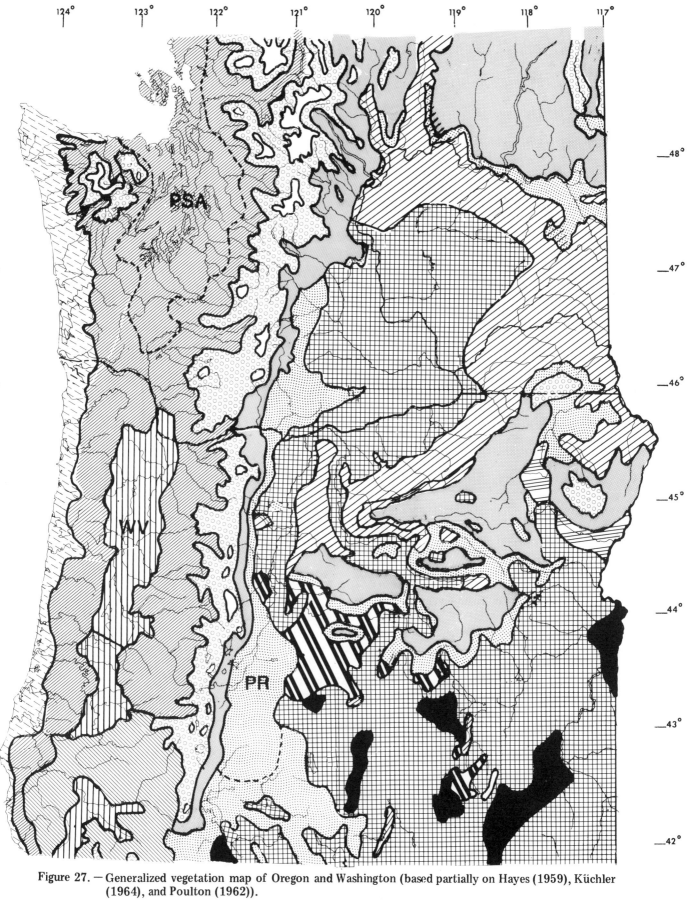

Figure 27. — Generalized vegetation map of Oregon and Washington (based partially on Hayes (1959), Küchler (1964), and Poulton (1962)).

We have used zones as basic organizational units for the communities within the subdivisions just outlined. Zones of various types have been used for many years in biogeography, particularly in mountain regions. They have been defined by many criteria, such as climate, existing vegetation, and potential (climax) vegetation (Daubenmire 1946). Perhaps a zone is most usefully defined as the area in which one plant association is the climatic climax (Daubenmire 1968a). This delineates an area of essentially uniform macroclimate since the climax community on deep, loamy soils and undulating topography[3] is primarily a product of macroclimate. Such zones tend to occur sequentially along moisture and temperature gradients which extend through broad regions or up mountain slopes.

Vegetational zones based on climax vegetation are the organizational basis for much of this paper. Our scheme is imperfect, however, for many reasons. In some areas, data are insufficient for construction of such a system and a more typological approach must be used (e.g., in southwestern Oregon). In other cases, zonal or modal habitats (loamy soils and undulating topography) are rare or absent, such as in most of the mountainous areas and the major part of eastern Oregon. Consequently, our forested zones are based not on climatic climax communities but rather on areas in which a single tree species is the major climax dominant; e.g., *Pinus ponderosa* or *Tsuga heterophylla*.[4] These areas do tend to occur as zonal sequences. Typological systems are inadequate since they depend on existing vegetation and frequently emphasize widespread seral types that span widely varying environments or economically valuable species rather than biologic features.

Someday, northwestern communities might be organized along the lines of climax series as recently carried out by Daubenmire and Daubenmire (1968) in the forests of eastern Washington and northern Idaho, but available data are inadequate. In such a system, phytogeographic data are organized by dominants in climax communities, e.g., *Tsuga heterophylla*, *Abies lasiocarpa*, and *Pinus ponderosa* without reference to zones, thereby avoiding the many difficulties posed by rugged mountain topography (Daubenmire 1968a).

Since we have elected to use zones, the reader should consider some attributes of the zonal scheme as he views the landscape:

1. Zones may occur as sequential belts on mountain slopes, but more often they interfinger, with each attaining its lower elevational limits in valleys and its highest limits on ridges; as a consequence, the zones along the slopes of a narrow valley can be reversed from their otherwise normal altitudinal relationship (Daubenmire 1946) (fig. 28).
2. Related to this phenomenon is the tendency for species or associations, occupying modal sites in one zone, to occur on moist, cool habitats in the adjacent warmer and drier zone and on warm, dry habitats in the adjacent cooler and moister zone. For example, east of the Cascade crest, *Pseudotsuga menziesii* can occur as a climax species on relatively moist habitats in the *Pinus ponderosa* Zone or on relatively dry ridges within the *Abies grandis* Zone. In the Washington steppe, the zonal *Agropyron-Festuca* association may occur as a topographic climax on steep north slopes in the drier *Artemisia/Festuca* Zone or on steep south slopes in the moister *Festuca/Symphoricarpos* Zone.

---

[3] By definition, deep loamy soils and undulating topography constitute the zonal or modal habitat. In essence, neither soils nor topography significantly modify the macroclimatic factor in development of plant communities on these sites.

[4] In two cases, we have named zones after major seral species characteristic of only that zone—*Picea sitchensis* and *Tsuga mertensiana*.

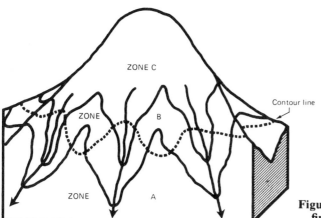

Figure 28.—Schematic diagram illustrating interfingering of zones in mountainous topography.

3. Disturbance and the resulting seral vegetation may obscure zonal sequences. Pioneer species often range through several vegetational zones. Many of the seral dominants in a given zone tend to be climax species in adjacent warmer and drier zones. Hence, *Pinus ponderosa* and *Pseudotsuga menziesii* are common pioneers on disturbed sites in the *Abies grandis* Zone. The relative abundance of understory and reproducing tree species often indicates trends in such areas.

4. Zonal schemes reflect plant responses to strong macroclimatic gradients in temperature and moisture. Unusual physical or chemical soil properties sometimes override climatic factors to severely modify zonation patterns. Serpentine areas and the pumice region east of Crater Lake, Oregon, are examples.

Before proceeding, we would like to discuss further the features which differentiate and are common to our geographic subdivisions of the forests and steppes. The interior valleys of western Oregon and timberline and alpine regions are treated in individual chapters and require little additional elaboration here. The interior valleys are potentially forested regions; for a variety of reasons, including the mosaic of forested and nonforested communities which characterizes the valleys in the past and present, we have treated them separately. Geographic diversity in timberline (subalpine parklands) and alpine regions exists but will be discussed in Chapter X.

Forests

In Oregon and Washington, forests dominate the landscapes west of the Cascade Range and the mountain slopes to the east (fig. 27). A great many tree species are endemic, but with rare exception, the dominants are conifers. In fact, the finest coniferous forests in the world occur in this region and adjacent parts of California and British Columbia. Some of the major tree species are listed in table 3 along with data on the sizes and longevity typically achieved. The absence of major hardwood dominants, a unique phenomenon in temperate zone forests of the world, is discussed at length in Chapter IV. There is extensive literature on the physiology and life history of many of these tree species, much of this autecological knowledge having been generated to provide data needed for forest management. We have cited little of this literature due to limitations of time and space; comprehensive publications which introduce and summarize many autecological data are Fowells (1965) and Krajina (1969). Readers wishing an introduction to silvicultural systems appropriate to the major forest types should see USDA Forest Service (1973).

Table 3. — Ages and dimensions typically attained by forest trees on better sites in the Pacific Northwest and their relative tolerances[1]

Species	Age	Diameter	Height	Tolerance[2]
	Years	Centimeters	Meters	
Abies amabilis	400+	90-110	45-55	VTOL
Abies concolor	300+	100-150	40-55	TOL
Abies grandis	300+	75-125	40-60	TOL
Abies lasiocarpa	250+	50-60	25-35	TOL
Abies magnifica	300+	100-125	40-50	INTER
Abies procera	400+	100-150	45-70	INTOL
Chamaecyparis lawsoniana	500+	120-180	60	TOL
Chamaecyparis nootkatensis	1,000+	100-150	30-40	TOL
Larix occidentalis	700+	140	50	INTOL
Libocedrus decurrens	500+	90-120	45	INTER
Picea engelmannii	500+	100+	45-50	TOL
Picea sitchensis	800+	180-230	70-75	TOL
Pinus contorta	250+	50	25-35	INTOL
Pinus lambertiana	400+	100-125	45-55	INTER
Pinus monticola	400+	110	60	INTER
Pinus ponderosa	600+	75-125	30-50	INTOL
Pseudotsuga menziesii	750+	150-220	70-80	INTOL
Sequoia sempervirens	1,000+	150-380	75-100	TOL
Thuja plicata	1,000+	150-300	60+	TOL
Tsuga heterophylla	400+	90-120	50-65	VTOL
Tsuga mertensiana	400+	75-100	25-35	TOL
Acer macrophyllum	300+	50	15	TOL
Alnus rubra	100	55-75	30-40	INTOL
Lithocarpus densiflorus	180	25-125	15-30	TOL
Populus trichocarpa	200+	75-90	25-35	INTOL
Quercus garryana	500	60-90	15-25	INTOL

[1] Developed from a variety of sources, the most important being Fowells (1965). Maximum ages and sizes for species are generally much greater than those indicated here.

[2] Tolerance scale: VTOL = very tolerant of shade, TOL = tolerant, INTER = intermediate shade tolerance (greater in youth, lesser at maturity), and INTOL = intolerant.

The primarily forested areas are divisible into three major subregions: (1) western Washington and the Cascade and Coast Ranges of northern Oregon (but including a coastal strip extending into California); (2) interior southwestern Oregon, i.e., the Cascade Range and portions of the Coast Ranges inland from the *Tsuga* and *Picea* forests found as a narrow coastal strip (fig. 27); and (3) eastern Washington and Oregon, including the eastern slopes of the Cascade Range.

The forests of western Washington and northwestern Oregon are the archetype of mesic temperate coniferous forests in the world. Many of the dominants are endemic to this distinctive coastal forest region[5] of *Pseudotsuga menziesii*, *Tsuga heterophylla*, and *Thuja plicata*, and many others find their center of distribution and attain maximum development here. The environment is mild and extremely favorable for forest development.

As one moves south from this region, the climate becomes increasingly warmer and drier. More and more California species are added to the flora and give the forest region of interior southwestern Oregon much of its character (e.g., *Pinus lambertiana*, *Libocedrus decurrens*, and *Lithocarpus densiflorus*). The forest zones of this region clearly represent a northern extension of the mixed-conifer and mixed sclerophyll forests characteristic of the Sierra Nevada and California Coast Ranges, respectively.

The eastern Washington and Oregon forests are primarily Rocky Mountain forest types. *Pinus ponderosa*, as much as any species, characterizes the forests at lower elevations, and *Abies lasiocarpa*, those at higher elevations. Pacific coastal elements mix with the Rocky Mountain elements in these interior forests, however, particularly on the eastern slopes of the Cascade Range and in extreme northeastern Washington.

The zones in the three subregions can be related in terms of the factor primarily responsible for a given part of the zonal sequence (table 4). The relationship between some

[5] Known as the "Pacific Coastal Forest" (W.S. Cooper, personal communication), "Northern Pacific Coast Rainy Western Hemlock Forest Biome" (Shelford 1963), and by an endless variety of other designations attempting to recognize its distinctive character.

Table 4. — Forested zones in different parts of Oregon and Washington

Zonal groups	Interior southwestern Oregon	Eastern Washington and Oregon	Western Washington
Xerophytic	Interior Valley	*Juniperus occidentalis* *Pinus ponderosa* *Pseudotsuga menziesii*	([1])
	—— Mixed Evergreen ——		
Temperate	Mixed Conifer *Abies concolor*	*Abies grandis* *Tsuga heterophylla*	*Tsuga heterophylla* *Picea sitchensis* —*Abies amabilis*—
Subalpine	*Abies magnifica shastensis* *Tsuga mertensiana*	*Abies lasiocarpa*	*Tsuga mertensiana*

[1] There are xerophytic areas in the Puget Trough.

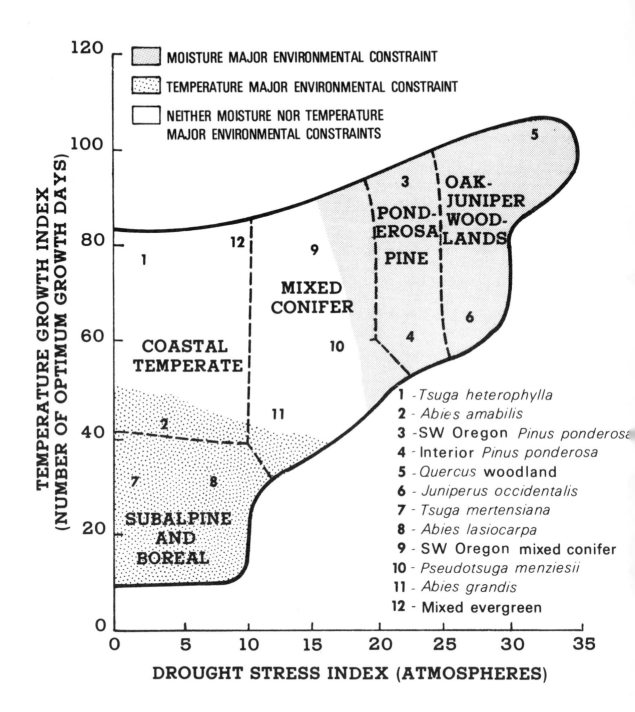

Figure 29.—Tentative distribution of some of the major forest zones within an environmental field based on moisture (maximum plant moisture stress during the dry season) and temperature (optimum growth days computed by the procedure of Cleary and Waring (1969)) *(courtesy R. H. Waring and Coniferous Forest Biome, U.S. International Biological Program).*

50

of these zones and environment can also be graphically portrayed (fig. 29). In the xerophytic zones, moisture is limiting for many species; each step up the zonal sequence indicates a more favorable moisture regime. Communities are very responsive to small differences in environment affecting plant moisture stresses (e.g., in exposure and soil depth). Within the temperate zones, neither moisture nor temperature conditions are severely limiting for the forest species. Forests grow and develop best in these zones, and species composition does not shift so markedly in response to local differences in site conditions. Temperature is the major factor separating subalpine types from temperate, with the related phenomenon of snowpack also profoundly influencing variation in community mosaics. Both factors appear limiting for many lower elevation species.

Steppes

Steppe and shrub-steppe occupy the basins in the rain shadow east of the Cascade Range (fig. 27). This is a region characterized by bunchgrasses (e.g., *Agropyron spicatum*, *Festuca idahoensis*, *Poa sandbergii*) and sagebrushes (e.g., *Artemisia tridentata*, *A. arbuscula*, and *A. rigida*). The vegetation of all or portions of this area is often referred to as desert, high desert, northern desert shrub, Great Basin desert, desert scrub, or by similar designations; as Daubenmire (1970) points out, "a combination of hot dry summers . . . rattlesnakes, horned lizards, tarantulas, and cacti seem to evoke this [a desert] classification. . . ." Daubenmire (1970) goes on to suggest that steppe is the more appropriate term based on existence of an appreciable cover of perennial grasses on zonal soils. Shrub steppe and meadow steppe are physiognomic subdivisions of steppe (perennial grassland) in which there are conspicuous (but discontinuous) layers of shrubs and a high proportion of broad-leaved forbs, respectively.

In figure 27, we have divided the steppes into three major units (plus a fourth unit of forest-steppe transition—the *Juniperus occidentalis*): (1) steppe, i.e., without *Artemisia tridentata* as a component; (2) shrub-steppe in which *Artemisia tridentata* and perennial grass codominate; and (3) desert shrub. Earlier, Franklin and Dyrness (1969) referred to "steppe (without shrubs)" but, as Daubenmire (personal communication) has pointed out, almost all the perennial grasslands in the Northwest have at least some shrub component. So defined, steppe is found primarily around the eastern rim of the Columbia basin (fig. 27). Shrub-steppe occupies the center of the Columbia basin and most of arid central and southeastern Oregon. Desert shrub is found only in isolated localities in southeastern Oregon.

Because of the information available, we have broken our treatment of the steppe and shrub-steppe into geographic units: the Columbia Basin Province (Chapter VIII) and central and southeastern Oregon (Chapter IX). General information on both regions is found in Chapter VIII.

Figure 30.—Old-growth *Pseudotsuga menziesii* **stand, with abundant** *Tsuga heterophylla* **reproduction, typical of the dense coniferous forests clothing the Cascade and Coast Ranges in western Washington and northwestern Oregon.**

CHAPTER IV. FOREST ZONES OF WESTERN WASHINGTON AND NORTHWESTERN OREGON

Western Washington and northwestern Oregon comprise the most densely forested region in the United States. These forests represent the maximal development of temperate coniferous forests in the world in terms of extent and size (fig. 30). For this reason, we would like to begin this chapter with a consideration of their outstanding features. This will be followed by an overview of their present condition and environment and detailed descriptions of the four major zones encountered in this area—the temperate *Picea sitchensis* and *Tsuga heterophylla* Zones, cool temperate *Abies amabilis* Zone, and subalpine *Tsuga mertensiana* Zone.[1]

Uniqueness of These Temperate Coniferous Forests

Western Washington and northwestern Oregon are the locale of the classic coniferous forests of the world, i.e., the popularly known region of *Pseudotsuga menziesii* dominance, of *Tsuga heterophylla-Thuja plicata* climaxes, and coastal *Picea sitchensis* "rain forests" (Weaver and Clements 1938, Shelford 1963, Heusser 1960). Few have appreciated how truly unique these forests are among the mesic temperate forests of the world, however. This includes many features including composition and productivity.

One of the most outstanding features of these forests is the nearly total dominance of coniferous species. Küchler (1946) reports the ratio of hardwoods to conifers in this region as 1:1,000 based on timber volume. In nearly all other mesic temperate zone regions, certainly in the northern hemisphere, deciduous hardwood or mixed hardwood-coniferous forests are the major natural forest formation. Coniferous species tend to be concentrated on more stressful habitats within a hardwood or mixed-forest matrix or to function primarily as pioneer or seral species which give way to a hardwood-dominated climax forest. Exceptions to such broad generalizations can be found; but the afore-mentioned pattern can be observed repeatedly in the mesic, moderate temperate zones of eastern Asia (particularly Japan), eastern United States, and western Europe. In the wet, mild climate of western Washington and Oregon, the patterns are reversed with deciduous hardwoods playing minor roles in mature forests and generally being concen-trated upon stressful habitats (e.g., *Quercus* woodlands) or functioning as pioneer species (e.g., *Alnus rubra*). Plant geographers have repeatedly commented upon the coniferous dominance of this region and offered hypothetical explanations; we will consider possible explanations shortly.

A second outstanding feature of the coniferous forests of this region is the size and longevity of the dominant species (table 3). In other parts of North America or the world, there are individual species which rival in height and diameter the northwestern American species or which exceed them in longevity. In no region is found a group of dominant trees which are the overall match of these, however. Every single coniferous genus

---

[1] The following discussion focuses on the unique features of the mesic temperate coniferous forests found in this region. Coniferous forests also dominate the slopes of interior mountains east of the Cascade Range and the southern Cascade Range and Sierra Nevada. However, as one gradually moves away from western Washington and Oregon into more continental or Mediterranean climates, the distinctive features of the mesic temperate forests under consideration are gradually lost; and coniferous forests are encountered which more clearly have analogs elsewhere in the north temperate zone.

represented finds its largest (and often longest lived) specific representative here—and sometimes its second and third largest as well: *Abies, Picea, Pseudotsuga, Pinus, Chamaecyparis, Thuja, Sequoia, Larix, Libocedrus,* and *Tsuga.*

These large, long-lived species do not occur as scattered individuals but dominate dense forests which clothe the landscape. These forests have produced the greatest biomass accumulations of any plant formations in the temperate zone and, possibly, the world. Annual productivity of young forests can be very great on favorable habitats; Fujimori (1971) reports an annual net productivity of 36.2 metric tons of biomass per hectare in a 26-year-old coastal *Tsuga heterophylla* stand. Annual productivity is substantially less on the average (perhaps 15 to 25 metric tons per hectare in fully stocked stands on better than average sites). Other mesic temperate zone forests may equal or exceed the northwestern forests in annual productivity in early years, but the coniferous forests in this region continue to grow substantially in height and accumulated biomass for decades or even centuries after others have essentially reached a state of equilibrium. As a consequence, it is not unusual to find closed mature forests in the *Tsuga heterophylla, Picea sitchensis,* and *Abies amabilis* Zones which have per-hectare values in excess of 100 square meters of basal area, 800 or 900 metric tons of live aboveground biomass, and 2,000 cubic meters of wood volume. Total nondestructive biomass analyses of superlative stands of *Sequoia sempervirens, Abies procera,* and *Pseudotsuga menziesii* conducted jointly by the Coniferous Forest Biome, U.S. International Biological Program and Japanese International Biological Program,[2] show maximum values far in excess of these. Earlier, Fujimori (1972) had made some tentative estimates of maximum biomass accumulation based on mensurational data from "record" plots reported in the literature:

Forest type	Biomass (metric tons per hectare)
Sequoia sempervirens	4,525
Pseudotsuga menziesii	2,437

After these estimates are adjusted drastically downward to take account of small plots placed within the densest parts of the stands, the values (2,300 and 1,600 metric tons for *Sequoia* and *Pseudotsuga,* respectively) still exceed any biomass figures ever reported.

What factors are responsible for this dominance of coniferous species and the high stand productivities and biomass accumulations? Küchler (1946) has probably given the compositional question greater consideration than any other author. He concludes that conifer dominance is mainly due to the history of climatic events over geologic time (evolution of the forests since the Miocene) and not primarily a consequence of the prevailing climate.

Certainly, conditions over geologic time, particularly during the Pleistocene, have been important elements in selecting a coniferous-dominated forest from the mixed Arcto-Tertiary forests of the Miocene. Evidence accumulated largely through the U.S. International Biological Program's Coniferous Forest Biome project increasingly suggests that the present climate is particularly favorable to coniferous species, however, and holds at least one key to coniferous dominance. Two climatic factors are notable: (1) the region has high total precipitation, but most occurs during the winter, and summers are relatively

---

[2] Data collected by Takao Fujimori and Charles Grier on file at the U.S. Forest Service, Forestry Sciences Laboratory, Corvallis, Oregon. Data reduction and analysis were not complete at the time this manuscript was prepared, but basal area per hectare was about 340 square meters in the *Sequoia* stand and 150 square meters in the *Abies* stand. Although *Sequoia sempervirens* forests of this size are found in coastal northern California and not in Oregon or Washington, it seems appropriate to include them in this discussion of the mesic temperate coniferous forests of which they are clearly one element.

dry (table 2); and (2) winters are very mild. Either one or both phenomena are absent or much less pronounced in other mesic temperate regions. As a consequence, coniferous species in western Washington and Oregon carry on substantial amounts of their yearly assimilation during the fall, winter, and spring. Furthermore, moisture stress, a significant factor on many forest sites during the summer, reduces the amount of assimilation possible during this season. Both the mild winters and dry summers would appear to give a definite advantage to evergreen species, i.e., to conifers over deciduous hardwoods. These are by no means the only factors apparently favoring coniferous forests but appear to be major elements in the complex explanation of coniferous dominance which is evolving.

The preceding discussion helps explain the high productivities encountered in many of the forests of western Washington and Oregon (e.g., see Fujimori 1971) but offers little enlightenment as to why trees grow to such large sizes and why stands have such large biomass accumulations. We will not explore this question here other than to point out two requirements for such a phenomenon. First, it is necessary to have species with the genetic potential for sustained height growth and longevity. The appropriate genotypes are obviously present in this area and are generally absent from other north temperate zone forests. What factors have favored these genotypes in the Pacific Northwest but not in other temperate regions? An absence of gene pool depletion in the Pacific Northwest during the Pleistocene may be one contrast with many other temperate regions (Silen 1962). Second, it is necessary to have an environment which will allow the species to attain their genetic potential, i.e., to grow dense and tall and to survive for many centuries. Climatic conditions may again prove to be a key in explanation. For example, Fujimori (1972) points out that strong winds which disturb or weaken forest communities in many other temperate regions (the typhoons of Asia and hurricanes of the eastern United States) are relatively uncommon in the Pacific Northwest.

Present Forest Conditions

At the time of the first settlers, conifer stands clothed almost the entire area of western Washington and northwestern Oregon from ocean shore to timberline except for the Willamette valley and some prairies in the Puget Sound trough. Presently, 82 percent of western Washington and Oregon is still classed as forest land (Barrett 1962), a total of 11,764,000 hectares.

As mentioned, forests grow rapidly in the region, the size, density, and longevity of virgin old-growth stands being practically unparalled (fig. 30). Dominant tree species typically reach heights of 50 to 75 meters at maturity (table 3). Some species often live well beyond 500 years, a significant feature from a successional standpoint. For example, in this area, *Pseudotsuga menziesii* is mainly a pioneer or seral species that reproduces after fire or other disturbances. Its long lifespan enables it to persist during extended periods of stability and to reseed an area following the next disturbance.

Much forest land in western Washington and Oregon is occupied today by relatively young seral stands that have followed clearing, logging, and wildfire. Such stands are typically referred to as "second growth" regardless of origin. Clearing away the obstructing forest was, of course, the first order of business for settlers. Much of this cleared acreage has since been covered by towns, cities, and farms. The lumber industry began almost simultaneously and grew rapidly in importance about the turn of the century. Activity was initially most intense in the lowlands of western Washington but slowly shifted southward and higher into the mountains as the virgin forests were cut over. Carelessness, often associated with clearing and logging, has resulted in extensive fires during the dry, warm summers and falls (fig. 31). Single burns usually reforest well from individual and/or grouped trees left by the fast-moving fires (fig. 32); areas burned repeatedly often remain treeless for many years (fig. 33).

Figure 31.—Wildfires of both natural and human origin have been responsible for the extensive seral forests of *Pseudotsuga menziesii.*

Figure 32.—Single wildfires typically leave individual trees or groups of trees which reforest the burned area quickly.

Figure 33.—Repeated wildfires can produce extensive tracts which remain deforested for decades without human intervention; Tillamook Burn, Oregon, first burned in 1933 and since has reburned several times.

Logging in the subregion is generally by clearcutting (Barrett 1962), partly for economic and partly for silvicultural reasons. *Pseudotsuga menziesii*, the preferred species, normally requires relatively open conditions for reproduction and rapid juvenile growth. In some areas, clearcuts are made as staggered patches (fig. 34), and in others, continuous clearcutting is practiced. Logging slash is typically broadcast-burned after cutting, and the site is seeded or planted, usually to *Pseudotsuga*.

Other silvicultural systems are equally or better suited to regeneration of *Pseudotsuga* forests, however, and are being used increasingly on sites where clearcutting is silviculturally unsatisfactory or esthetically unacceptable (USDA Forest Service 1973, Franklin and DeBell 1973).

Four vegetation zones can be recognized in western Washington and the Cascade and Coast Ranges of northern Oregon: the coastal *Picea sitchensis* Zone, the widespread *Tsuga heterophylla* Zone, *Abies amabilis* Zone, and the subalpine *Tsuga mertensiana* Zone. Typical tree species and their zonal occurrence in western Washington are listed in table 5.

Figure 34.—Old-growth forests in western Washington and Oregon are typically logged in "staggered-setting" clearcuts of 15 to 30 hectares (Santiam River drainage, Willamette National Forest, Oregon).

Table 5. — Representative tree species and their relative importance in both seral and climax communities in forested zones of western Washington[1]

Species	Zones			
	Picea sitchensis	*Tsuga heterophylla*	*Abies amabilis*	*Tsuga mertensiana*
Abies amabilis	m	m	M	M
Abies grandis	m	m	m	—
Abies lasiocarpa	—	—	m	M
Abies procera	—	—	M	m
Chamaecyparis nootkatensis	—	—	m	M
Picea sitchensis	M	m	—	—
Pinus monticola	—	[2] m	m	m
Pinus contorta	m	[2] m	m	m
Pseudotsuga menziesii	M	M	M	m
Tsuga heterophylla	M	M	M	m
Acer macrophyllum	m	m	—	—
Alnus rubra	M	M	—	—
Thuja plicata	M	M	m	—

[1] M = major species, m = minor species.
[2] Except major species in the Puget Trough Province.

Picea sitchensis Zone

Picea sitchensis characterizes this long narrow zone which stretches the length of Washington and Oregon's coast. It is, in fact, just part of a coastal forest zone which extends north into Alaska and south well into northern California where it grades, in part, into *Sequoia sempervirens* forests. The *Picea sitchensis* Zone is generally only a few kilometers in width, except where it extends up river valleys. On the west side of the Olympic Peninsula, where an extensive coastal plain exists, it is much broader. Although the zone is generally found below elevations of 150 meters, it goes to 600 meters when mountain masses are immediately adjacent to the ocean. This zone could be considered a variant of the *Tsuga heterophylla* Zone distinguished by *P. sitchensis*, frequent summer fogs, and proximity to the ocean. Perhaps for this reason Krajina (1965) has not recognized a similar zone in British Columbia; our *Picea sitchensis* Zone is comparable to the Coastal Subzone of the Humid Transition Life Zone recognized by D. R. M. Scott (Barrett 1962).

Environmental Features

The *Picea sitchensis* Zone has what could be considered the mildest climate of any northwestern vegetation zone (table 6). Extremes in moisture and temperature regimes are minimal; the climate is uniformly wet and mild. Precipitation averages 2,000 to 3,000 millimeters, but frequent fog and low clouds during the relatively drier summer months are probably as important in ensuring minimal moisture stresses. Fog drip adds precipitation as a consequence of condensation in tree crowns (Ruth 1954).

Some of the finest forest soils in the region are found in this zone—deep, relatively rich, and fine textured. Major great soil groups on upland forest sites are mainly Haplohumults (Brown Lateritics, Reddish Brown Lateritics, and Sols Bruns Acides). Surface soils are typically acid (e.g., pH 5.0 to 5.5), high in organic matter (e.g., 15 to 20 percent) and total nitrogen (e.g., 0.50 percent), and low in base saturation (e.g., 10 percent).

Table 6. — Climatic data from representative weather stations within the *Picea sitchensis* Zone

Station	Elevation	Latitude	Longitude	Temperature					Precipitation		
				Average annual	Average January	Average January minimum	Average July	Average July maximum	Average annual	June through August	Average annual snowfall
	Meters			- - - - - - - - - Degrees C. - - - - - - - - -					Millimeters		Centimeters
Quinault, Wash.	72	47°28'	123°51'	10.6	3.8	1.2	17.3	23.8	3,371	244	30
Astoria, Oreg.	66	46°11'	123°50'	10.6	4.7	2.1	16.0	20.6	1,967	140	--
Otis, Oreg.	49	45°02'	123°56'	10.3	5.3	2.2	15.3	20.9	2,496	163	--
Port Orford, Oreg.	96	42°44'	124°31'	11.3	7.9	4.3	14.9	19.2	1,780	81	--

Source: U. S. Weather Bureau (1956, 1965a, 1965b).

Forest Composition

The coniferous forest stands in this zone are typically dense, tall, and among the most productive in the world (Fujimori 1971) (fig. 35). Constituent tree species are *Picea sitchensis*, *Tsuga heterophylla*, *Thuja plicata*, *Pseudotsuga menziesii*, *Abies grandis*, and *A. amabilis* (in Washington). The first three are by far the most common. *Alnus rubra* is one of the most abundant trees on recently disturbed sites, and *Pinus contorta* is common along the ocean. *Sequoia sempervirens*, *Umbellularia californica*, and *Chamaecyparis lawsoniana* are found in this zone in southwestern Oregon.

Mature forests have lush understories with dense growths of shrubs, dicotyledonous herbs, ferns, and cryptogams (fig. 36). On sites modal in environmental conditions, *Polystichum munitum*, *Oxalis oregana*, *Maianthemum dilatatum*, *Montia sibirica*, *Tiarella trifoliata*, *Viola sempervirens*, *V. glabella*, *Disporum smithii*, *Vaccinium parvifolium*, and *Menziesia ferruginea* are common understory species. On less favorable sites, e.g., old sand dunes and steep slopes facing the ocean, dense understories dominated by ericads such as *Gaultheria shallon*, *Rhododendron macrophyllum*, and *Vaccinium ovatum* are

Figure 36.—*Polystichum munitum* **and** *Oxalis oregana* dominate the understory of this *Picea sitchensis* **stand; note the "prop" roots of the** *Picea* **which has developed on a rotting log (Neskowin Crest Research Natural Area, Siuslaw National Forest, Oregon).**

Figure 35.—A young, even-aged *Tsuga heterophylla* **stand typical of the dense, productive forests found in the** *Picea sitchensis* **Zone.**

common. Wetter forested sites also have dense understories where *Oplopanax horridum*, *Athyrium filix-femina*, *Blechnum spicant*, *Dryopteris austriaca*, and *Sambucus racemosa* var. *arborescens* are typical along with the "modal" site species mentioned earlier.

Cryptogams are extremely abundant and varied in the *Picea sitchensis* Zone (Sharpe 1956, Coleman et al. 1956, Harthill 1964). Some common ground cryptogams are *Eurhynchium oreganum*, *Hylocomium splendens*, *Hypnum circinale*, *Rhytidiadelphus loreus*, *Leucolepis menziesii*, and *Plagiomnium insigne*. *Isothecium stoloniferum*, *Ptilidium californicum*, *Porella navicularis*, and *Scapania bolanderi* are a small sample of the abundant epiphytes. *Alnus rubra* is an especially favorable host for epiphytic development (Pechanec and Franklin 1968, Coleman et al. 1956).

The limited community analyses available for typical *Picea-Tsuga* forests found in this zone confirm the generalized description. Hines (1971) recognizes two major community types along the northern Oregon coast: *Tsuga heterophylla-Picea sitchensis/Gaultheria*

shallon/Blechnum spicant and *Tsuga-Picea/Oplopanax horridum/Athyrium filix-femina*.[3]
The *Tsuga-Picea/Gaultheria/Blechnum* type is found in areas closest to the coast and
includes a dense shrub layer of *Gaultheria shallon* (42-percent cover), *Vaccinium ovalifolium*,
V. parvifolium, and *Menziesia ferruginea*. Major herbaceous species are *Blechnum spicant*
(48-percent cover), *Polystichum munitum*, and *Oxalis oregana*. *Thuja plicata*, *Pseudotsuga
menziesii*, and *Alnus rubra* are infrequently associated with the *Picea* and *Tsuga*. Ths
Tsuga-Picea/Oplopanax/Athyrium community is located farther inland. Many ecologists
would characterize the understory as a *Polystichum munitum-Oxalis oregana* type since
these two species have the greatest cover (49 and 50 percent, respectively). Other species
present include *Vaccinium ovalifolium*, *V. parvifolium*, *Menziesia ferruginea*, *Acer
circinatum*, *Oplopanax horridum*, and *Rubus spectabilis* in the shrub layer and *Blechnum
spicant*, *Disporum smithii*, *Athyrium filix-femina*, and *Maianthemum dilatatum* in the
herb layer. *Pseudotsuga menziesii*, particularly, but also *Thuja plicata* and *Alnus rubra*
are more frequent associates of the *Tsuga* and *Picea* in this community.

Successional Patterns

Early successional trends following fire or logging in the *Picea sitchensis* Zone are
similar to those encountered in the *Tsuga heterophylla* Zone (see next section). There is
a stronger tendency, however, toward development of dense shrub communities dominated
by *Rubus spectabilis*, *Sambucus racemosa* var. *arborescens*, and *Vaccinium* spp. because
of the favorable growing conditions. Several of these seral communities have been described
and mapped along the northern Oregon coast (Meurisse and Youngberg 1971, Meurisse
1972). Among the more common are the "Salal-Red Huckleberry" (*Gaultheria shallon-
Vaccinium parvifolium*), "Salmonberry-Sword Fern" (*Rubus spectabilis/Polystichum
munitum*), and "Vine Maple-Sword Fern" (*Acer circinatum/Polystichum munitum*) types.
The Salal-Red Huckleberry type has about 75-percent cover of the two dominants and
a very low herbaceous cover typified by *Polystichum munitum* and *Pteridium aquilinum*.
The "Salmonberry-Sword Fern" type has a high cover of *Rubus spectabilis* (average 65
percent) in association with other shrubs such as *Rubus parviflorus* and *Sambucus*, and
a relatively rich herbaceous layer characterized by *Polystichum munitum*, *Oxalis oregana*,
Maianthemum dilatatum, and *Blechnum spicant*. *Gaultheria shallon* is found only on
decaying wood in this type.

There are two major kinds of seral forest stands in the zone: (1) coniferous, containing
varying mixtures of *Picea*, *Tsuga*, and *Pseudotsuga*, and (2) the hardwood, *Alnus rubra*.
Alnus rubra reproduces abundantly and grows extremely fast on disturbed forest land
within the zone (Zavitkovski and Stevens 1972). In many cases, it overtops conifer regeneration, resulting in pure or nearly pure *Alnus* forest (fig. 37). Replacement of *A. rubra* by
other tree species is often very slow, even though it is a relatively short-lived species. This
is partially because of the dense shrubby understories of *Rubus spectabilis* and other
species typically associated with *Alnus* stands (Meurisse and Youngberg 1971). Successional
sequences have not been thoroughly studied, although it appears that *Alnus rubra* can
variously be replaced by semipermanent brushfields (Newton et al. 1968), by *Picea sitchensis*
released from a suppressed state (Franklin and Pechanec 1968), or by *Thuja plicata* or

---

[3] We use a slash between taxa in community names to indicate major layers (overstory tree, tall shrub,
and low shrub and herb) within plant groupings. A hyphen between taxa indicates they are in the same
layer.

Figure 37.—The prolific and fast-growing *Alnus rubra* often pioneers on logged or burned lands in the *Picea sitchensis* **Zone, offering severe competition for conifers; this typical 50-year-old** *Alnus rubra* **stand has an understory dominated by** *Rubus spectabilis* **(Siuslaw National Forest, Oregon).**

Tsuga heterophylla, the latter often invading via down logs. Fonda[4] has described a sere for river terraces on the western Olympic Peninsula in which *Alnus rubra* is replaced initially by *Picea sitchensis*, *Populus trichocarpa*, and *Acer macrophyllum*.

Henderson (1970) describes in more detail an age sequence in community composition of river-bottom *Alnus rubra* stands in the Coast Ranges of Oregon. Basically, succession in the understory proceeds from a grass-herb to shrub-fern understory over a 60- to 70-year period. Early understory dominants are *Holcus lanatus*, *Stellaria media*, *Montia sibirica*, *Digitalis purpurea*, *Rubus ursinus*, and *Stachys mexicana*. Species increasing with stand age are *Athyrium filix-femina*, *Polystichum munitum*, and *Rubus spectabilis*. Henderson (1970) tabulates the results from six studies which indicates the *Alnus rubra/ Rubus spectabilis* community is consistently associated with *Polystichum*, *Sambucus racemosa*, *Rubus ursinus*, *Oxalis oregana*, and *Galium triflorum*. In addition to these species which are present in all the descriptions, *Athyrium filix-femina*, *Stellaria media*, and *Montia sibirica* are usually present as well.

Alnus rubra is noteworthy for its soil-improving properties; this species fixes significant amounts of nitrogen in this region (Tarrant 1964, Tarrant and Trappe 1971) and can have other effects on nutrient cycling, soil chemistry, and microbiology as well (Franklin et al. 1968, Lu et al. 1968, Zavitkovski and Newton 1971).

Succession in most mature conifer forest types in this zone is toward replacement of mixed *Picea sitchensis*, *Thuja plicata*, *Tsuga heterophylla*, and *Pseudotsuga menziesii* forests by *Tsuga heterophylla* (see Neskowin Crest and Quinault Research Natural Areas in Franklin et al. (1972), for example). This species is apparently more tolerant than *Picea sitchensis* and dominates the reproduction in old-growth forests. Krajina (1969) has apparently reached a similar conclusion in coastal British Columbia. Hines (1971) found that *Picea sitchensis* was not reproducing at all in his *Tsuga-Picea/Gaultheria/Blechnum* community type and only about one-third of his *Tsuga-Picea/Oplopanax/Athyrium* stands had *Picea* seedlings, although the *Tsuga* was reproducing well. He concluded that the *Picea* is perpetuated, if at all, by natural openings created by windthrow or overstory mortality. Since *P. sitchensis*, *Thuja plicata*, and *Pseudotsuga menziesii* are all long-lived species, even very old stands usually retain at least some of the original representation of these species. On moist to wet sites, it appears *T. plicata* and, in some cases, *P. sitchensis* will be at least a part of the climax along with *T. heterophylla*.

Much of the forest regeneration in conifer stands takes place on rotting logs, "nurse logs," which often support hundreds of *Tsuga*, *Picea*, and *Thuja* seedlings (fig. 38) (Sharpe 1956, Hines 1971). Some of these survive, and their roots eventually reach mineral soil. The consequences are often readily visible in forests as lines of mature trees growing along the remains of the original nurse logs.

---

[4] R. W. Fonda. Forest vegetation in relation to river terrace development in the Hoh Valley, Olympic National Park, Washington. Unpublished manuscript available from author in Biology Department, Western Washington State College, Bellingham, Wash. 1972.

Figure 38.—*Tsuga, Picea,* **and** *Thuja* **seedlings developing on a rotting nurse log, a typical phenomenon in forests of the** *Picea sitchensis* **Zone (Quinault Research Natural Area, Olympic National Forest, Washington); the range pole in this and other photos, unless otherwise noted, is 1 m. in height and marked in dm. segments.**

Special Types

Including, as it does, the ocean strand, headland, and coastal plain environments, the *Picea sitchensis* Zone is the locale of a rich variety of specialized habitats. We will deal with the sand dune and strand communities in a later chapter on the vegetation of unique habitats, however, and confine this discussion of special types to considerations of the "Olympic rain forest," *Sequoia sempervirens* forest, forested swamps, and a coastal plain prairie.

It is important to note, however, that herb- or shrub-dominated communities which are largely confined to headland areas north of Coos Bay become increasingly common as ocean-front vegetation types south of Port Orford into northern California. Conversely, ocean-front forests become less common and discontinuous and tend to be found on headland areas south of Port Orford. This tendency for a reversal of vegetation types is increasingly evident and eventually results in a complete dominance of nonforested ocean-front communities south of about 40° north latitude.

Olympic Rain Forest

One element of the *Picea sitchensis* Zone which has been singled out for scientific and public attention is the "Olympic rain forest." This attention necessitates we make some mention of its characteristics and place within the zone. As generally recognized, the Olympic rain forests are old-growth forests, dominated by *Picea sitchensis* and *Tsuga heterophylla,* found in three or four major river valleys (Hoh, Quinault, Queets, and possibly Bogachiel Rivers) in the very heavy rainfall area on the western slopes of the Olympic Mountains (Kirk 1966). Among the distinctive characteristics of these forests found on older land surfaces (usually river terraces) are: (1) an abundance of *Acer macrophyllum* and *A. circinatum*; (2) conspicuous coverage of epiphytic plants, mostly cryptogams, but with one of the most abundant being *Selaginella oregana*, a club moss; (3) abundant nurse logs; and (4) heavy seasonal visitation by Roosevelt elk (*Cervis canadensis* var. *roosevelti*).

These *Picea-Tsuga* forests contrast with those normally found within the *Picea sitchensis* Zone (fig. 39). Trees are of massive size (dominant *Picea* are frequently 230- to 330-cm. or 90- to 130-inch d.b.h. and over 60 m. or 200 ft. tall), but the stands have relatively open canopies and low densities. Scattered through the forests are groves of large (75- to 100-cm. or 30- to 40-inch d.b.h.), epiphyte-draped *Acer macrophyllum*. The shrub layer is also relatively open except for large clumps of *Acer circinatum*. Other characteristic shrubs (all with low cover) are *Vaccinium ovalifolium*, *V. parvifolium*, *Rubus ursinus*, and *R. spectabilis*. *R. spectabilis* is relatively sparse compared with other coastal forest stands of comparable overstory density, however—a possible consequence of heavy grazing by elk (fig. 40). The major herbaceous species are *Oxalis oregana*, *Polystichum munitum*, *Tiarella unifoliata*, *Carex deweyana*, *Trisetum cernuum*, *Maianthemum dilatatum*, *Rubus pedatus*, *Montia sibirica*, *Athyrium filix-femina*, and *Gymnocarpium dryopteris*; *Polystichum* and *Oxalis* are clearly the most important. A heavy moss layer is typical including *Eurhynchium oreganum*, *Hypnum circinale*, *Rhytidiadelphus loreus*, *Leucolepis*

Figure 39.—Typical "Olympic rain forest" community in the Hoh River valley; note the large *Acer macrophyllum* **(left) and** *Picea sitchensis* **(upper right), open nature of the overstory, and heavy epiphyte cover on the** *Acer,* **all of which are characteristic of this** *Picea sitchensis* **Zone community (Twin Creek Research Natural Area, Olympic National Park, Washington).**

Figure 40.—Contrasting vegetation inside (left) and outside (right) a long-term elk exclosure, within an "Olympic rain forest"; within the exclosure is a dense tangle of shrubs (particularly *Rubus spectabilis*) and tree reproduction, although both are almost absent outside the exclosure (Bogachiel River valley, Olympic National Park, Washington).

menziesii, Plagiomnium insigne, and *Hylocomium splendens* as more common species. The heavy epiphyte coverage includes the cryptogams *Isothecium stoloniferum, Porella navicularis, Rhytidiadelphus loreus, Radula bolanderi, Frullania nisquallensis, Scapania bolanderi,* and *Ptilidium californicum,* and the vascular plants *Polypodium vulgare* and *Selaginella oregana.*

The forests appear to be in a near-climax condition, and although we consider *Picea sitchensis* to be a subclimax species elsewhere in the zone, this does not appear to be the case in these stands. *Picea* seedlings and saplings are often encountered, and their occurrence may be related to the open canopy and selective grazing of *Tsuga* seedlings by elk. In any case, almost all tree reproduction is on rotting nurse logs; lines of *Picea* reproduction can often be seen invading small stand openings along such down logs.

There are two general viewpoints of this so-called rain forest: (a) that it is a unique and distinctive type found only in selected valleys on the Olympic Peninsula or (b) that it is simply a variant of the normal coastal forests with maximal development of certain features and not appropriately referred to as "rain forest," considering the tropical connotations of this term. Kirk (1966) has probably most clearly stated the case for the uniqueness of the Olympic rain forest as well as the appropriateness of the term "rain forest," although her view is not unique (e.g., see Sharpe 1956). Arguments are based on the superabundant rainfall, massiveness of the trees, numerous canopy levels, evergreen habit of the dominants, profusion of epiphytes, and an intangible "overall quality of growth."

Fonda (see footnote 4) takes a very different view. First, based upon a study of long-term forest succession on river terraces, he concludes that the *Picea-Tsuga* forests which are the focus of attention are, in fact, a long-lived late successional stage typical of second-terrace levels. *Tsuga* forests would be the climax type in a sere which begins with *Alnus rubra* communities (flood plain) and progresses successively through *Picea sitchensis-Populus trichocarpa-Acer macrophyllum* (first terrace) and *Picea-Tsuga* (the classical "rain forest" on second terraces) stages, to a climax forest of *Tsuga heterophylla.* The *Acer macrophyllum* groves typically found in the *Picea-Tsuga* or second-terrace stage he

66

relates to local areas of shallow stony soil. A progression in soil development from unmodified colluvial materials to well-developed profiles is also associated with the vegetational sequence. He finds *Pseudotsuga menziesii* and *Thuja plicata* generally confined to the steep slopes or risers between the terraces and not generally found on the terraces themselves, although fire has created certain exceptions, at least in the case of *Pseudotsuga* (see Jackson Creek Research Natural Area in Franklin et al. 1972). Fonda enumerates the many differences between tropical rain forest and the Olympic forests and concludes that the term "rain forest" is inappropriate.

We agree with Fonda that the "Olympic rain forest" is, in fact, simply a variant of the *Picea sitchensis* Zone with its primary affinities to other northwestern coastal forests. The climate within these valleys and the presence of large elk herds are undoubtedly key contributors to its peculiar features, although time or successional stage may be another as Fonda suggests. *Tsuga heterophylla* would seem the likely climax species over a very long timespan; however, as long as the present open nature of the stands persists, and there is little evidence of change, *Picea sitchensis* will continue to perpetuate itself and can be considered a climax species.

Redwood Forests

Forests with significant amounts of *Sequoia sempervirens* extend 15 to 20 kilometers into coastal southwestern Oregon (fig. 41). As will be seen, it is difficult to relate these to any one vegetation zone, but it seems as appropriate to consider them in the discussion of the *Picea sitchensis* Zone as any other. Actually *S. sempervirens* forests in Oregon are typically found on slopes rather than in river bottoms and well inland from the narrow strip of *Picea sitchensis* Zone rather than within it. The latter situation is reported at least as far south as Del Norte County in northern California.[5] Such a pattern of distribution

---

[5] Personal communication, Dale Thornburgh, Humboldt State University, Arcata, California.

Figure 41.—Old-growth *Sequoia sempervirens* forest about 10 kilometers from the Pacific Ocean in southwestern Oregon; major associates are *Lithocarpus densiflorus, Pseudotsuga menziesii, Rhododendron macrophyllum,* and *Vaccinium ovatum* (Wheeler Creek Research Natural Area, Siskiyou National Forest, Oregon).

67

suggests that the *Sequoia* and *Picea* forests are not strict analogs as many have supposed.

Evidence in southwestern Oregon suggests that distribution of *Sequoia sempervirens* actually overlaps three vegetation zones—extending from the *Picea sitchensis* Zone, across the narrow strip of *Tsuga heterophylla* Zone present in this area, into the Mixed Evergreen (*Pseudotsuga*-sclerophyll) Zone (Dyrness et al. 1973). Its occurrence in the Mixed Evergreen Zone may be questionable, however, as Thornburgh (personal communication) feels that mixed evergreen stands where *Sequoia* occurs are probably climax *Tsuga* sites from which fire has excluded the *Tsuga*. The major community associates in southwestern Oregon are *Pseudotsuga menziesii*, *Lithocarpus densiflorus*, *Rhododendron macrophyllum*, and *Vaccinium ovatum* on dry sites, and *Pseudotsuga*, *Umbellularia californica*, *Acer macrophyllum*, *Vaccinium ovatum*, *Polystichum munitum*, and *Oxalis oregana* on moist sites.

The successional status of *Sequoia sempervirens* is presently uncertain. It appears to us its role and the climax associates vary from habitat to habitat. In one area (Dyrness et al. 1973), reproduction of *S. sempervirens* and *Tsuga heterophylla* was sporadic and only that of *Lithocarpus densiflorus* was abundant. Almost all the large *S. sempervirens* bore massive fire scars suggesting it may be a seral species dependent upon fire for reproduction. In California, some scientists feel *S. sempervirens* forests are seral and strongly dependent upon periodic fires and flooding for their perpetuation (Stone et al. 1972). *Tsuga heterophylla* and *Lithocarpus densiflorus* are typically identified as climax species. Nevertheless, Thornburgh (see footnote 5) proposes that *S. sempervirens* will remain a dominant in climax forests because its long-lived habit requires only very low replacement rates—rates which can be met by establishment of new trees or sprouts in small canopy openings. Waring[6] concurs in a climax designation for *S. sempervirens* based on its relatively high shade tolerance; at the same time, he acknowledges the significance of flooding and fire in reproduction of *S. sempervirens* on alluvial flats and on mountain slopes, respectively.

Forested Swamps

Cedar and alder swamps are another specialized series of communities which are probably more common in the *Picea sitchensis* Zone than in any other. Swamps of this type are found in the *Tsuga heterophylla* Zone as well (figs. 42 and 43), and even, occasionally, in the *Abies amabilis* Zone. However, they are most common on the coastal plains and portions of glacial drift adjacent to Puget Sound. The constant habitat characteristic is a high water table, or even standing surface water, for all or a portion of the year.

The chief tree species on these sites are *Thuja plicata* or *Alnus rubra* or both. In fact, it is in some of these swamp communities that *A. rubra* appears to be a climax species. *Picea sitchensis*, *Pinus monticola*, *P. contorta*, and *Tsuga heterophylla* may also be present. Although the understory is often dominated by only one or two species, such as *Lysichitum americanum* or *Carex obnupta*, a great variety of shrubby or herbaceous species may be present. Some of the more characteristic are *Blechnum spicant*, *Athyrium filix-femina*, *Oenanthe sarmentosa*, *Stachys mexicana*, *Mitella* spp., *Tolmiea menziesii*, *Spiraea douglasii*, *Salix hookeriana*, and *Rubus spectabilis*.

Among the best developed coastal swamps are those on the coastal plain of the western Olympic Peninsula. All of the above-mentioned tree species may be present as well as *Abies amabilis*; *Thuja plicata* is most conspicuous because of its large diameter and "candelabra" tops, although *Pinus monticola* is typically the tallest tree when present.

---

[6] Personal communication, R. H. Waring, School of Forestry, Oregon State University, Corvallis, December 18, 1972.

Figure 42.—Swamp community of *Alnus rubra* and *Carex obnupta* in winter aspect; communities of this type in which *Alnus, Thuja plicata, Lysichitum americanum,* and *Carex* are dominants are scattered throughout the *Picea sitchensis* and *Tsuga heterophylla* Zones (Cedar Flats Research Natural Area, Gifford Pinchot National Forest, Washington).

Figure 43.—Cedar swamp created by beaver activity; despite its considerable tolerance of high water tables, the *Thuja plicata* in areas of deeper water have been killed (Wind River Research Natural Area, Gifford Pinchot National Forest, Washington).

A survey of water table depths and forest composition in this area shows that *Thuja* and *Alnus rubra* are the most common tree species present, are little affected by water table depth, and grow reasonably well even when stagnant water occurs at the soil surface during the winter (Minore and Smith 1971). *Picea sitchensis* and *Tsuga heterophylla* are also common but are somewhat less tolerant of high water tables. The understory in many of these swamps can only be described as rank with very dense shrub layers dominated by *Gaultheria shallon* (2 m. or more in height), *Menziesia ferruginea, Vaccinium alaskaense, V. ovatum, V. parvifolium,* and *Rubus spectabilis.* Typical herbs are *Blechnum spicant, Maianthemum dilatatum, Lysichitum americanum,* and *Cornus canadensis.*

Prairies

These treeless areas in an otherwise heavily forested locale are scattered through the *Picea sitchensis* Zone. We exclude from consideration here the grassy or shrubby headland or ocean-front communities (such as the one described by Davidson 1967); they will be dealt with in a later chapter.

Only a single prairie has been a subject of study to date—the Quillayute Prairie on the coastal plain of the western Olympic Peninsula (Lotspeich et al. 1961). This prairie is dominated by *Pteridium aquilinum* var. *pubescens*; other common species are *Achillea millefolium, Anthoxanthum odoratum, Eriophyllum lanatum, Fragaria vesca* var. *bracteata, Holcus lanatus, Hypericum perforatum, Prunella vulgaris* var. *lanceolata,* and *Spiraea douglasii* var. *menziesii.* Encroachment of the forest on this prairie was believed to be extremely slow prior to human disturbance. Rapid invasion of the prairie by *Picea sitchensis* has followed plowing, burning, or grazing of the original vegetation.

Tsuga heterophylla Zone

The *Tsuga heterophylla* Zone is the most extensive vegetation zone in western Washington and Oregon and the most important in terms of timber production. This is the region famous for subclimax *Pseudotsuga menziesii* and climax *Tsuga heterophylla-Thuja plicata* formations (Weaver and Clements 1938, Oosting 1956, Cooper 1957). The *Tsuga heterophylla* Zone extends south from British Columbia through the Olympic Peninsula, Coast Ranges, Puget Trough, and both Cascade physiographic provinces in western Washington (fig. 44).

Figure 44.—*Pseudotsuga menziesii* **is often the sole dominant tree in forests of the** *Tsuga heterophylla* **Zone, even in old-growth stands, although it is almost invariably seral (along Cave Creek, Gifford Pinchot National Forest, Washington).**

70

In Oregon, it is split into two major segments—located in the Coast Ranges and in the Western and High Cascades Provinces—by the Willamette and other dry interior valleys. The southern limits are the Klamath Mountains on the coast (except for a narrow coastal strip) and the divide between the North and South Umpqua Rivers in the Cascade Range (about 43°15′N. lat.). Elevational range in the Cascade Range varies from essentially sea level to 600 or 700 meters at 49° north latitude and from 150 to 1,000 meters at 45° north latitude. In the Olympic Mountains, the *Tsuga heterophylla* Zone occurs between 150 and 550 meters on the western slopes and from nearly sea level to 1,125 meters on the drier eastern slopes (Fonda 1967). Where conditions are favorable, elements of this zone appear on the east side of the Cascade Range (discussed in a later chapter) and in the northern Rocky Mountains (Daubenmire 1952, Daubenmire and Daubenmire 1968).

The *Tsuga heterophylla* Zone as recognized here is analogous to Krajina's (1965) Coastal Western Hemlock Zone, except for our separation of the coastal *Picea sitchensis* Zone. In western Washington, it is essentially in agreement with Scott's definition of Merriam's Humid Transition Zone (Barrett 1962); however, in southwestern Oregon, the *Tsuga heterophylla* Zone does not include much of the area considered Humid Transition Zone.

It is emphasized that although this is called the *Tsuga heterophylla* Zone, based on the potential climax species, large areas are dominated by forests of *Pseudotsuga menziesii*. Much of the zone has been logged or burned, or both, during the last 150 years, and *Pseudotsuga* is usually a dominant (often a sole dominant) in the seral stands which have developed (Munger 1930, 1940). Even old-growth stands (typically 400 to 600 years old) frequently retain a major component of *Pseudotsuga* (fig. 44).

Environmental Features

The *Tsuga heterophylla* Zone has a wet, mild, maritime climate (table 7). Since the zone lies farther from the ocean, moisture and temperature extremes are greater than in the *Picea sitchensis* Zone. Also, there is a great deal of climatic variation in this widespread zone associated with latitude, elevation, and location in relation to mountain massifs. Precipitation averages 1,500 to 3,000 millimeters and occurs mainly during the winter; Orloci (1965) has suggested 1,650 centimeters precipitation is the lower limit of the Coastal Western Hemlock Zone in British Columbia. Summers are relatively dry with only 6 to 9 percent of the total precipitation. Moisture stresses are sufficient to result in distinctive community spectra along moisture gradients (McMinn 1960). Mean annual temperatures average 8° to 9° C., and neither January nor July temperatures are extreme (table 7).

Despite the fact soils in the *Tsuga heterophylla* Zone are derived from a wide variety of parent rocks, they tend to have some general features in common. Soil profiles are generally at least moderately deep and of medium acidity. Surface horizons are well aggregated and porous. Organic matter content ranges from moderate in the Cascades to high in portions of the Coast Ranges and Olympic Peninsula, where thick, very dark A1 horizons are especially common. Forest floor depths are generally less than 7 centimeters, except at higher elevations where they may reach 15 centimeters in thickness. Depending on degree of profile development, amounts of clay accumulation in the B horizon vary from low to medium. Most soils in the zone are of medium texture, ranging from sandy loam to clay loam. In some areas, well-developed soils are limited to moderate slope positions, but on the steeper slopes, poorly developed, often shallow, soils are encountered most often. Great soil groups, characteristic of the *Tsuga heterophylla* Zone, include Dystrochrepts (Sols Bruns Acides), Haplumbrepts (Western Brown Forest soils), and Haplohumults (Reddish Brown Lateritic soils) in the Coast Ranges and Olympic Peninsula Provinces, and Haplorthods (Brown Podzolic soils), and Xerumbrepts and Vitrandepts (Regosols) in the Cascade Range.

Table 7. — Climatic data from representative stations within the *Tsuga heterophylla* Zone

Station	Eleva-tion	Lati-tude	Longi-tude	Temperature					Precipitation		
				Average annual	Average January	Average January minimum	Average July	Average July maximum	Average annual	June through August	Average annual snowfall
	Meters			- - - - - - - - - Degrees C. - - - - - - - - -					Millimeters		Centi-meters
Darrington, Wash.	168	48°15′	121°36′	9.7	1.0	−3.3	17.4	25.7	2,045	154	120
Greenwater, Wash.	521	47°09′	121°39′	7.4	−1.2	−3.7	15.8	22.6	1,487	138	198
Castle Rock, Wash.	36	46°17′	122°54′	10.4	2.9	−.2	17.6	26.7	1,453	109	27
Wind River, Wash.	351	45°48′	121°56′	8.8	0	−3.7	17.5	26.9	2,528	119	233
Detroit, Oreg.	485	44°44′	122°09′	9.3	.9	−3.2	17.9	27.5	1,929	110	156
McKenzie Bridge, Oreg.	419	44°10′	122°10′	10.1	1.6	−2.6	18.9	29.4	1,789	106	--
Valsetz, Oreg.	346	44°50′	123°40′	9.6	2.4	−.7	16.6	25.6	3,207	144	38

Source: U. S. Weather Bureau (1956, 1965a, 1965b).

Forest Composition

Major forest tree species in this zone are *Pseudotsuga menziesii*, *Tsuga heterophylla*, and *Thuja plicata*. *Abies grandis*, *Picea sitchensis* (near the coast), and *Pinus monticola* occur sporadically. Both *Pinus monticola* and *Pinus contorta* are common on glacial drift in the Puget Sound area. In Oregon, especially near the southern limits of the zone, *Libocedrus decurrens*, *Pinus lambertiana*, or even *Pinus ponderosa* may be encountered. *Abies amabilis* is common near the upper altitudinal limits or even well within the *Tsuga heterophylla* Zone in the northern Cascade Range and Olympic Mountains. *Chamaecyparis lawsoniana* is a major element of the forests in a portion of the southern Oregon Coast Ranges. *Taxus brevifolia* is found throughout the zone, but always as a subordinant tree.

Hardwoods are not common in forests of the *Tsuga heterophylla* Zone and, except on recently disturbed sites or specialized habitats (e.g., riparian sites), are almost always subordinant. *Alnus rubra*, *Acer macrophyllum*, and *Castanopsis chrysophylla* are the most widespread. *Populus trichocarpa* and *Fraxinus latifolia* with *Acer macrophyllum* and *Alnus rubra* are found along major water courses. *Arbutus menziesii* and *Quercus garryana* may be found on drier, lower elevation sites anywhere in the zone, but are not characteristic. *Umbellularia californica* and *Lithocarpus densiflorus* are found in the southern Oregon Coast Ranges.

The forest communities of the *Tsuga heterophylla* Zone have been studied in detail at many locations. Excluding strictly successional studies, these include (1) community classifications of seral *Pseudotsuga menziesii* stands (Spilsbury and Smith 1947; Becking 1954, 1956), (2) community descriptions for limited areas (Dirks-Edmunds 1947, Macnab 1958, Neiland 1958, Merkle 1951, Anderson 1967, Roemer 1972, Hawk 1973), and (3) investigations of the entire spectrum of forest communities (Bailey 1966, Orloci 1965, Corliss and Dyrness 1965, Rothacher et al. 1967, Fonda 1967, Fonda and Bliss 1969, Meurisse and Youngberg 1971, and Hines 1971). Bailey and Poulton (1968), Mueller-Dombois (1965), and Bailey and Hines (1971) concentrated upon seral communities, McMinn (1960) upon community-moisture relationships, and Eis (1962) upon community correlations with environment and productivity. Cryptogamic components of forests in the *Tsuga heterophylla* Zone have been reported by Pechanec (1961), Higinbotham and Higinbotham (1954), Spilsbury and Smith (1947), Orloci (1965), and Becking (1954).

The earliest comprehensive studies of vegetation within the *Tsuga heterophylla* Zone stressed the usefulness of understory species as indicators of *Pseudotsuga menziesii* site quality. For example, Spilsbury and Smith (1947) recognized five site types in western British Columbia, Washington, and Oregon and related them to growth rate of *Pseudotsuga*. These site types, from best to poorest sites, are: (1) "*Polystichum munitum*," (2) "*Polystichum-Gaultheria shallon*," (3) "*Gaultheria*," (4) "*Gaultheria-Parmelia*" (a "pale green" lichen), and (5) "*Gaultheria-Usnea* (a "bearded" lichen). Later, Becking (1954), recognizing a similar basic dichotomy, classified *Pseudotsuga menziesii* stands in western Washington and Oregon into two main groups—"*Polystichum-Pseudotsuga* forest type group" and "*Gaultheria-Pseudotsuga* forest type group." *Pseudotsuga* site index (height in feet at 100 years) averaged approximately 165 for *Polystichum* stands and 115 for *Gaultheria* stands.

Subsequent studies have shown a similar spectrum of communities arranged along moisture gradients. On dry sites, understories are characterized by *Holodiscus discolor* or *Gaultheria shallon*, or both. At the opposite end of the gradient, very moist sites are typified by *Polystichum munitum* and *Oxalis oregana*, with the wettest forested sites indicated by *Lysichitum americanum*. Intermediate mesic sites are typified by *Rhododendron macrophyllum* and *Berberis nervosa* in some areas or by codominance of *Polystichum* and *Gaultheria*. Although the details of community composition and nomenclature vary with the investigation and the locale, this same basic pattern—*Gaultheria* at the dry end of the scale to *Polystichum* on moist sites—is repeated throughout the *Tsuga heterophylla* Zone.

Six old-growth associations are representative of this moisture spectrum on the western slopes of Oregon's Cascade Range at about 45° N. latitude (table 8).[7] This classification is based on computer-assisted computation of similarity index values and a two-dimensional ordination technique (Franklin et al. 1970). These associations, listed in order from dry to wet, are: *Pseudotsuga menziesii/Holodiscus discolor, Tsuga heterophylla/ Castanopsis chrysophylla, Tsuga heterophylla/Rhododendron macrophyllum/Gaultheria shallon, Tsuga heterophylla/Rhododendron macrophyllum/Berberis nervosa, Tsuga heterophylla/Polystichum munitum,* and *Tsuga heterophylla/Polystichum munitum-Oxalis oregana.* Although coverage and constancy data contained in table 8 are from old-growth associations, the same types of understories are found in much younger forests, including seral stands dominated completely by *Pseudotsuga menziesii* (e.g., see Spilsbury and Smith 1947).

The *Pseudotsuga menziesii/Holodiscus discolor* association typifies the driest forested sites. The overstory is relatively open (fig. 45) and consists primarily of *Pseudotsuga menziesii,* although species such as *Libocedrus decurrens, Arbutus menziesii,* and *Acer macrophyllum* are often present. All age classes of *Pseudotsuga* from seedlings to veterans

---

[7] Source for this discussion is an unpublished manuscript, "A preliminary classification of forest communities in the central portion of the western Cascades in Oregon," by C. T. Dyrness, J. F. Franklin, and W. H. Moir, 70 p., 1972, on file at Forestry Sciences Laboratory, Pacific Northwest Forest and Range Experiment Station, Corvallis, Oregon.

Table 8. – Cover and constancy of important species in six associations found in the *Tsuga heterophylla* Zone of the western Oregon Cascade Range[1]

Species and stratum	Pseudotsuga/ Holodiscus		Tsuga/ Castanopsis		Tsuga/ Rhododendron/ Gaultheria		Tsuga/ Rhododendron/ Berberis		Tsuga/ Polystichum		Tsuga/ Polystichum-Oxalis	
	Cover	Constancy	Cover	Constancy	Cover	Constancy	Cover	Constancy	Cover	Constancy	Cover	Constancy
	-- *Percent* --											
Overstory trees:												
Tsuga heterophylla	–	–	7	56	20	76	43	100	44	100	29	100
Pseudotsuga menziesii	41	100	36	100	45	100	45	100	42	100	38	100
Thuja plicata	–	–	tr	12	3	47	13	72	16	80	13	75
Libocedrus decurrens	6	50	tr	19	1	12	–	–	–	–	–	–
Pinus lambertiana	1	50	tr	12	2	12	–	–	–	–	–	–
Acer macrophyllum	2	25	tr	6	1	12	–	–	2	47	7	62
Arbutus menziesii	2	38	1	25	–	–	–	–	–	–	–	–
Tree regeneration:												
Tsuga heterophylla	tr	25	4	81	8	100	8	100	9	100	11	100
Pseudotsuga menziesii	8	100	5	75	–	–	–	–	–	–	–	–
Thuja plicata	–	–	1	31	1	29	2	56	3	53	2	38
Libocedrus decurrens	3	38	tr	6	–	–	–	–	–	–	–	–
Pinus lambertiana	tr	38	1	38	tr	6	–	–	–	–	–	–
Acer macrophyllum	tr	12	–	–	tr	6	tr	11	tr	13	–	–
Arbutus menziesii	tr	12	tr	12	–	–	–	–	–	–	–	–
Shrubs:												
Acer circinatum	19	88	18	88	21	88	9	83	2	87	6	88
Rhododendron macrophyllum	tr	12	40	100	40	100	13	89	1	67	tr	12
Castanopsis chrysophylla	1	50	23	100	2	82	2	78	tr	13	tr	25
Holodiscus discolor	5	88	tr	12	–	–	–	–	–	–	–	–
Corylus cornuta var. californica	7	88	1	56	tr	12	tr	11	tr	33	tr	12
Taxus brevifolia	4	38	5	81	6	59	7	78	4	47	1	50
Cornus nuttallii	2	50	5	94	3	53	2	50	1	33	1	38
Vaccinium parvifolium	1	62	1	75	2	71	1	83	2	87	3	100
Berberis nervosa	16	100	10	100	14	100	11	100	8	100	13	100
Gaultheria shallon	7	62	40	100	40	100	4	89	2	53	4	75
Rubus ursinus	1	75	1	75	1	65	2	83	1	67	1	75
Symphoricarpos mollis	2	88	tr	6	–	–	2	17	–	–	–	–
Herbs:												
Achlys triphylla	tr	50	1	56	tr	29	tr	33	1	47	2	75
Viola sempervirens	tr	12	1	62	1	65	2	83	2	73	1	62
Trillium ovatum	–	–	tr	19	tr	29	1	56	1	87	1	62
Polystichum munitum	4	100	1	75	1	65	4	94	26	100	27	100
Linnaea borealis	3	75	5	100	5	82	13	100	13	80	11	50
Vancouveria hexandra	tr	25	tr	38	tr	6	tr	39	tr	27	4	88
Galium triflorum	tr	38	–		tr	18	tr	11	1	60	1	38
Trientalis latifolia	1	100	1	69	tr	29	tr	44	tr	27	tr	12
Lathyrus polyphyllus	3	38	–	–	–	–	–	–	–	–	–	–
Madia gracilis	1	50	–	–	–	–	–	–	–	–	–	–
Collomia heterophylla	1	38	tr	6	–	–	–	–	–	–	–	–
Hieracium albiflorum	1	62	tr	38	tr	12	tr	28	tr	20	tr	25
Synthyris reniformis	4	75	tr	31	tr	12	tr	17	–	–	–	–
Xerophyllum tenax	2	25	10	81	2	53	2	50	tr	7	–	–
Iris tenax	1	62	tr	25	tr	6	–	–	–	–	–	–
Festuca occidentalis	1	88	tr	6	tr	6	–	–	tr	7	–	–
Whipplea modesta	8	100	tr	38	1	29	1	17	tr	20	–	–
Chimaphila umbellata	1	88	1	69	2	82	4	83	tr	53	1	38
Coptis laciniata	–	–	1	25	1	53	4	89	3	73	1	25
Tiarella unifoliata	–	–	–	–	tr	12	tr	28	4	73	tr	12
Disporum hookeri	–	–	–	–	tr	18	tr	17	tr	27	1	50
Asarum caudatum	–	–	–	–	–	–	–	–	tr	27	1	25
Athyrium filix-femina	–	–	–	–	–	–	–	–	tr	13	tr	12
Blechnum spicant	–	–	–	–	–	–	–	–	1	27	1	38
Oxalis oregana	–	–	–	–	–	–	–	–	tr	7	38	100

[1] tr = trace (less than 0.5 percent cover); a dash means the species wasn't found.

Figure 45. —A *Pseudotsuga menziesii/ Holodiscus discolor* **community typical of dry forest sites in the** *Tsuga heterophylla* **Zone; note the open nature of the stand and reproduction of** *Pseudotsuga,* **which is climax here (H. J. Andrews Experimental Forest, Oregon).**

are present in these stands (fig. 45), indicating it appears as the major climax species. *Holodiscus discolor, Corylus cornuta* var. *californica, Symphoricarpos mollis,* and *Gaultheria shallon* typify the shrub layer. A number of herbs not common on mesic sites find their forest optimum here; e.g., *Synthyris reniformis, Whipplea modesta, Hieracium albiflorum, Festuca occidentalis* and *Iris tenax* (fig. 46). Similar communities are much more common outside the *Tsuga heterophylla* Zone; e.g., in southwestern Oregon.

Figure 46.—**Herb layer within the** *Pseudotsuga menziesii/Holodiscus discolor* **association. Herbs visible here are** *Synthyris reniformis, Trientalis latifolia, Whipplea modesta,* **and** *Iris tenax;* **principal low shrubs are** *Symphoricarpos mollis* **and** *Berberis nervosa;* **grasses are** *Festuca occidentalis* **and** *Bromus* **sp.**

The *Tsuga heterophylla/Castanopsis chrysophylla* association is generally located on dry, exposed ridgetops. Its slightly more mesic position than the *Pseudotsuga/Holodiscus* is indicated by the presence of *Tsuga heterophylla*, both in the overstory and understory (table 8). The *Tsuga/Castanopsis* association typically has a rather open overstory tree canopy and a very dense shrub layer dominated by *Rhododendron macrophyllum* and *Castanopsis chrysophylla* (fig. 47). Although both *Tsuga* and *Pseudotsuga* seedlings and saplings are commonly present, *Tsuga* would be expected to be the eventual climax dominant if these sites were protected from disturbance. Because of the complete dominance of shrubs, the herb layer is very poorly developed and consists largely of scattered clumps of *Xerophyllum tenax* and *Linnaea borealis*.

The *Tsuga heterophylla/Rhododendron macrophyllum/Gaultheria shallon* association occupies cooler and moister sites than the *Tsuga/Castanopsis*. Although *Pseudotsuga menziesii* may still be dominant in the overstory, most old-growth stands also include a considerable component of mature *Tsuga heterophylla*. Tree regeneration is clearly dominated by *Tsuga*, and *Pseudotsuga* regeneration is completely lacking (table 8). *Thuja plicata* is the only other commonly occurring tree species. The *Tsuga/Rhododendron/Gaultheria* association is characterized by moderately dense tree overstory and shrub layers (fig. 48). Dominant shrub species are *Rhododendron macrophyllum* and *Gaultheria shallon*. As is the case with the *Tsuga/Castanopsis* association, the herb layer is very poorly developed (table 8).

Figure 47.—A stand representative of the *Tsuga heterophylla/Castanopsis chrysophylla* association typical of low elevation ridgetops in the *Tsuga heterophylla* Zone; note the dense tall shrub layer dominated by *Castanopsis* and *Rhododendron macrophyllum* (H. J. Andrews Experimental Forest, Oregon).

Figure 48.—A stand representative of the *Tsuga heterophylla/Rhododendron macrophyllum/Gaultheria shallon* association showing the typically dense shrub layer (H. J. Andrews Experimental Forest, Oregon).

The *Tsuga heterophylla/Rhododendron macrophyllum/Berberis nervosa* association and its relatives typify the climatic climax for the *Tsuga heterophylla* Zone (figs. 49 and 50). Trees in old-growth stands consist primarily of *Pseudotsuga menziesii*, *Tsuga heterophylla*, and *Thuja plicata*. *Tsuga heterophylla* is, theoretically, the sole climax species based on size-class analyses. However, long-lived *Pseudotsuga menziesii* and *Thuja plicata* are often present in stands undisturbed for 500 years or more. The understory is generally balanced between layers in the *Tsuga/Rhododendron/Berberis* association (table 8). Major shrubs are *Berberis nervosa*, *Acer circinatum*, *Vaccinium parvifolium*, *Rubus ursinus*, and *Rhododendron macrophyllum*. Typical herbs are *Linnaea borealis*, *Viola sempervirens*, *Coptis laciniata*, and *Goodyera oblongifolia*. The most common moss is *Eurhynchium oreganum*. *Polystichum munitum* and *Gaultheria shallon* are often present, but not as understory dominants.

The *Tsuga heterophylla/Polystichum munitum* association occupies moist sites, e.g., toe slopes and north aspects, within the *Tsuga heterophylla* Zone. Stands typically are made up of old-growth *Pseudotsuga menziesii* and *Tsuga heterophylla*, providing a relatively dense overstory canopy (table 8). The sparse shrub layer is dominated by *Berberis nervosa*; other common shrub species are *Acer circinatum* and *Vaccinium parvifolium*. The most outstanding characteristic of the understory vegetation is the robust growth of *Polystichum munitum* (fig. 51). Less noticeable, but generally present herbaceous species include *Linnaea borealis*, *Tiarella unifoliata*, and *Coptis laciniata*.

Figure 49.—A stand representative of the *Tsuga heterophylla/Rhododendron macrophyllum/Berberis nervosa* association which is the regional climatic climax for the *Tsuga heterophylla* Zone in the Cascades of Oregon (H. J. Andrews Experimental Forest).

Figure 50.—Shrub and herb layers within the *Tsuga heterophylla/Rhododendron macrophyllum/Berberis nervosa* association. Visible here are *Rhododendron* and *Berberis* in the shrub layer, and *Linnaea borealis* and *Viola sempervirens* in the herb layer.

The moist end of the community spectrum in the *Tsuga heterophylla* Zone of the Oregon Cascades is occupied by the *Tsuga heterophylla/Polystichum munitum-Oxalis oregana* association. Stands typifying this association are generally found on very moist streamside slopes. The overstory usually includes *Pseudotsuga menziesii*, *Tsuga heterophylla*, and *Thuja plicata* (table 8). Very large *Pseudotsuga* are encountered in old-growth stands, as growth conditions are near optimum for the species. Size-class analyses indicate *Tsuga heterophylla* will be the major climax species, but *Thuja plicata* also reproduces in sufficient quantity to perpetuate itself in some stands. The understory is dominated by a lush growth of herbs (fig. 52), including *Polystichum munitum*, *Oxalis oregana*, *Vancouveria hexandra*, and *Achlys triphylla*. Frequently, such moisture-loving herbs as *Asarum caudatum*, *Blechnum spicant*, and *Athyrium filix-femina* are present in at least small amounts. Mosses and liverworts are also common, including *Eurhynchium oreganum* and *Scapania bolanderi*.

Recent studies of forest vegetation on terraces and flood plains along the McKenzie River in the western Cascades of Oregon have resulted in the description of two climax associations on alluvial soils (Hawk 1973). The distribution of these two units, which are closely related to the associations just discussed, is largely controlled by soil characteristics. Deep, fine-textured soils with abundant moisture support a *Tsuga heterophylla/Acer circinatum/Polystichum munitum-Oxalis oregana* association. The tree layer in this

Figure 51.—*Polystichum munitum* is the major understory dominant in stands representative of the *Tsuga heterophylla/Polystichum munitum* association (H. J. Andrews Experimental Forest, Oregon).

Figure 52.—*Polystichum munitum* and *Oxalis oregana* dominate the lush herbaceous understory on moist sites in the *Tsuga heterophylla* Zone (*Tsuga heterophylla/Polystichum munitum-Oxalis oregana* association in the H. J. Andrews Experimental Forest, Oregon).

association includes *Pseudotsuga menziesii, Thuja plicata, Libocedrus decurrens,* and scattered *Abies grandis,* as well as *Tsuga heterophylla.* However, dominance of *Tsuga* in the understory suggests its eventual climax role. Riverside sites with coarse-textured, stony soils support a more drought-tolerant assemblage of species—the *Tsuga heterophylla/ Berberis nervosa-Gaultheria shallon/Linnaea borealis* association. An outstanding feature of stands characteristic of this association is extensive areas of very shallow soil and exposed stones which, although virtually devoid of vascular species, are covered with a dense moss layer made up of *Eurhynchium oreganum, Hylocomium splendens,* and *Rhytidiadelphus triquetrus.*

The forest communities within the *Tsuga heterophylla* Zone of the Oregon Coast Ranges have been intensively studied and classified in four areas ranging from Coos County on the south to Clatsop County to the north (Corliss and Dyrness 1965, Bailey 1966, Hines 1971, Meurisse and Youngberg 1971, and Meurisse 1972). The spectrum of climax forest associations which emerges from these studies is very similar to that encountered in the western Cascades of Oregon (table 9). At the dry end of the moisture gradient, Bailey (1966) and Corliss and Dyrness (1965) have identified a *Pseudotsuga menziesii/Holodiscus discolor/Gaultheria shallon* association which is characterized by climax *Pseudotsuga.* The species composition of this unit is similar to the *Pseudotsuga/ Holodiscus* association of the Oregon Cascades, except for substantially greater abundance

Table 9. — Climax associations within the *Tsuga heterophylla* Zone in the Oregon Coast Ranges arranged along a moisture gradient[1]

Generalized spectrum	Coos County (Bailey 1966)	Benton and Lincoln Counties (Corliss and Dyrness 1965)	Tillamook County (Meurisse and Youngberg 1971)	Clatsop, Tillamook, and Lincoln Counties (Hines 1971)
Pseudotsuga menziesii/ Holodiscus discolor/ Gaultheria shallon	Pseudotsuga/ Holodiscus/ Gaultheria	Pseudotsuga/ Holodiscus/ Gaultheria		
Tsuga heterophylla- Pseudotsuga menziesii/ Rhododendron macrophyllum/ Berberis nervosa	Tsuga-Pseudotsuga/ Rhododendron/ Berberis			
Tsuga heterophylla/ Vaccinium membranaceum/ Xerophyllum tenax				Tsuga/Vaccinium/Xerophyllum
Tsuga heterophylla/ Acer circinatum/ Gaultheria shallon	Tsuga heterophylla/ Acer circinatum/ Berberis nervosa	Tsuga/Acer/Gaultheria		Tsuga heterophylla/ Berberis nervosa/ Trientalis latifolia
Tsuga heterophylla/ Gaultheria shallon- Polystichum munitum		Tsuga/Gaultheria- Polystichum	Tsuga/Gaultheria- Polystichum	Tsuga heterophylla/ Vaccinium ovalifolium/ Polystichum munitum
Tsuga heterophylla/ Polystichum munitum	Tsuga heterophylla/ Polystichum munitum- Oxalis oregana	Tsuga/Polystichum	Tsuga/Polystichum	
Tsuga heterophylla/ Polystichum munitum- Oxalis oregana	Thuja plicata/ Adiantum pedatum- Athyrium filix-femina	Tsuga/Polystichum-Oxalis	Tsuga/Polystichum-Oxalis	Tsuga heterophylla/ Polystichum munitum- Adiantum pedatum

[1] From dry to wet (top to bottom).

of *Gaultheria shallon*. The *Tsuga heterophylla-Pseudotsuga/Rhododendron macrophyllum/ Berberis nervosa* association described by Bailey (1966) (table 9) appears to be equivalent to the *Tsuga heterophylla/Castanopsis chrysophylla* association in the Cascades (table 8). Both are typified by sizable amounts of *Pseudotsuga* regeneration as well as *Tsuga*, and in both units, *Castanopsis chrysophylla* and *Rhododendron macrophyllum* are diagnostic species. The *Tsuga heterophylla/Vaccinium membranaceum/Xerophyllum tenax* association (Hines 1971) occurs on shallow, stony soils at high elevations in the headwater areas of the Trask and Wilson Rivers. *Vaccinium membranaceum* is actually a minor component of these stands, and the species composition of the association closely resembles that of the *Tsuga heterophylla/Acer circinatum/Gaultheria shallon* association. Both are characterized by a well-developed shrub layer dominated by *Acer circinatum*, *Gaultheria shallon*, and *Berberis nervosa*.

The regional climatic climax for the *Tsuga heterophylla* Zone in the Oregon Coast Ranges is thought to be the *Tsuga heterophylla/Polystichum munitum* association. On this basis, we can conclude that modal sites in the Coast Ranges tend to have more favorable moisture regimes than comparable areas in the Cascades which are occupied by the *Tsuga heterophylla/Rhododendron macrophyllum/Berberis nervosa* association (table 8). Coastal stands typical of the *Tsuga/Polystichum* association generally include considerable quantities of *Oxalis oregana* and variable amounts of such shrubs as *Acer circinatum*, *Vaccinium parvifolium*, and *V. ovatum* (fig. 53). The wet end of the moisture gradient in the Coast Ranges is occupied by the *Tsuga heterophylla/Polystichum munitum-Oxalis oregana* association (table 9). As a result of wetter site conditions in the Coast Ranges, this unit differs appreciably from typical *Tsuga/Polystichum-Oxalis* stands in the Oregon Cascades. Perhaps the most noticeable difference is the greater occurrence of such moisture-loving ferns as *Adiantum pedatum*, *Athyrium filix-femina*, and *Blechnum spicant* in the coastal stands.

Figure 53.—A stand representative of the *Tsuga heterophylla/Polystichum munitum* **association in the southern Oregon Coast Ranges with the shrub layer dominated by** *Vaccinium ovatum* **(Cherry Creek Research Natural Area).**

None of the work discussed thus far has included consideration of *Thuja plicata* communities on wet sites. Such vegetation types are frequently encountered in the *Tsuga heterophylla* Zone of western Washington. The *Thuja plicata-Tsuga heterophylla/Oplopanax horridum/Athyrium filix-femina* association is typical of very wet lower slopes and stream terraces and is characterized by an extremely lush understory of shrubs, dicotyledonous herbs, and ferns (fig. 54). Although *Pseudotsuga menziesii* may be present, tree size-class analyses indicate eventual codominance of *Tsuga heterophylla* and *Thuja plicata* in climax stands. Understory dominants include *Oplopanax horridum*, *Athyrium filix-femina*, *Blechnum spicant*, *Vaccinium* spp., *Gymnocarpium dryopteris*, and *Dryopteris austriaca*, *Trautvetteria caroliniensis*, *Anemone deltoidea*, *Viola glabella*, *Streptopus* spp., *Smilacina* spp., *Tiarella trifoliata*, and *Achlys triphylla*. Another community, found on swampy sites, is typified by *Thuja plicata* and *Lysichitum americanum* plus a wide variety of semiaquatic species.

Slightly drier habitats in the Washington Cascades support a *Thuja plicata/Acer*

Figure 54.—*Thuja plicata* **and dense understories of shrubs and herbs, such as** *Oplopanax horridum* **and** *Athyrium filix-femina*, **are typical of old-growth stands on wet benches and stream terraces in the** *Tsuga heterophylla* **Zone (Mount Baker National Forest, Washington).**

81

circinatum/herb community.[8] On these sites, competitive interrelationships apparently favor the dominance of *Thuja plicata* over *Tsuga heterophylla*. However, climax stands generally contain at least a small component of *Tsuga heterophylla*. The most conspicuous understory species in this community are *Acer circinatum* and *Tiarella unifoliata*. Other important species are *Oplopanax horridum*, *Rubus parviflorus*, *Clintonia uniflora*, *Smilacina stellata*, *Athyrium filix-femina*, *Galium triflorum*, *Osmorhiza chilensis*, *Gymnocarpium dryopteris*, and *Disporum smithii*.

There are many variations of the general community pattern throughout the *Tsuga heterophylla* Zone. There is a general response to decreasing moisture (or, conversely, increasing moisture stress) from north to south, for example. In Washington, sites sufficiently dry to develop the *Pseudotsuga*/*Holodiscus* community are relatively rare. Instead, *Pseudotsuga menziesii*/*Gaultheria shallon* community, in which *Tsuga heterophylla* is the probable climax, is usually found on the poorest quality sites. Similarly, communities with *Polystichum*-type understories are more widespread in Washington. There is a rather regular increase of *Pseudotsuga* importance from north to south within the zone, given stands of similar age. A notable difference between communities described in Oregon and in Washington is in the importance of *Rhododendron macrophyllum*. In many *Tsuga heterophylla* Zone communities in Oregon, *Rhododendron macrophyllum* is a dominant; in Washington, it occurs sporadically.

Some of the few detailed descriptions of forest communities available for the *Tsuga heterophylla* Zone of western Washington have been provided by Fonda and Bliss (1969). Their work centered in the eastern portion of the Olympic Peninsula in what they termed "the eastern montane zone." The driest community encountered was located on south- or west-facing slopes at 550 to 900 meters in rain-shadow areas (e.g., the Elwha and Dungeness drainages). Stands are now dominated by 270- to 300-year-old *Pseudotsuga menziesii* and, because of the dry nature of the sites, *Tsuga heterophylla* is only very slowly extending its range. In these areas, the shrub layer is well developed and completely dominated by *Gaultheria shallon*. Other common shrubs include *Rosa gymnocarpa*, *Berberis nervosa*, and *Symphoricarpos albus*. The distinctive herb layer is made up of such species as *Chimaphila umbellata*, *Adenocaulon bicolor*, *Achlys triphylla*, *Campanula scouleri*, *Vicia americana*, *Trisetum cernuum*, *Hieracium albiflorum*, and *Bromus* sp. On moister sites, away from the influence of the rain shadow, stands contain abundant *Tsuga heterophylla* regeneration even though overstories are often dominated by *Pseudotsuga menziesii*. Such stands, situated at 600- to 1,100-meter elevation, have two principal understory types. On the drier west- and south-facing slopes, understories are dominated by dense *Gaultheria shallon*, accompanied by such herbs as *Chimaphila umbellata*, *C. menziesii*, and *Xerophyllum tenax*. On the other hand, stands occupying north- and east-facing slopes have a more poorly developed shrub layer consisting of *Vaccinium parvifolium*, *V. ovalifolium*, *Gaultheria shallon*, and *Berberis nervosa*, and a herb layer containing *Polystichum munitum*, *Linnaea borealis*, *Tiarella unifoliata*, and *Listera caurina*.

Successional Patterns

The early stages of secondary succession following logging and burning in this zone have been the subject of a number of studies. Unfortunately, however, much of this research has been limited to the first 5 to 8 years after complete tree removal, and, therefore, detailed

---

[8] Unpublished report, "Phytosociological reconnaissance of western redcedar stands in four valleys of the North Cascades Park complex," by Joseph W. Miller and Margaret M. Miller, 50 p., Dec. 1970. On file at Forestry Sciences Laboratory, Corvallis, Oregon.

successional patterns for the entire period of forest reestablishment have not been worked out. Timber harvesting operations in the *Tsuga heterophylla* Zone generally involve clearcut logging, followed by controlled burning of the logging slash to reduce wildfire hazard. During the first growing season after burning, the sparse plant cover is made up of residual species from the original stand, plus small amounts of invading herbaceous species such as *Senecio sylvaticus, Epilobium angustifolium*, and *E. paniculatum* (fig. 55) (Dyrness 1965, 1973). A moss-liverwort stage has also been noted during the first year (Isaac 1940, Ingram 1931).

Vegetation the second year is dominated by invading annual herbaceous species, which

Figure 55.—Herbaceous stages in secondary succession in the *Tsuga heterophylla* **Zone; range poles are 4 feet tall and marked in 6-inch segments (H. J. Andrews Experimental Forest, Oregon).**

A. **Virtually bare condition during the first growing season following slash burning.**

B. **The second year after slash burning, invading** *Senecio sylvaticus* **dominates the site.**

D. **By the fifth growing season after slash burning, the invading shrub,** *Ceanothus velutinus*, **is beginning to gain dominance over the herbaceous invaders (here, mainly** *Epilobium angustifolium* **and** *E. paniculatum*).

C. **During the third year,** *Epilobium angustifolium* **replaces** *Senecio*, **which has almost completely dropped out of the stand.**

produce large numbers of small, windborne seeds (fig. 55). Over much of the zone, a very high proportion of the second-year cover is made up of *Senecio sylvaticus*, a species which is present in only very small amounts in subsequent years (Brown 1963, Dyrness 1973, Isaac 1940). West and Chilcote (1968) have shown this short-term dominance is related to high nutrient requirements which are generally satisfied only on recently burned sites. Perennial invading herbaceous species, such as *Epilobium angustifolium*, *Cirsium vulgare*, and *Pteridium aquilinum*, rapidly build up their populations until the fourth or fifth year when their rate of increase slackens (fig. 55) (Ingram 1931, Isaac 1940, Yerkes 1960, Gashwiler 1970).

This successional stage, sometimes called the weed stage, gradually gives way to a shrub-dominated period (fig. 56). These shrubs, including residual species such as *Acer circinatum*, *Rubus ursinus*, *Berberis nervosa*, *Rhododendron macrophyllum*, and *Gaultheria shallon*, as well as invaders such as *Ceanothus velutinus* and *Salix* spp., then dominate the site until they are overtopped by tree saplings, generally *Pseudotsuga menziesii* (Kienholz 1929, Ingram 1931, Isaac 1940).

Herbaceous species surviving from the original forest stand (e.g., *Polystichum munitum*, *Trientalis latifolia*, and *Oxalis oregana*) are reported to be of only minor importance in early successional stages in the Cascade Range (Kienholz 1929, Isaac 1940, Yerkes 1960). However, in the Coast Ranges of Oregon, residual species, especially *Polystichum munitum*, often constitute an important component of the seral vegetation following logging. In addition, both shrub and total plant cover tend to be substantially greater on Coast Ranges cutting units than in the Cascades (Morris 1958). Typical shrub species near the coast include *Rubus spectabilis*, *Rubus parviflorus*, and *Gaultheria shallon*.

The vegetation in early stages of succession following logging and burning is characteristically very heterogeneous. Much of this variability is attributable to site differences caused by a wide range of types of logging disturbance and degrees of burning severity (Dyrness 1965, 1973). The effects of burning in control of species composition are perhaps the most notable. For example, annual invading herbaceous species, such as *Senecio sylvaticus* and *Epilobium paniculatum*, show a marked preference for burned areas;

Figure 56.—A stage of shrub dominance typically follows herbaceous stage in *Tsuga heterophylla* **Zone secondary successions;** *Rhododendron macrophyllum,* **residual from the original forest stand, dominates this 10-year-old clearcut (Santiam River drainage, Willamette National Forest, Oregon).**

whereas residual species such as *Polystichum munitum* and *Trientalis latifolia* occur more frequently on unburned sites. Of the shrubs, *Ceanothus* spp. are often almost entirely restricted to burned areas, and residual species (e.g., *Rhododendron macrophyllum* and *Vaccinium parvifolium*) generally are much more common on unburned sites (Morris 1958, Steen 1966).

The relationships between early stages of secondary succession and plant communities, present in the original *Pseudotsuga menziesii-Tsuga heterophylla* stands, have been studied on Vancouver Island by Mueller-Dombois (1965) and in the southern Oregon Coast Ranges by Bailey (1966). Both workers found that in these coastal areas the characteristic forest plants were present in sufficient quantities after logging and burning to permit successful identification of the preexisting communities. Mueller-Dombois concluded that the cutover vegetation is denser than that found under the original stand due to spreading forest weeds, semitolerant of shade, and invading shade-intolerant weeds. Both weed groups compete for the same vacant spaces, but in general, the intolerant weeds appear to be more successful in burned areas and the semitolerant forest weeds in unburned localities.

Kellman (1969), working in southern British Columbia, found that species present in stands before logging maintained themselves in logged areas, although with reduced frequency, and gradually increased in importance as secondary succession advanced. Invading species, responding mainly to tree canopy removal, were largely concentrated in severely disturbed areas. In his study areas, patterns of species occurrence showed considerable variation from stand to stand; thus, he was unable to find consistent differences in species distributions which could be related to successional stages. Kellman concluded "that no discrete communities, as measured by the methods employed, existed in the vegetation studied. It is proposed that these unique groupings were determined either by chance or by correlations between species' propagule availabilities within the stand, with the species' distributions being controlled only weakly by environmental factors."

Although much work remains to be done in classification and description of seral communities in the *Tsuga heterophylla* Zone, sufficient information is available for western Oregon to suggest broad successional trends (table 10). Seral communities on dry sites most often contain conspicuous quantities of *Gaultheria shallon*, a species which is also very characteristic of the climax associations. The most important seral tree species is *Pseudotsuga menziesii*, although in some areas of the Coast Ranges *Acer macrophyllum* may also be dominant. On intermediate sites, early seral communities are often shrub dominated and include such species as *Vaccinium parvifolium*, *V. ovatum*, and *Rubus spectabilis*, as well as *Gaultheria shallon*. The most widely distributed seral tree species on moist to wet sites in the Coast Ranges is *Alnus rubra*. Understory vegetation in stands of *Alnus rubra* is most frequently dominated by *Rubus spectabilis* in the shrub layer and *Polystichum munitum* in the herb layer. Other species characteristic of seral communities on moist sites include *Acer circinatum* and *Lotus crassifolius*.

The composition and density of the seral forest stands are dependent on the type of disturbance, available seed source, and environmental conditions. A very common occurrence is the development of dense, nearly pure, essentially even-aged stands of *Pseudotsuga menziesii* (fig. 57). This tendency is encouraged by the extensive planting and seeding of this species after logging or wildfires. These stands are often dense enough to eliminate most of the understory vegetation (fig. 58). Reestablishment of the characteristic understory species and invasion of western hemlock then take place as mortality begins to open up the stand at 100 to 150 years of age. On the other hand, stands of *Pseudotsuga* may be relatively open, resulting in persistent understories dominated by *Gaultheria shallon* or *Acer circinatum*. Other common types of young stands in the zone are (1) those dominated

Table 10. – Seral communities and hypothesized equivalent climax associations reported for the *Tsuga heterophylla* Zone in western Oregon arranged along a moisture gradient from dry (top) to wet (bottom)

Early seral communities	Late seral communities	Climax association
Pteridium aquilinum-Gaultheria shallon (Corliss and Dyrness 1965)		*Pseudotsuga menziesii/Holodiscus discolor/Gaultheria shallon*
Pteridium aquilinum-Gaultheria shallon (Corliss and Dyrness 1965)	*Pseudotsuga menziesii/Gaultheria shallon* (Corliss and Dyrness 1965)	*Tsuga heterophylla/Acer circinatum/ Gaultheria shallon*
Vaccinium parvifolium/ Gaultheria shallon, Rubus parviflorus/Trientalis latifolia (Bailey and Poulton 1968, Bailey and Hines 1971, Meurisse and Youngberg 1971)	*Acer macrophyllum/Symphoricarpos mollis* (Bailey and Poulton 1968, Bailey and Hines 1971)	
	Pseudotsuga menziesii/Acer circinatum/Gaultheria shallon (Dyrness, Franklin, and Moir, unpublished)	*Tsuga heterophylla/Rhododendron macrophyllum/Gaultheria shallon*
	Pseudotsuga menziesii/Acer circinatum/Berberis nervosa (Dyrness, Franklin, and Moir unpublished)	*Tsuga heterophylla/Rhododendron macrophyllum/Berberis nervosa*
Vaccinium ovatum-Rubus spectabilis, Vaccinium parvifolium/Gaultheria shallon (Meurisse and Youngberg 1971)		*Tsuga heterophylla/Gaultheria shallon-Polystichum munitum*
Vaccinium parvifolium/Gaultheria shallon (Bailey and Poulton 1968)		
Alnus rubra/Rubus spectabilis/ Polystichum munitum (Corliss and Dyrness 1965)	*Pseudotsuga menziesii/Acer circinatum/Polystichum munitum* (Corliss and Dyrness 1965)	*Tsuga heterophylla/Polystichum munitum*
Acer circinatum/Polystichum munitum (Bailey and Hines 1971)		
Alnus rubra/Acer circinatum, Alnus rubra/Polystichum munitum (Bailey and Poulton 1968)		
Alnus rubra/Rubus spectabilis/ Polystichum munitum (Corliss and Dyrness 1965)		*Tsuga heterophylla/Polystichum munitum-Oxalis oregana*
Pteridium aquilinum-Lotus crassifolius, Alnus rubra/Rubus spectabilis/ Polystichum munitum (Meurisse and Youngberg 1971)		
Pteridium aquilinum-Lotus crassifolius, Alnus rubra/Rubus parviflorus (Bailey and Hines 1971)		

Figure 57.—Dense, nearly pure, essentially even-aged *Pseudotsuga menziesii* stands typically develop on cutover areas in western Washington and northwestern Oregon by natural seeding or planting.

Figure 58. — Young conifer stands in the *Tsuga heterophylla* Zone are often dense enough to completely eliminate most of the understory; dense 66-year-old *Pseudotsuga menziesii* stand near Cottage Grove, Oregon.

by *Alnus rubra*, particularly on *Polystichum munitum*-characterized habitats (Bailey and Poulton 1968), and (2) stands in which *Tsuga heterophylla* or *Thuja plicata* are major components right from the beginning of secondary forest development. The latter two situations are most commonly found in wetter parts of the *Tsuga heterophylla* Zone; i.e., the northern Washington Cascade Range, the west side of the Olympic Mountains, and in the Coast Ranges.

Truly climax forests are rare, but examples of old-growth forests which have been undisturbed for 400 to 600 years are relatively common. From these, we can draw some conclusion about the potential climax species. The eventual replacement of *Pseudotsuga* by *Tsuga heterophylla* in the absence of disturbance has been described by many authors (e.g., by Munger 1940, Hansen 1947, Cooper 1957, and Barrett 1962, in general terms; and by Bailey 1966, Fonda 1967, and Orloci 1965, for specific areas). As mentioned, *Tsuga heterophylla* may appear with the *Pseudotsuga* and a mixed stand develop. On other sites, significant *Tsuga* invasion may not occur for 50 or 100 years after disturbance. The point is that on most sites in the zone, *Tsuga heterophylla* is able to reproduce itself beneath the forest canopy and the relatively intolerant *Pseudotsuga* is not (fig. 59). Numerous examples can be found of mixed old-growth *Pseudotsuga-Tsuga* forest in which abundant seedlings, saplings, and poles of *Tsuga heterophylla* are present and those of *Pseudotsuga* are completely lacking.

However, there is some variation within the zone regarding the climax tree species. On environmentally median sites, *Tsuga heterophylla* appears to be essentially the sole climax species. On very dry sites, *Tsuga heterophylla* is absent and, consequently, *Pseudotsuga menziesii* attains a climax role. On the wet to very wet sites, *Thuja plicata* will certainly be a part of any climax forest; size-class analyses do not support climax status for this species on modal or dry sites, however, even though many ecologists have hypothesized a *Tsuga-Thuja* climax for most of the region. The tendency for *Thuja plicata* to be a part of the mixed stands, successionally intermediate between the pioneer *Pseudotsuga* forests and the climax *Tsuga heterophylla* forests, has perhaps misled some ecologists in interpreting its successional role.

Figure 59.—The shade-tolerant, climax *Tsuga heterophylla* is capable of repro-ducing under a forest canopy, but the intolerant *Pseudotsuga menziesii* is not; *Tsuga* saplings developing under a *Pseudo-tsuga* overstory (Santiam River drainage, Willamette National Forest, Oregon).

Special Types

Puget Sound Area

Here, large areas differ from the surrounding *Tsuga heterophylla* Zone in community types. Prairie, oak woodland, and pine forest are encountered, for example. Climate and soil are both major factors in these differences. The area lies in the rain shadow of the Olympic Mountains. Precipitation is typically between 800 and 900 millimeters in the Puget lowlands, although it drops as low as 460 millimeters on the northeastern side of the Olympic Peninsula and in the San Juan Islands (U.S. Weather Bureau 1960b). And, as pointed out earlier, a continental ice sheet covered the Puget Trough during the Pleistocene epoch (Vashon glaciation). Consequently, most soils have developed in glacial drift and outwash. Such soils are often coarse textured, poor in nutrients, and excessively drained.

Although plant communities around Puget Sound are similar to others in the *Tsuga heterophylla* Zone, there are many notable features which are not common elsewhere: (1) stands with *Pinus contorta*, *Pinus monticola*, and even *Pinus ponderosa* as major constituents along with *Pseudotsuga menziesii,* and with *Gaultheria shallon* as an extremely common understory species; (2) *Quercus garryana* groves—many being actively invaded by *Pseudotsuga menziesii;* (3) extensive prairies often being invaded by *Pseudotsuga menziesii* and associated with groves of *Quercus;* (4) abundant poorly drained sites with swamp or bog communities; and (5) occurrence of species rarely or never found elsewhere in western Washington or northwestern Oregon, e.g., *Juniperus scopulorum, Populus tremuloides, Pinus ponderosa,* and *Betula papyrifera.* None of these communities or community mosaics, except the bogs (Rigg 1917, 1919, 1922a, 1922c, 1958; Fitzgerald 1966), have been studied in detail, although generalized accounts of one or more may be found in Lang (1961), Barrett (1962), Hansen (1947), and Rigg (1913). Roemer (1972) describes comparable communities from the east coast of Vancouver Island. The San Juan Islands will be discussed in Chapter XI.

Prairies are a conspicuous feature of the landscape south of Puget Sound. These grassy openings include the Tacoma Prairies (Hansen 1947) and Wier Prairie (Lang 1961) near Olympia, Washington. The development of prairie (grassland) soil profiles in many of the openings indicates they have been free of forest for many years. The origin and maintenance of these prairie openings are thought to have been due to two main factors: (1) the occurrence of droughty, gravelly soils derived from glacial outwash materials coupled with low summer precipitation and (2) frequent burning of the prairies by natural causes, Indians, and possibly early white settlers. Lang (1961) has described the composition of nearly undisturbed portions of Wier Prairie as consisting primarily of bunches of *Festuca idahoensis* with the intervening space covered by a thick moss layer, principally *Rhacomitrium canescens.* Species penetrating the moss layer include *Carex pensylvanica, Dodecatheon hendersonii, Camassia quamash, Saxifraga integrifolia, Sisyrinchium angustifolium, Armeria maritima, Viola adunca, V. nuttallii, Zigadenus venenosus,* and *Balsamorhiza deltoidea.* Many of these species clearly suggest high soil water contents during the winter and early spring.

Since settlement, the extent of these prairies has been rapidly diminishing as a result of invasion by *Pseudotsuga menziesii* and *Quercus garryana* (fig. 60). Accelerated tree invasion has been attributed to the advent of fire protection, grazing, and other means. *Quercus garryana* stands within the prairie-woodland mosaic generally include the following understory species: *Cytisus scoparius, Symphoricarpos albus, Arctostaphylos uva-ursi, Carex pensylvanica, Festuca idahoensis,* and *Fragaria vesca* var. *bracteata* (Lang 1961). *Pseudotsuga menziesii* stands recently established on what were formerly prairie sites are characteristically dense with sparse understory layers made up of such species as *Symphoricarpos albus, Osmaronia cerasiformis, Goodyera oblongifolia, Corallorhiza maculata,* and *Sanicula crassicaulis* (Lang 1961). *Pinus contorta* (generally) and relict stands of *Pinus ponderosa* are also associated with the gravelly soils and prairies.

"Mima mounds" are found on some of these prairies (fig. 60). These mounds range from swellings on the prairie surface to a maximum of about 2.1 meters in height and average about 12.2 meters in diameter. Several different theories of origin have been proposed, two of which are seriously considered. Dalquest and Scheffer (1942) attributed them to gopher activity (*Thomomys talpoides*). Newcomb (1952) related them to the glacial climate by proposing the mounds were due to thrusting action of ground ice wedges.

Eventually, the Puget lowlands may be recognized as a separate vegetative zone similar to the Coastal Douglas-fir Zone found in British Columbia (Krajina 1965). Its close relations with this zone are obvious. Many of the Puget lowland communities also appear related to those found in the Willamette valley; possibly some ecologists would group them together. However, both coastal British Columbia and the Puget Sound areas were glaciated and do involve large bodies of ocean water which significantly affect the local climate. Neither of these circumstances applies to the Willamette and other interior valleys of western Oregon.

Figure 60. —Mima Prairie near Olympia, Washington, showing (1) distribution of the mima mounds, which average 12 meters in diameter and 2 meters in height, and (2) invasion of a prairie area by *Pseudotsuga menziesii* (*photo courtesy V. B. Scheffer*).

Grass Balds in the Oregon Coast Ranges

Scattered peaks in the Oregon Coast Ranges, most often composed of intrusive igneous bedrock, support meadow vegetation generally described as grass balds. These openings are conspicuous features in otherwise uninterrupted forest vegetation. The most outstanding examples of grass balds are located on or near the summits of the following mountains: Marys Peak, Monmouth Peak, Grass Mountain, Prairie Peak, Roman Nose Mountain, Tyee Mountain, and Saddle Mountain. Elevations range from 779 meters for Tyee Mountain to 1,249 meters for Marys Peak. Merkle (1951) and Detling (1954) have provided lists of species occurring in grass balds situated on Marys Peak and Saddle Mountain, respectively. However, the only phytosociologically oriented study of these bald communities is that of Aldrich (1972). Aldrich, in his study of six grass balds, identified and described two climax associations, the *Lomatium martindalei* and *Elymus glaucus* associations, and three seral communities designated as the *Carex rossii*, *Cynosurus echinatus*, and *Viola adunca* communities.

According to Aldrich (1972), the *Lomatium martindalei* association is limited to shallow soils adjacent to rock outcrop areas and is made up of low-growing mosses, grasses, and herbs. Important characteristic species are *Lomatium martindalei*, *Aira praecox*, and *Hypochaeris radicata*. Other species often present include *Lupinus lepidus*, *Aira caryophyllea*, *Koeleria cristata*, and *Agoseris heterophylla*. The *Elymus glaucus* association occurs on deeper soils and is characterized by *Elymus glaucus*, *Bromus carinatus*, *Galium aparine*, *Carex pachystachya*, and *Poa pratensis*. The *Carex rossii* and *Cynosurus echinatus* units are interpreted as seral communities within the *Elymus glaucus* habitat type. The dominant plant in the *Carex rossii* community is *Pteridium aquilinum*. Other important species include *Collinsia parviflora, Delphinium menziesii, Thermopsis montana* var. *venosa,* and *Galium oreganum.* Important species within the *Cynosurus echinatus* community include, in addition to *Cynosurus, Taraxacum officinale, Bromus rigidus, Geranium molle, Bromus commutatus,* and *Arrhenatherum elatius.* The *Viola adunca* community is characterized by high cover of *Carex californica* and *Festuca rubra* and the occurrence of *Viola adunca* and *Iris tenax.*

Talus Communities

Nonforested talus or scree slopes occur in many parts of the *Tsuga heterophylla* Zone. Very often the communities on these sites are dominated by the shrub *Acer circinatum.* These should probably be described as a series of communities because of the great variety in stand composition and structure, depending on substrate (size, arrangement, and chemistry of rocks, moisture conditions), elevation, exposure, etc. Some typical associates of *Acer circinatum* on more xeric talus include *Cryptogramma crispa, Festuca occidentalis, Holodiscus discolor, Symphoricarpos mollis, Corylus cornuta* var. *californica, Cheilanthes gracillima, Selaginella wallacei, Xerophyllum tenax, Synthyris reniformis, Rhacomitrium canescens* var. *ericoides, Aira caryophyllea,* and *Ceanothus sanguineus. Acer circinatum*-dominated talus communities may be found well into the *Abies amabilis* Zone and be intergraded with the *Alnus sinuata* communities described later in this chapter.

Talus communities dominated by *Acer macrophyllum* occur in the Oregon Coast Ranges (Bailey and Poulton 1968). Similar communities have been noted in northern Washington (fig. 61).

Figure 61.—Talus communities dominated by *Acer macrophyllum* **(background) or** *Acer circinatum* **(foreground) are widespread in the** *Tsuga heterophylla* **Zone (Lake Twentytwo Research Natural Area, Mount Baker National Forest, Washington).**

Port-Orford-cedar Variant

Near its southern edge in the Oregon Coast Ranges, *Chamaecyparis lawsoniana* is added to *Tsuga heterophylla* Zone forests. In these forests, *Chamaecyparis* is associated with species such as *Pseudotsuga menziesii, Tsuga heterophylla, Abies grandis, Thuja plicata* (only on wetter sites), *Lithocarpus densiflorus, Rhododendron macrophyllum, Arbutus menziesii,* and *Pinus lambertiana* (fig. 62). It appears to have the same ecological role as *Pseudotsuga menziesii,* a long-lived but seral dominant. Old-growth specimens of *Chamaecyparis lawsoniana* develop thick bark and are quite resistant to fire. Old but vigorous specimens frequently have numerous fire scars (fig. 63). Structural analyses of old-growth stands indicate *Chamaecyparis* is not capable of reproducing under closed

Figure 62.—*Chamaecyparis lawsoniana* **is found in the southern part of the** *Tsuga heterophylla* **Zone and attains optimal development there; note the dense understory of evergreen shrubs (Siskiyou National Forest, Oregon).**

Figure 63.—Old-growth *Chamaecyparis lawsoniana* **are quite fire resistant, and vigorous specimens frequently have deep basal fire scars (Coquille River Falls Research Natural Area, Siskiyou National Forest, Oregon).**

forest conditions and is replaced by more tolerant associates—*Abies grandis* and *Tsuga heterophylla.*[9]

Chamaecyparis lawsoniana is considered in this discussion of the *Tsuga heterophylla* Zone because it attains optimum development there. In fact, it grows in several different zones and on a wide variety of sites in southwestern Oregon—from sand dunes along the coastal strip to over 1,500 meters in the Siskiyou Mountains and down into the interior valleys; from swampy sites to dry, rocky ridges; and even on serpentine soils (Whittaker 1960, Fowells 1965). Indeed, it is difficult to understand why the natural range of the species is so restricted geographically. *Chamaecyparis lawsoniana* is seriously threatened by a killing root disease, *Phytophthora lateralis,* which was recently introduced into its natural range. This disease has already decimated *Chamaecyparis lawsoniana* in the coastal region and could possibly eliminate it from most of its natural range.

Abies amabilis Zone

The *Abies amabilis* Zone lies between the temperate mesophytic *Tsuga heterophylla* Zone of the lowlands and the subalpine *Tsuga mertensiana* Zone. It occurs on the western slopes of the Cascade Range from British Columbia south to about 44° north latitude, generally at elevations from 1,000 to 1,500 meters in Oregon, 900 to 1,300 meters in southern Washington (Franklin and Bishop 1969), and 600 to 1,300 meters in northern Washington. The *Abies amabilis* Zone is also conspicuous in the Olympic Mountains, except in the rain shadow on the northeastern slopes of the peninsula (Fonda 1967, Fonda and Bliss 1969). Where local conditions are favorable, comparable communities are found (1) on eastern slopes of Washington's Cascade Range, (2) south to 43° north latitude in Oregon's Western Cascades Province, (3) on wet, cool sites (streamsides and benches) in the *Tsuga heterophylla* Zone, and (4) in isolated locales in the northern Oregon Coast Ranges (Hines 1971).

We are uncertain as to whether the *Abies amabilis* Zone should be considered a subalpine or cool temperate formation. In earlier work, the subalpine designation is more common (e.g., Franklin 1966). More recent work in connection with the Coniferous Forest Biome, U.S. International Biological Program (fig. 29), suggests the cool temperate designa-

---

[9] Data on file at Forestry Sciences Laboratory, Pacific Northwest Forest and Range Experiment Station, Corvallis, Oregon.

tion may be more proper. Vegetative analyses often show a preponderance of temperate over subalpine elements although many of the environmental features (e.g., short, cool growing season and significant winter snowpack) are more characteristic of subalpine areas. Fonda and Bliss (1969) designate their *Abies amabilis-Tsuga heterophylla* forests as "montane" rather than subalpine, and Krajina (1965) refers to his zonal analog as the "Wet Subzone, Coastal Western Hemlock Zone." In any case, our *Abies amabilis* Zone includes most of the poorly defined Canadian Zone (Barrett 1962) as well as the upper edges of the Humid Transition Zone (Franklin and Bishop 1969).

Environmental Features

The *Abies amabilis* Zone is wetter and cooler than the adjacent *Tsuga heterophylla* Zone and receives considerably more precipitation in the form of snow (table 11), much of which accumulates in winter snowpacks as deep as 1 to 3 meters. The complex of soil-forming processes leads toward podzolization. This trend is less pronounced in the south where Haplorthods (Brown Podzolic soils) are the rule, and most pronounced in the north where Cryorthods (Podzols) are typical. Organic matter accumulations are of a mor or duff-mull type. These generally average only 3 to 7 centimeters thick (Williams and Dyrness 1967), except in northern Washington where accumulations up to 30 centimeters or more in thickness are encountered along with well-developed bleicherde (A2) and ortstein (B2ir) mineral horizons.

Table 11.— Climatic data from representative weather stations within the *Abies amabilis* Zone

Station	Eleva-tion	Lati-tude	Longi-tude	Temperature					Precipitation		
				Average annual	Average January	Average January minimum	Average July	Average July maximum	Average annual	June through August	Average annual snowfall
	Meters			- - - - - - - - - Degrees C. - - - - - - - - - -					Millimeters		Centi-meters
Snoqualmie Pass, Wash.	991	47°25'	121°25'	5.5	−3.2	−6.6	14.4	21.1	2,656	227	982
Spirit Lake, Wash.	1,063	46°16'	122°09'	5.6	−2.0	−4.4	14.9	22.3	2,253	140	718
Government Camp, Oreg.	1,280	45°18'	121°45'	5.6	−1.7	−4.9	14.0	20.8	2,190	190	792

Source: U. S. Weather Bureau (1956, 1965a).

Forest Composition

Forest composition in the *Abies amabilis* Zone varies widely, depending upon stand age, history, and locale (Franklin 1965a, 1965b). Typical tree species are *Abies amabilis*, *Tsuga heterophylla*, *Abies procera*, *Pseudotsuga menziesii*, *Thuja plicata*, and *Pinus monticola* (fig. 64). Around Mount Adams and in Oregon, *Abies lasiocarpa*, *Abies grandis*, *Picea engelmannii*, *Pinus contorta*, and *Larix occidentalis* may also occur in the zone. At its upper margin, *Tsuga mertensiana* and *Chamaecyparis nootkatensis* appear. Under-stories are usually dominated by ericaceous genera, such as *Vaccinium*, *Menziesia*,

Figure 64.—Mixed stands are typical of the *Abies amabilis* Zone; overstory dominants in this stand are *Abies procera, Pseudotsuga menziesii,* and *Tsuga heterophylla* (background), but poles (foreground) and reproduction are the climax *Abies amabilis* (Wildcat Mountain Research Natural Area, Willamette National Forest, Oregon).

Gaultheria, Chimaphila, Rhododendron, and *Pyrola. Cornus canadensis, Clintonia uniflora, Rubus lasiococcus, R. pedatus, Linnaea borealis, Xerophyllum tenax,* and *Viola sempervirens* are also common species. *Rhytidiopsis robusta* is the most constant and conspicuous bryophyte.

Composition of forest communities within the *Abies amabilis* Zone varies markedly on both geographic and local scales. Most synecological studies have recognized a series of community types within a study area (e.g., Franklin 1966, Fonda and Bliss 1969, Hines 1971), although others (Thornburgh 1969, Del Moral 1973) have found a continuum viewpoint useful. In southern Washington and northern Oregon, much of the variation in community composition is associated with moisture regimes (Franklin 1966). Some understory constancy and coverage data for an exemplary group of associations along a moisture gradient within the Mount Rainier Province (Franklin 1965b) are provided in table 12. The *Abies amabilis/Vaccinium alaskaense* association is the zonal climax in this area. It is characterized by well-developed shrub, herb, and moss layers (fig. 65). *Vaccinium alaskaense, V. ovalifolium, Cornus canadensis, Clintonia uniflora, Linnaea borealis,* and *Rhytidiopsis robusta* typify these modal communities. Drier sites are occupied by associations with depauperate understories dominated by low shrubs such as *Gaultheria shallon* and *Berberis nervosa* or, on lithosolic sites, by the coarse liliaceous *Xerophyllum tenax* (*Abies amabilis/Xerophyllum tenax*/Lithosol and *Abies amabilis/Gaultheria* associations in table 12). More mesic sites have communities with herb-rich understories dominated by species such as *Tiarella unifoliata*, *Streptopus roseus*, *Achlys triphylla*, *Gymnocarpium dryopteris*, *Vancouveria hexandra*, and *Smilacina stellata* (*Abies amabilis/Streptopus roseus* association in table 12) (figs. 66 and 67). On some habitats, *Oxalis oregana* and *Blechnum spicant* are also characteristic herbs. Communities on the wettest forested sites have dense, lush understories typified by a variety of succulent herbs, *Athyrium filix-femina*, and *Oplopanax horridum* (*Abies amabilis/Oplopanax horridum* association in table 12).

Table 12. — Constancy and coverage of selected plants species in five plant associations within the *Abies amabilis* Zone representing a moisture gradient from dry (on left) to wet (on right) (from Franklin 1966)

Stratum and species	*Abies amabilis/ Xerophyllum tenax/* Lithosol		*Abies amabilis/ Gaultheria shallon*		*Abies amabilis/ Vaccinium alaskaense*		*Abies amabilis/ Streptopus roseus*		*Abies amabilis/ Oplopanax horridum*	
	Constancy	Cover	Constancy	Cover	Constancy	Cover	Constancy	Cover	Constancy	Cover
	--------------------------------- Percent ---------------------------------									
Shrubs:										
Acer circinatum	100	13	80	13	65	6	50	2	83	4
Gaultheria shallon	100	9	100	40	38	1	—	—	—	—
Berberis nervosa	100	8	100	13	54	3	—	—	—	—
Vaccinium parvifolium	100	7	100	2	65	2	—	—	33	1
Vaccinium membranaceum	100	4	60	tr[1]	73	4	100	2	17	tr
Vaccinium alaskaense[2]	100	tr	80	2	100	33	100	10	83	1
Oplopanax horridum	—	—	—	—	—	—	50	tr	100	31
Shrub total	—	47	—	76	—	48	—	16	—	41
Herbs:										
Xerophyllum tenax	100	50	100	1	65	5	—	—	—	—
Linnaea borealis	100	3	100	2	81	3	—	—	17	tr
Chimaphila umbellata	67	1	80	tr	77	4	—	—	—	—
Cornus canadensis	67	1	40	tr	88	6	67	1	—	—
Achlys triphylla	67	1	60	1	62	1	100	14	100	18
Clintonia uniflora	33	1	60	1	73	2	100	6	83	5
Rubus lasiococcus	—	—	—	—	23	1	83	4	83	tr
Smilacina stellata	—	—	—	—	50	tr	83	3	67	3
Tiarella unifoliata	—	—	—	—	77	tr	100	18	100	12
Trillium ovatum	—	—	80	tr	—	tr	100	1	100	1
Gymnocarpium dryopteris	—	—	—	—	—	—	33	8	100	42
Athyrium filix-femina	—	—	—	—	—	—	67	tr	100	19
Herb total	—	58	—	8	—	23	—	88	—	128
Mosses:										
Eurhynchium oreganum	67	1	100	11	50	1	17	tr	50	1
Rhytidiopsis robusta	100	12	100	7	100	14	100	1	83	2
Homalothecium megaptilum	33	tr	100	2	35	tr	—	—	—	—
Brachythecium velutinum	—	—	20	tr	35	tr	83	tr	100	1
Moss total	—	19	—	26	—	17	—	2	—	10

[1] tr represents trace (less than 0.5 percent cover).
[2] Includes *Vaccinium ovalifolium* when present.

Similar types of communities are found in other regions of the *Abies amabilis* Zone. In the Olympic Mountains, herb-rich communities characterized by *Oxalis oregana* and *Blechnum spicant* are much more common than in southern Washington (Fonda 1967, Fonda and Bliss 1969). In the northern Washington Cascade Range, the climatic climax communities are similar to the *Abies amabilis/Vaccinium alaskaense* association except for a somewhat greater abundance of mesic herbs and of *Rubus pedatus*. In general, the community mosaic in the northern Cascade Range is shifted toward the moist end of the spectrum because of the cooler, moister climate. The *Abies amabilis* Zone in the High

Figure 65.—Understory in an *Abies amabilis/Vaccinium alaskaense* community, the zonal climax in the southern Washington Cascade Range; *Vaccinium alaskaense, Xerophyllum tenax, Cornus canadensis, Clintonia uniflora,* and *Rhytidiopsis robusta* are notable species (Wind River Research Natural Area, Gifford Pinchot National Forest, Washington).

Figure 66.—*Abies procera* attains maximum development on mesic sites in the *Abies amabilis* Zone; these sites are characterized by herb-rich understories (near Mount St. Helens, Gifford Pinchot National Forest, Washington).

Figure 67.—Lush herbaceous understories typify forest understories on moist sites in the *Abies amabilis* Zone; *Achlys triphylla, Vancouveria hexandra, Gymnocarpium dryopteris,* and *Streptopus roseus* on *Abies amabilis/Streptopus* habitat type (Gifford Pinchot National Forest, Washington).

Cascades around Mount Adams and in Oregon is characterized by communities with depauperate understories dominated by *Vaccinium membranaceum* and *Xerophyllum tenax* (Franklin 1966). In the Western Cascades of Oregon, many *Abies amabilis* Zone communities have added *Rhododendron macrophyllum* as an understory dominant (fig. 68) (see footnote 7). Herb-rich and *Vaccinium alaskaense-Cornus canadensis* associations are also present south to at least 44° north latitude. One distinctive community dominated by *Picea engelmannii* and *Abies amabilis* with a lush herbaceous understory is characteristic of wet, frosty habitats throughout the northern Oregon Cascade Range. Generally, the proportion of communities with an understory dominated by herbs and by *Vaccinium* spp. (except *V. membranaceum*) appears to increase and decrease, respectively, as one moves south in the Cascade Range in Oregon.

Communities in which *Abies amabilis* is a major climax component occur sporadically and not as an identifiable zone in high rainfall areas (over 2,000 mm. annually) in the northern Oregon Coast Ranges (Hines 1971). The community types present include herb-rich (*Tsuga heterophylla-Abies amabilis/Vaccinium ovalifolium/Oxalis oregana*) and *Rhododendron*-dominated types (*Tsuga heterophylla-Abies amabilis/Rhododendron macrophyllum/Cornus canadensis*). An anomaly in this area is a frequent occurrence of stands of *Abies procera* at higher elevations (750 to 1,000 meters) from which *Abies amabilis* is absent (Hines 1971, Merkle 1951, Neiland 1958, Franklin 1964).

Figure 68.—**Understory of community on** *Abies amabilis/Rhododendron macrophyllum-Vaccinium alaskaense/Cornus canadensis* **habitat type (H. J. Andrews Experimental Forest, Oregon).**

Successional Patterns

The major climax species throughout the zone is *Abies amabilis* (Thornburgh 1969, Fonda and Bliss 1969, Franklin 1966); size and age class analyses of many stands illustrate this clearly (fig. 69). A typical successional sequence begins with invasion of the site by

Figure 69.—*Abies amabilis* **is the major climax species in this zone dominating the reproduction in old-growth stands;** *Abies amabilis* **saplings in an** *Abies procera* **stand (Wildcat Mountain Research Natural Area, Willamette National Forest, Oregon).**

Pseudotsuga menziesii or *Abies procera*, or both. Western hemlock may be established simultaneously or develop later under a forest canopy, or both. In any case, *Pseudotsuga* and *Abies procera* fail to reproduce. The heavy-seeded, fire-sensitive *Abies amabilis* is usually last to invade the site (Schmidt 1957), coming in under the mixed canopy of *Pseudotsuga menziesii*, *Abies procera*, and *Tsuga heterophylla*; if seed sources are available, *Abies amabilis* can also function as a pioneer species, however.

Four hundred to 500 years after disturbance, a typical mixed stand includes scattered, large (100- to 150-cm d.b.h.) *Pseudotsuga menziesii*, more abundant but somewhat smaller (70- to 100-cm. d.b.h.) *Tsuga heterophylla*, and abundant seedlings, saplings, and poles of *Abies amabilis*. Many stands with this or a similar structure are encountered in the Cascade Range.

Stand structure analyses indicate *Tsuga heterophylla* is a minor climax species along with *Abies amabilis*, especially at lower elevations within the zone. Climax species on wet sites are not easily predicted since seedlings and saplings of all species are relatively scarce. It does appear that both *Tsuga heterophylla* and *Thuja plicata* are at least minor climax species.

Replacement of the shade-tolerant *Tsuga heterophylla* by *Abies amabilis* on most sites is probably a consequence of mechanical factors (Thornburgh 1969). Fragile *Tsuga* seedlings are unable to survive winter accumulations of litter and snow characteristic of forest floors in the zone, whereas those of *Abies amabilis* are (Thornburgh 1969, Kotor 1972). Surviving *Tsuga heterophylla* reproduction is invariably confined to down logs and mounds of rotten wood. *Tsuga heterophylla* seedlings do conduct photosynthesis more efficiently at low light intensities, and reportedly they grow faster than those of *Abies amabilis* (Thornburgh 1969).

Abies amabilis sometimes occurs within the *Tsuga heterophylla* Zone and grows to large size there (e.g., see Hades Creek in Franklin et al. 1972). Fire, slow growth rates, and a sensitivity to moisture stress have been mentioned as possible factors restricting

Abies amabilis at lower elevations (Schmidt 1957, Thornburgh 1969, Kotor 1972). Kotor (1972), in particular, has proposed that *Abies amabilis* is physiologically unsuited for the generally drier habitats found in the lowlands and has provided some evidence to substantiate this hypothesis. Successional relationships between *Tsuga heterophylla* and *Abies amabilis* are reversed at lower elevations with *Tsuga* playing the climax role. The lesser shade tolerance and slower growth rate of *Abies* seedlings (Thornburgh 1969) and greater drought tolerance of *Tsuga* seedlings (Kotor 1972) are certainly key factors in the seral role of *Abies amabilis* in such habitats.

Early stages in forest succession (from disturbance through development of a closed forest canopy) have not been studied in detail. However, major dominants during the preforest stage of succession are generally species present in the forest stands before disturbance. Some examples are all the *Vaccinium* spp., *Xerophyllum tenax*, and *Sorbus* spp. *Pteridium aquilinum* is often a major invader not present before disturbance. An exception to these generalizations is on wet sites where disturbance may produce dense shrub communities of *Sambucus*, *Rubus spectabilis*, and *Ribes* spp.

Special Types

Alnus sinuata Communities

Alnus sinuata characterizes a community type found most often in this zone. These shrub communities occupy sites subject to deep winter snow accumulations and extensive snow creep; they often suffer recurrent snow avalanches as well. Consequently, *Alnus sinuata* individuals, 3 to 5 meters tall, have strongly bowed stems (fig. 70). In the Washington Cascade Range, repeated avalanching is at least partially responsible for creation and maintenance of these communities (fig. 71). In Oregon, *Alnus sinuata* communities

Figure 70.—*Alnus sinuata* **communities are on sites subject to heavy snow accumulations and snow creep or avalanching; hence, the 3- to 5-meter-tall** *Alnus* **have strongly bowed stems (H. J. Andrews Experimental Forest, Oregon).**

Figure 71.—*Alnus sinuata* **communities are permanently maintained on some sites by recurrent snow avalanches (Mount Rainier National Park, Washington).**

appear on sites that are not avalanche tracks but do have heavy snow accumulations and abundant seepage water. High water tables have been ascribed to a nearly impervious subsoil in some areas,[10] but in others there appears to be no difference in substrate between forest and shrub communities (Daubenmire and Daubenmire 1968, Aller 1956). *Alnus sinuata* communities are to all appearances a stable community type or are only very slowly encroached upon by forest vegetation. The only conifer capable of surviving and reproducing on sites with recurrent avalanches is *Chamaecyparis nootkatensis*. Since the sites occupied by *Alnus sinuata* communities are normally very wet, the understory is typically rich in dicotyledonous herbs, such as *Montia* spp., ferns, and Carices and usually includes *Oplopanax horridum*. Similar communities occur in higher forested zones on the east slopes of the Cascade Range, in the northern Rocky Mountains (Daubenmire and Daubenmire 1968), and in the Blue Mountains where *Alnus incana* dominates (personal communication, Dr. F. C. Hall). Avalanche tracks on which *Populus tremuloides* is the major shrub are also encountered in the northeastern Washington Cascade Range and Okanogan Highlands.

Meadow Communities

There are many types of mountain meadow and other nonforested communities associated with the *Abies amabilis* Zone. A mesic meadow type, dominated by *Pteridium aquilinum* and *Rubus parviflorus*, is extremely widespread and is often contiguous with *Alnus sinuata* communities and may intergrade with subalpine meadow types at higher elevations (personal communication, Mr. G. W. Douglas). Hickman (1968) categorized and listed constituent species for the different vegetation types found in forest openings along the ridges and peaks of Oregon's Western Cascades Province; the majority of these are associated with the *Abies amabilis* Zone. Included in Hickman's lists are three different meadow types, and seep, talus, and outcrop communities.

Tsuga mertensiana Zone

The *Tsuga mertensiana* Zone is the highest forested zone along the western slopes and crest of the Cascade Range and in the Olympic and Klamath Mountains. Elevational limits of the *Tsuga mertensiana* Zone are generally between 1,300 and 1,700 meters in northern Washington, 1,250 and 1,850 meters around Mount Rainier (Franklin and Bishop 1969), and 1,700 and 2,000 meters in the southern Oregon Cascades. The zone extends varying distances east of the Cascade crest until it is gradually replaced by the *Abies lasiocarpa* Zone more typical of interior subalpine environments. A similar replacement of *Tsuga mertensiana* forests by those of *Abies lasiocarpa* occurs in the rain shadow portion of the Olympic Mountains (Fonda 1967, Fonda and Bliss 1969). *Tsuga mertensiana*-dominated forests reappear in portions of the northern Rocky Mountains in association with those of *Abies lasiocarpa* and *Picea engelmannii* (Daubenmire 1952, Daubenmire and Daubenmire 1968, Habeck 1967).

The *Tsuga mertensiana* Zone can be divided into major subzones—a lower subzone of closed forest and an upper parkland subzone. In the lower subzone, there is essentially continuous forest cover of *Tsuga mertensiana* and its associates. The upper subzone is a mosaic of forest patches and tree groups interspersed with shrubby or herbaceous subalpine communities. In this section, we will be concerned only with the lower subzone; the

---

[10] Unpublished soil survey data from the H. J. Andrews Experimental Forest on file at Forestry Sciences Laboratory, Pacific Northwest Forest and Range Experiment Station, Corvallis, Oregon.

subalpine meadow-forest mosaic will be considered during the discussion of timberline and alpine regions.

As defined here, the *Tsuga mertensiana* Zone is comparable to the Mountain Hemlock Zone of Krajina (1965); his parkland and forest subzones are also identical. The *Tsuga mertensiana* Zone also includes most of the Hudsonian Life Zone (Barrett 1962) and, perhaps, the upper part of the Canadian Zone as well. Douglas (1970, 1972) refers to the parkland subzone of the *Tsuga mertensiana* Zone as the "subalpine zone" and considers the closed forest subzone simply a component of a midelevation "Pacific silver fir zone" (personal communication). This view is partially justified by the occurrence of *Abies amabilis* as a major climax species in the closed *Tsuga mertensiana* forests in most of Washington and northern Oregon. Similar *Tsuga mertensiana* forests, without *Abies amabilis*, extend north to Alaska, south to California, and east to Montana, however, and the group constitutes a vegetationally and environmentally distinctive formation. Consequently, it seems parochial to eliminate one segment of the formation on the basis of *Abies amabilis*, and we have elected to recognize a closed forest subzone of the *Tsuga mertensiana* Zone.

Environmental Features

The *Tsuga mertensiana* Zone is wet and is the coolest of the forested zones in western Oregon and Washington (table 13). Annual precipitation ranges from about 1,600 to 2,800 millimeters. This includes 400 to 1,400 centimeters of snowfall which accumulates in snowpacks up to 7.5 meters deep. The sharp increase in snow accumulation within this zone is a consequence of the elevation of the freezing isotherm during the winter months (Peterson 1969, Brooke et al. 1970). Detailed micro- and macroclimatic data collected in British Columbia (Brooke et al. 1970) are certainly representative of *Tsuga mertensiana* Zone in northern Washington and probably beyond. These data suggest a snow duration of at least 6 months but seldom more than 8 months in the closed forest subzone.

Soils within the zone are podzolic, although the degree of podzolization varies greatly. In the north, Cryorthods (Podzols and Gley Podzolic soils) are common, with well-developed mor or duff-mull humus layers 5 to 10 centimeters thick. In central and southern Oregon, Cryorthods (Podzols) are rare; and weakly developed Haplorthods (Brown Podzolic soils) with a relatively thin (2 to 5 centimeters), but densely matted, felty mor humus layer are the rule (Williams and Dyrness 1967).

Table 13. — Climatic data from representative weather stations within the *Tsuga mertensiana* Zone

Station	Elevation	Latitude	Longitude	Temperature					Precipitation		
				Average annual	Average January	Average January minimum	Average July	Average July maximum	Average annual	June through August	Average annual snowfall
	Meters			- - - - - - - - - Degrees C. - - - - - - - - - -					Millimeters		Centimeters
Mount Baker Lodge, Wash.	1,362	48°52'	121°40'	4.5	−2.6	−5.7	12.1	17.5	2,821	313	1,398
Paradise Ranger Station, Wash.	1,821	46°47'	121°44'	3.4	−3.4	−7.0	11.6	17.4	2,635	226	1,362
Crater Lake National Park Headquarters, Oreg.	2,124	42°54'	122°08'	3.8	−3.7	−8.4	13.4	21.9	1,643	99	1,324

Source: U. S. Weather Bureau (1956, 1965a, 1965b).

Forest Composition

Forest composition within the zone varies considerably with locale (table 14). Relatively few species are found as dominants—*Tsuga mertensiana* in old-growth forests throughout the zone (fig. 72) and *Abies lasiocarpa* or *Pinus contorta*, or both, in seral stands in drier portions of the zone. *Abies amabilis* is conspicuous in the zone in Washington and northern Oregon but drops out completely in the southern Oregon High Cascades and Siskiyou Mountains. *Abies amabilis* and *Chamaecyparis nootkatensis* are major associates, and *Pseudotsuga menziesii*, *Abies lasiocarpa*, and *Pinus monticola* are minor associates in wetter portions of the *Tsuga mertensiana* Zone in the Olympic Mountains (Fonda and Bliss 1969). There are a wide variety of understory species of which many are Ericaceae, Rosaceae, and Compositae.

The spectrum of communities found within the *Tsuga mertensiana* Zone varies locally with gradients in temperature, moisture, and accumulation and duration of snow, as well as geographically (Brooke et al. 1970, Franklin 1966). Some of the associations which

Figure 72.—**Pure, old-growth stands of** *Tsuga mertensiana* **are common throughout the** *Tsuga mertensiana* **Zone; note the clumping tendency of the trees in this stand near timberline (Gifford Pinchot National Forest, Washington).**

Table 14. — Relative importance of tree species in the closed-forest subzone of the *Tsuga mertensiana* **Zone.**

	Washington			Oregon	
Species	Western Cascades	Eastern Cascades	Olympic Mountains	Northern Cascades	Southern Cascades
Abies amabilis	[1] M	m	M	M	—
Abies lasiocarpa	m	M	m	M	—
Abies magnifica var. *shastensis*	—	—	—	—	M
Abies procera	m	—	—	m	—
Chamaecyparis nootkatensis	M	m	M	m	—
Picea engelmannii	—	m	—	m	m
Pinus albicaulis	—	m	—	m	m
Pinus contorta	m	M	m	M	M
Pinus monticola	m	m	m	m	m
Pseudotsuga menziesii	m	m	m	m	—
Thuja plicata	m	—	—	—	—
Tsuga heterophylla	m	m	m	m	—
Tsuga mertensiana	M	M	M	M	M

[1] M indicates a species is a major constituent of either seral or climax forests or both; m indicates a species is a minor component.

have been described are (the latinized association names of Brooke et al. (1970) have been altered to our format):

Association	Location	Author
Abies amabilis-Tsuga mertensiana/ Xerophyllum tenax	Central Oregon Western Cascades	[11]
Abies amabilis-Tsuga mertensiana/ Vaccinium membranaceum	Southern Washington-northern Oregon	Franklin 1966
Abies amabilis/Veratrum viride	Southern Washington-northern Oregon	Franklin 1966
Abies amabilis/ Menziesia ferruginea	Southern Washington-northern Oregon	Franklin 1966
Chamaecyparis nootkatensis/ Rhododendron albiflorum	Southern Washington-northern Oregon	Franklin 1966
Tsuga mertensiana/Vaccinium membranaceum	British Columbia	Brooke et al. 1970
Tsuga mertensiana/Cladothamnus pyrolaeflorus	" "	" "
Tsuga mertensiana-Abies amabilis/ Vaccinium alaskaense	" "	" "
Abies amabilis/Streptopus roseus	" "	" "
Thuja plicata/Oplopanax horridum	" "	" "
Chamaecyparis nootkatensis/ Lysichitum americanum	" "	" "

[11] See footnote 7.

One of the most widespread associations is the *Tsuga-Abies/Vaccinium membranaceum* type which occurs from British Columbia to central Oregon. In southern Washington, this community has a depauperate understory composed of *Vaccinium membranaceum, Xerophyllum tenax, Pyrola secunda, Rubus lasiococcus,* and perhaps a few additional species (Franklin 1966). The understory is frequently enriched as the upper limits of closed forest are reached by species such as *Vaccinium deliciosum* and *Phyllodoce empetriformis.* A variant of this community, in which the understory is completely dominated by *Xerophyllum tenax* is common in the southern Washington and northern Oregon Cascade Range (fig. 73). These communities dominated by *Vaccinium membranaceum* and *Xerophyllum tenax* are obviously closely related to Daubenmire and Daubenmire's (1968) *Abies lasiocarpa/Xerophyllum* association in the Rocky Mountains as well as Fonda and Bliss' (1969) *Abies lasiocarpa* type in drier portions of the Olympic Mountains.

On wetter, cooler habitats, e.g., north slopes and along drainages, communities have dense shrub understories 1 to 1½ meters in height. The most extreme example in southern Washington is a *Chamaecyparis nootkatensis/Rhododendron albiflorum* association (fig. 74) (Franklin 1966). A tangle of *Vaccinium ovalifolium, V. membranaceum, Rhododendron*

Figure 73.—**Communities with understories dominated by** *Xerophyllum tenax* **and** *Vaccinium membranaceum* **are very widespread on poorer sites in the** *Tsuga mertensiana* **Zone; a** *Tsuga mertensiana-Abies amabilis/Xerophyllum tenax* **community on a ridgetop with shallow soil (Wildcat Mountain Research Natural Area, Willamette National Forest, Oregon).**

Figure 74.—**Representative stand of the** *Chamaecyparis nootkatensis/Rhododendron albiflorum* **association;** *Abies amabilis* **and** *Tsuga mertensiana* **are associated with** *Chamaecyparis* **in the overstory and** *Menziesia ferruginea* **and** *Vaccinium* **spp. with** *Rhododendron* **in the dense, tangled shrub layer (Snoqualmie National Forest, Washington).**

albiflorum, Menziesia ferruginea, and *Sorbus* spp. is characteristic of this shrub layer. Typical herbs include *Rubus pedatus, R. lasiococcus, Valeriana sitchensis, Viola sempervirens, Listera caurina,* and *Erythronium montanum.*

Interestingly, the *Tsuga mertensiana/Cladothamnus pyrolaeflorus* association described from British Columbia has a similar dense tangle of shrubs in the understory but is reportedly found on habitats drier than normal (Brooke et al. 1970). Communities on wet sites in that area are characterized by understory species such as *Lysichitum americanum* and *Coptis asplenifolia* where seepage is slow-moving or stagnant, and *Oplopanax horridum, Athyrium filix-femina,* and a variety of other herbs where seepage is fast flowing.

Herb-rich understories are characteristic of habitats with a temporary seepage influence and deep soils. Associations of this type include the *Abies amabilis/Veratrum viride* (Franklin 1966) and *Abies amabilis/Streptopus roseus* (Brooke et al. 1970). Modal habitats in British Columbia and the central Cascade Range, particularly toward the lower edge of the *Tsuga mertensiana* Zone, are occupied by communities with understories dominated by shrubs such as *Vaccinium alaskaense, Menziesia ferruginea, V. membranaceum,* a few herbs such as *Rubus pedatus* or *R. lasiococcus* and the moss *Rhytidiopsis robusta.* Associations of this type include the *Abies amabilis/Menziesia ferruginea* (Franklin 1966) and *Tsuga mertensiana-Abies amabilis/Vaccinium alaskaense* (Brooke et al. 1970).

The most extensive *Tsuga mertensiana* forests are those found along the crest of the Cascade Range in central and southern Oregon, and these have not yet been the subject of published reports. These forests are generally much simpler structurally and floristically than those mentioned above. The sparse understories are usually dominated by *Vaccinium membranaceum* or *V. scoparium* along with minor amounts of other species such as *Chimaphila umbellata* and *Arctostaphylos nevadensis.* The absence of data on these forests is a major gap in our knowledge of the forests of this zone. Some suggestions about the nature of some of these types can be found in the discussion of the *Abies magnifica shastensis* Zone in Chapter VI.

Successional Patterns

Large acreages of the *Tsuga mertensiana* Zone have been burned over during the last 150 years in southern Washington and Oregon. Consequently, seral communities are often conspicuous. Early stages in succession have not been studied, but they often involve domination of the site by fire-resistant species conspicuous in the understory—*Vaccinium membranaceum* and *Xerophyllum tenax,* for example.

Forest development on burned areas is often very slow, as would be expected in the severe environment of this zone. In some cases, semipermanent communities of *Vaccinium* spp., *Xerophyllum, Sorbus* spp., and *Spiraea* have been created by repeated burning. Indians used this method to perpetuate fields of *Vaccinium membranaceum* from which they collected berries for food. Many of these areas are valuable recreational resources which will be managed to perpetuate the *Vaccinium* resource (Minore 1972b) (fig. 75).

Successional sequences of tree species vary geographically. On moist sites, *Tsuga mertensiana* and *Abies amabilis* can function as pioneer species. On dry sites, and particularly in Oregon's High Cascades Province, seral forests of *Pinus contorta* (fig. 76) or *Abies lasiocarpa* (fig. 77) are often the first to develop. The distributional pattern and successional role of *Abies lasiocarpa* suggest it has a greater tolerance of drought than *Abies amabilis* and *Tsuga mertensiana,* and this hypothesis has been partially verified experimentally by Lowery (1972). Studies of seedling root growth, root-shoot ratios, and

Figure 75.—Fields of *Vaccinium membranaceum* and other shrubs have been created and maintained by repeated burning in the *Tsuga mertensiana* and *Abies amabilis* Zones (near Mount Adams, Washington).

Figure 76.—*Pinus contorta* is a major pioneer species in High Cascades portions of the *Tsuga mertensiana* Zone; here, *Tsuga mertensiana* is developing in a stand of dead and dying *Pinus contorta* (near Olallie Butte, Mount Hood National Forest, Oregon).

Figure 77.—Seral forest of nearly pure *Abies lasiocarpa* developed on a warm, south slope within the *Tsuga mertensiana* Zone; similar stands are common throughout much of the Washington Cascade Range (proposed Steamboat Mountain Research Natural Area, Gifford Pinchot National Forest, Washington).

survival under different levels of moisture stress all indicate *Abies lasiocarpa* is more drought tolerant. In any case, these pioneer forests are gradually replaced by *Tsuga mertensiana* and, except in southern Oregon, by *Abies amabilis* (Franklin and Mitchell 1967).

In Washington and northern Oregon, *Abies amabilis* appears to be the major climax species in closed-forest portions of the *Tsuga mertensiana* Zone (Franklin 1966, Thornburgh 1969). Although nearly all old-growth stands do contain mature *Tsuga mertensiana*, reproduction is often almost absent; a full range of *Abies amabilis* size classes is usually present. *Tsuga mertensiana* and *Chamaecyparis nootkatensis* may be minor climax species on some habitats. In southern Oregon, more tolerant arborescent associates are absent from *Tsuga mertensiana* stands and *Tsuga* is the climax species. Apparently, regeneration develops after the old-growth stands begin to break up, since significant *Tsuga mertensiana* regeneration is not common under closed-forest canopies.

Special Types

As mentioned earlier, most of the subalpine nonforested communities will be discussed in Chapter X on timberline and alpine vegetation. There is one class or type of "meadow" community which is more typically found in the forested subzone of the *Tsuga mertensiana* Zone and, less frequently, in the *Abies amabilis* Zone. These are the marshes, boggy meadows, or moors which are found in poorly drained depressions or on wet, poorly aerated soils on gentle slopes. Brooke et al. (1970) describe such areas as the *Eriophorum-Sphagnum* association which includes as character species *Carex aquatilis, Eriophorum polystachion, Equisetum palustre, Viola palustris, Sphagnum squarrosum, S. magellanicum, Scirpus cespitosus,* and *Trientalis arctica.* They indicate a wide range of variation in this association which develops on open stagnant water or where drainage is greatly impeded. Accumulations of organic matter may rise above the ground surface with scattered small pools of open water interspersed.

Almost no descriptions of these mountain wet meadows, bogs, or moors are available for Oregon or Washington. However, they do occur and often are conspicuous features

Figure 78.—**Boggy meadows, marshes, or moors are characteristic of poorly drained areas in the forest subzone of the** *Tsuga mertensiana* **Zone and** *Abies amabilis* **Zone, particularly in the Washington Cascade Range. Left:** *Carex aquatilis/Salix pedicellaris* **community showing marginal shrub belt dominated by** *Spiraea, Lonicera, Alnus,* **and** *Salix* **(proposed Steamboat Mountain Research Natural Area, Gifford Pinchot National Forest, Washington). Right: Marshes and bogs dominated by Carices;** *Menyanthes trifoliata* **is conspicuous in the small foreground pond (proposed Goat Marsh Research Natural Area, Gifford Pinchot National Forest, Washington).**

in some subalpine forest landscapes. One such area near Mount Adams[12] (fig. 78) has *Carex aquatilis, Salix pedicellaris, Eriophorum polystachion, Vaccinium occidentale, Agrostis thurberiana, Aster occidentalis, Carex luzulina,* and *Kalmia polifolia* as dominants. There are many distinctive associated species including *Equisetum* sp., *Cicuta douglasii, Carex muricata, Dodecatheon jeffreyi, Camassia quamash, Epilobium glandulosum, Pedicularis groenlandica, Senecio subnudus, Juncus ensifolius* var. *montanus, Tofieldia glutinosa, Carex disperma,* and *Saxifraga oregana.* An ecotonal belt of shrubs 1 to 2 meters high is characteristic between forest and moor, and in this area, *Spiraea douglasii, Lonicera caerulea, Alnus sinuata, Salix phylicifolia,* and *Viburnum edule* dominated the ecotone (fig. 78). In an extensive marshy landscape near Mount St. Helens[13] (fig. 78), *Carex aquatilis, C. rostrata, C. lenticularis, Scirpus microcarpus, Eriophorum polystachion, Calamagrostis canadensis, Glyceria elata,* and *Juncus ensifolius* var. *montanus* are major species in one or more of the varied community types. *Menyanthes trifoliata* characterizes some shallow ponds. Marshes and moors of these types frequently provide habitat for unusual plants such as *Drosera* spp. and *Utricularia* spp. (see Gold Lake Bog Research Natural Area in Franklin et al. 1972).

---

[12] Data from proposed Steamboat Mountain Research Natural Area on file at USDA Forest Service, Forestry Sciences Laboratory, Corvallis, Oreg.

[13] Data from proposed Goat Marsh Research Natural Area on file at USDA Forest Service, Forestry Sciences Laboratory, Corvallis, Oreg.

CHAPTER V. INTERIOR VALLEYS OF WESTERN OREGON

"Interior" valleys refer to the valley bottoms and lowlands enclosed by the Cascade Range on the east and the Coast Ranges or Siskiyou Mountains on the west. The major units of this type are the Umpqua, Rogue River, and Willamette valleys. Because of their location they are relatively warm, dry regions, too dry for many of the mesic species found on adjacent mountain slopes, such as *Tsuga heterophylla*. These areas were settled relatively early (during the middle of the 19th century) and have been subjected to extensive human influences. Cities, farmlands, and other developments dominate the landscape (fig. 8) and even where "natural" vegetation remains, it has been largely molded by human activities— fire control programs, clearing, logging, and grazing, for example, or a combination of these. Even the presettlement Indians had substantially influenced the vegetational mosaic by their extensive use of wildfire in hunting activities (Johannessen et al. 1971).

The vegetational mosaic today includes *Quercus* woodlands, coniferous forests, grasslands, sclerophyllous shrub communities (sometimes referred to as chaparral), and riparian forests. All must be considered seminatural in character because of human activities. We can refer to these valley mosaics as the "Interior Valley" or *"Pinus-Quercus-Pseudotsuga"* Zone, recognizing that this is a typological unit rather than one based on climax community types. As will be seen, it is not yet clear what the climax types are in the various valleys except that the zonal or climatic climaxes are almost certainly forest communities and that they vary in each of the valleys.

The reader should recognize that even though these three major interior valleys share many features in common, there are significant differences between them. Further, this treatment is heavily weighted toward the most mesic, the Willamette valley, because of the near absence of data for the Rogue River and Umpqua valleys. Detling (1968) has taken a different approach, typifying vegetation of the Willamette valley as "Pine-oak Forest" and that of the main Umpqua and Rogue valleys as "Chaparral" with a peripheral belt of Pine-oak Forest in the latter.

Environmental Features

The interior valleys of western Oregon are the warmest and driest regions west of the Cascade Range (table 15). This is primarily because they are in the rain shadow of the Coast Ranges or Siskiyou Mountains. This effect is accentuated by their location at the dry end of the general north to south decrease in precipitation and increase in temperature found in western Washington and Oregon. Summers are hot and dry, potential evapotranspiration far exceeding the moisture buildup during the mild, wet winters. There is, of course, a climatic gradient across these valleys; precipitation gradually decreases moving east and down the eastern slopes of the Coast Ranges to a minimum within the valleys and increases again ascending the western foothills of the Cascade Range. Also, the valleys become increasingly hot and dry from north to south. For example, the Willamette valley is the wettest and the upper Rogue River valley (around Medford, Oregon) is the driest. More locally, the Willamette valley is hotter and drier in the upper valley (around Eugene) than it is 100 miles to the north at its mouth (Portland).

Soils in the Interior Valley Zone include many types found infrequently elsewhere in the Pacific Northwest. Some of the more important soil great groups are Albiqualfs, Argixerolls, Haplohumults, and Haploxerolls (Prairie, Planosol, Alluvial, Reddish Brown

Table 15. — Climatic data from representative stations within the Interior Valley Zone

Area and Station	Eleva- tion	Lati- tude	Longi- tude	Temperature					Precipitation		
				Average annual	Average January	Average January minimum	Average July	Average July maximum	Average annual	June through August	Average annual snowfall
	Meters			----------Degrees C.----------					---Millimeters---		Centi- meters
alem, Oreg.	60	44°55′	123°01′	11.4	3.6	−0.1	19.2	28.0	1,038	55	20
orvallis, Oreg.	62	44°38′	123°12′	11.3	4.0	.6	18.9	27.1	1,004	47	--
ugene, Oreg.	110	44°07′	123°13′	11.2	3.7	.1	19.1	27.9	1,040	53	18
oseburg, Oreg.	154	43°14′	123°22′	12.1	5.2	1.6	19.9	28.0	830	47	17
rants Pass, Oreg.	282	42°26′	123°19′	12.1	3.9	−.3	21.2	32.3	767	31	17
edford, Oreg.	400	42°22′	122°52′	12.2	3.0	−1.2	22.3	31.8	497	35	19

Source: U. S. Weather Bureau (1965a).

Lateritic, and Sols Bruns Acide great soil groups). Chromoxererts (Grumusols), swelling, clayey soils dominated by montmorillonite clay, are found in these interior valleys. Litho-solic soils (Xerumbrepts) are common on many of the valley foothills, particularly in the Umpqua and upper Rogue River valleys. These shallow rocky soils accentuate the diffi-culties for vegetation during the droughty summer months.

Quercus Woodland

Forest stands, groves, and savannas dominated by the deciduous oaks, *Quercus garryana* and *Quercus kelloggii*, and by the evergreen *Arbutus menziesii* are a conspicuous feature of the Interior Valley Zone (fig. 79). Thilenius (1964) indicates there are over 400,000 hectares of oak woodlands in northwestern Oregon alone. *Quercus garryana* is often the sole dominant in the Willamette valley, although *Acer macrophyllum*, *Pseudotsuga menziesii*, or *Arbutus menziesii* may be present. *Quercus kelloggii* is found from the southern Willamette valley south. In the Umpqua valley, mixed stands of both *Quercus* spp. and *Arbutus menziesii* are common, often with scattered *Pseudotsuga menziesii* (Gratkowski 1961a) (fig. 80). In the Rogue River valley, *Quercus garryana* appears to be more important than *Quercus kelloggii* on the most xeric habitats.

Willamette Valley

Quercus communities have been studied in detail only in the Willamette valley. Thilenius (1964, 1968) recognized four major *Quercus garryana* communities in the central valley area and named them after understory dominants: *Corylus cornuta-Polystichum munitum*, *Amelanchier alnifolia-Symphoricarpos albus*, *Prunus avium-Symphoricarpos albus*, and

Figure 79.—Woodlands of *Quercus garryanna* and *Q. kelloggii* are typical of the Interior Valley Zone; oak savanna in the Rogue River Valley.

Figure 80.—Mixed stands of *Quercus garryana* and *Arbutus menziesii* with scattered *Pseudotsuga menziesii* are typical of many low hills in the Umpqua valley; note the *Rosa eglanteria* and *Rhus diversiloba* in the foreground pasture.

Rhus diversiloba. Characteristic understory species for these communities are listed in table 16. The *Corylus cornuta-Polystichum munitum* and *Amelanchier alnifolia-Symphoricarpos albus* are types found on the least disturbed sites. *Prunus avium*, an extremely tolerant introduced shrub or tree, dominates the understory in the third community type. *Prunus avium* saplings often produce nearly impenetrable thickets.

The *Quercus garryana/Rhus diversiloba* community is most widespread and occupies the most xeric habitats (Thilenius 1964, 1968). It is also found on sites which have been or are heavily grazed and can be the consequence of this grazing. *Rhus diversiloba* occurs in two growth forms—a ground cover and a liana—which are usually connected by a dense, shallow root system (fig. 81). It is generally less palatable to livestock than any of its associates. Thilenius (1964) hypothesized that even if the ground cover *Rhus diversiloba* is grazed, it can draw on liana *Rhus*, which is out of reach of grazing animals, for photosynthate. He further suggests that once *Rhus diversiloba* has replaced its associates, it maintains dominance even when grazing is stopped.

Thilenius (1964) recognized a *Rhus*-Gramineae variant of the *Rhus diversiloba* community which includes an abundance of herbs (fig. 82) (introduced species are indicated by an asterisk): *Poa pratensis**, *Dactylis glomerata**, *Agrostis idahoensis*, *Festuca rubra*, *Torilis arvensis**, *Holcus lanatus**, *Elymus glaucus*, *Danthonia californica*, *Plantago lanceolata**, *Bromus rigidus**, *Hypericum perforatum**, and *Sanicula crassicaulis*.

Table 16. — Characteristic species of major *Quercus garryana* communities in the central Willamette valley

Layer	*Corylus cornuta/Polystichum munitum*	*Amelanchier alnifolia/Symphoricarpos albus*	*Prunus avium/Symphoricarpos albus*	*Rhus diversiloba*
Tree	*Quercus garryana* *Acer macrophyllum* *Pseudotsuga menziesii* *Abies grandis*	*Quercus garryana* *Pseudotsuga menziesii* *Acer macrophyllum*	*Quercus garryana* *Pseudotsuga menziesii* *Acer macrophyllum*	*Quercus garryana* *Pseudotsuga menziesii* *Acer macrophyllum*
Tall Shrub	*Corylus cornuta* *Prunus avium*[1] *Amelanchier alnifolia* *Holodiscus discolor* *Osmaronia cerasiformis* *Crataegus douglasii*	*Amelanchier alnifolia* *Osmaronia cerasiformis* *Prunus avium*[1]	*Prunus avium*[1] *Corylus cornuta* *Amelanchier alnifolia*	
Low Shrub	*Polystichum munitum* *Symphoricarpos albus* *Rubus parviflorus* *Rhus diversiloba* *Pteridium aquilinum* *Rubus ursinus* *Rosa eglanteria*[1] *Rosa gymnocarpa*	*Symphoricarpos albus* *Rhus diversiloba* *Polystichum munitum* *Rosa nutkana* *Rosa eglanteria*[1]	*Symphoricarpos albus* *Rhus diversiloba* *Polystichum munitum* *Rubus ursinus*	*Rhus diversiloba* *Rosa eglanteria*[1] *Symphoricarpos albus* *Rubus ursinus*
Herb	*Tellima grandiflora* *Galium* spp.	*Galium* spp.	*Galium* spp.	*Poa pratensis*[1] *Torilis arvensis*[1] *Galium* spp. *Osmorhiza chilensis* *Satureja douglasii* *Fragaria virginiana* var. *platypetala* *Elymus glaucus*

Source: Thilenius (1964).
[1] Exotic species.

Figure 81.—*Rhus diversiloba* **occurs as interconnected liana and shrub forms in** *Quercus* **woodlands which is believed to be one reason for its increase after it is grazed; both forms are visible here and are probably connected by their root systems (Pigeon Butte Research Natural Area, William L. Finley National Wildlife Refuge, Oregon).**

Hall (1956) also provides some data on composition of a *Quercus garryana* woodland. Typical species in a relatively undisturbed stand were *Rhus diversiloba*, *Amelanchier alnifolia*, *Ligusticum apiifolium*, *Elymus glaucus*, *Bromus vulgaris*, *Osmorhiza chilensis*, *Vicia americana*, and *Trisetum cernuum*. Thinning the *Quercus* resulted in an increase in most of these species as well as addition of *Sanicula crassicaulis* and *Fragaria* sp. as important species. Clearing resulted in abundant *Quercus* coppice but *Rhus diversiloba* and *Elymus glaucus* were the only understory increasers.

Umpqua and Rogue Valleys

Whittaker (1960) described Oak Woodland as the driest forested formation in his transect across the Siskiyou Mountains. In his driest exemplary locale (adjacent to but not in the Rogue River valley), oak woodlands dominated by an overstory of *Quercus kelloggii* and *garryana* and an understory of grasses occupied east and southeast slopes; *Arbutus menziesii* was absent, but *Ceanothus integerrimus*, *Arctostaphylos viscida*, and *Cercocarpus betuloides* were important high shrubs. Northeastern slopes (more mesic sites) were occupied by open stands of *Pseudotsuga menziesii*, *Pinus ponderosa*, and *Libocedrus decurrens* over a well-developed lower canopy of *Quercus garryana*. *Quercus* coverage decreased and grass coverage increased on increasingly xeric sites, and south or southwest slopes were grassland with only widely scattered *Quercus* spp. or none at all.

Waring (1969) described a Black Oak Type (*Quercus kelloggii*) in the foothills of the eastern Siskiyou Mountains which commonly included *Pseudotsuga menziesii*, *Pinus ponderosa*, *Arbutus menziesii*, and occasional *Quercus garryana*. Typical understory species were *Rhus diversiloba*, *Apocynum pumilum*, *Lonicera hispidula* var. *vacillans*, *Balsamorhiza deltoidea*, *Arctostaphylos viscida*, *Festuca californica*, *Lupinus albifrons*, *Brodiaea multiflora*, *Boschniakia strobilacea*, *Castilleja* spp., *Gilia capitata*, and *Plectritis macrocera*.

Obviously, a number of diverse entities are here lumped under "*Quercus* Woodland." They range from very open savannas with grass understories to dense forest stands and

Figure 82.—*Quercus garryana* **communities with grass-***Rhus diversiloba* **understories are widespread in the Willamette valley;** *Corylus cornuta* **var.** *californica* **(left foreground) is more common than** *Rhus* **in this 80-year-old stand near Albany, Oregon.**

from pure *Quercus* types to communities with an abundance of conifer associates, particularly *Pseudotsuga menziesii* and *Pinus ponderosa*. Obviously included are communities both seral and climax. Some of these successional relationships will be considered later in this paper.

Conifer Forests

Conifer forests occupy large areas of the foothill regions in and around the interior valleys. *Pseudotsuga menziesii* is probably the most common species. *Abies grandis* is also common, particularly on more mesic sites. *Pinus ponderosa* increases in frequency from north to south. In the central Willamette valley, it is found primarily on specialized habitats—coarse alluvial deposits along river channels which are subject to high water tables in winter and drought in summer—rather than on the uplands. In the southern Willamette valley and in the Umpqua and Rogue River valleys, it is more widespread on uplands in association with other conifers and hardwoods (Gratkowski 1961a). *Libocedrus decurrens* appears from the southern Willamette valley south and is conspicuous in some valley forests. *Pinus lambertiana* and *P. jeffreyi* (on serpentine) are characteristic of the Rogue River valley (Gratkowski 1961a). *Tsuga heterophylla* is effectively excluded from all but the peripheries of the most mesic valley, the Willamette. *Thuja plicata*, which is somewhat more tolerant of dry climates, is sporadically present on mesic habitats (ravines, streamsides, etc.) in the Willamette valley particularly, but not exclusively, at the northern end around Portland.

Hardwoods are typical conifer associates and include *Acer macrophyllum, Quercus garryana, Q. kelloggii,* and *Arbutus menziesii.* Several other valley hardwoods are primarily riparian and will be discussed later. The intergradation possible between the previously discussed *Quercus*-dominated communities and those dominated by conifers should be obvious. Hardwoods often function as pioneer species which produce, in time, various hardwood-conifer mixtures.

Willamette Valley

Coniferous stands in the Willamette valley are composed mainly of *Pseudotsuga menziesii* (Hansen 1947, Sprague and Hansen 1946, Habeck 1961, Anderson 1967). *Abies grandis* and *Acer macrophyllum* are also widespread components. *Quercus garryana* and *Arbutus menziesii* are usually present as remnants of pioneer stands which prepared the way for the conifers.

Total community descriptions are sparse in the literature but are available for two areas near Corvallis in the west-central valley (Sprague and Hansen 1946, Anderson 1967). Anderson (1967) recognized at least two community types which typify some of the foothill coniferous forests: the *Corylus cornuta californica/Bromus vulgaris* and *Acer circinatum/Gaultheria shallon (Corylus cornuta californica-Holodiscus* subtype) communities. Compositional data for these communities are provided in table 17. It is apparent from a comparison of Anderson (1967) and Thilenius (1964) that most of the understory dominants can be found under both *Quercus garryana* and *Pseudotsuga menziesii* canopies. Some additional comments on understory composition in Willamette valley conifer stands can be found in Sprague and Hansen (1946) and Sabhasri and Ferrell (1960).

Some interesting coniferous communities have been examined in the foothills of the upper Willamette valley southwest of Eugene.[1] In one area, *Pseudotsuga menziesii* dominates with *Libocedrus decurrens* as the major associate. Both species are typically of the same age, but the *Libocedrus* forms a lower stratum below the dominant *Pseudotsuga* canopy. *Acer macrophyllum* is a constant associate, *Abies grandis* is confined to moister habitats, and *Pinus ponderosa* is conspicuous along some streams. Major understory species are *Acer circinatum, Gaultheria shallon, Corylus cornuta* var. *californica, Holodiscus*

---

[1] Data on proposed Camas Swale, Fox Hollow, and Mohawk Research Natural Areas, on file at USDA Forest Service Forestry Sciences Laboratory, Corvallis, Oregon.

Table 17. — Characteristic species in two coniferous forest communities found in the foothills of the Willamette valley

Layer	Corylus cornuta californica/Bromus vulgaris	Acer circinatum/Gaultheria shallon (Corylus cornuta californica-Holodiscus subtype)
Tree	Dominant — *Pseudotsuga menziesii, Abies grandis, Acer macrophyllum* Occasional — *Quercus garryana*	Dominant — *Pseudotsuga menziesii* Occasional — *Abies grandis, Acer macrophyllum, Cornus nuttallii*
Shrub	*Corylus cornuta* var. *californica, Holodiscus discolor, Rosa gymnocarpa, Symphoricarpos albus, Rhus diversiloba*	*Acer circinatum, Corylus cornuta* var. *californica, Holodiscus discolor, Rosa gymnocarpa*
Herb	*Bromus vulgaris, Aster radulinus, Fragaria vesca* var. *bracteata, Satureja douglasii, Vicia americana* var. *truncata, Berberis nervosa, Synthyris reniformis, Madia madioides, Osmorhiza chilensis*	*Gaultheria shallon, Berberis nervosa, Adenocaulon bicolor, Polystichum munitum, Anemone deltoidea, Galium triflorum, Festuca occidentalis*

Source: Anderson (1967).

Figure 83.—**Mixed forest of** *Pseudotsuga menziesii* **and** *Pinus ponderosa* **which is characteristic of many coniferous stands in the interior valleys from the upper Willamette valley south; this dry, ridgetop stand has a shrub layer dominated by** *Holodiscus discolor* **and** *Rhus diversiloba* **(proposed Fox Hollow Research Natural Area near Eugene, Oregon).**

discolor, *Taxus brevifolia, Rhus diversiloba, Philadelphus lewisii, Polystichum munitum, Synthyris reniformis, Cardamine pulcherrima* var. *tenella, Calypso bulbosa, Hieracium albiflorum, Fragaria* sp., *Festuca* sp., and *Adenocaulon bicolor*. A series of east-west oriented ridges crosses a second area, producing an alternation of forest stands on the north and south aspects. South slopes and ridgetops have a mixture of *Pseudotsuga* and *Pinus ponderosa* with minor amounts of *Libocedrus, Quercus garryana, Q. kelloggii,* and

Arbutus menziesii, especially near the ridgetop (fig. 83). The understory is characterized by *Holodiscus discolor*, *Rhus diversiloba*, *Gaultheria shallon*, *Corylus cornuta* var. *californica*, *Synthyris reniformis*, and mosses. The *Pseudotsuga/Acer circinatum/Polystichum munitum* community on the north slopes offers a sharp contrast; *Gaultheria shallon* (especially on upper north slopes), *Berberis nervosa*, *Holodiscus discolor*, *Abies grandis*, and *Acer macrophyllum* are often additional associates. Communities with understories dominated by herbs, particularly *Polystichum munitum*, are not unusual in valley coniferous forests, but they are confined to distinctly cooler and moister habitats.

Umpqua and Rogue Valleys

Very little is known about the composition of coniferous stands in the Umpqua and Rogue River valleys other than the major tree dominants. Gratkowski (1961a) reports *Pseudotsuga menziesii* forest as common on hill crests and more mesic slopes in the Umpqua valley, often in association with *Pinus ponderosa* and *Libocedrus decurrens*. The Rogue River valley forests are perhaps the most diverse found in the three valleys, with *Pseudotsuga* still important but *Pinus ponderosa*, *P. lambertiana*, and *Libocedrus decurrens* also common (Gratkowski 1961a). *Pinus jeffreyi* woodlands, which are found on serpentine substrate, are also present but are discussed in Chapter XI.

The only other coniferous forest data are from the foothills around the Rogue River valley (Whittaker 1960, Waring 1969). Whittaker's (1960) data are hard to interpret, but it appears that he classes low-elevation conifer forests in the western Siskiyou Mountains with the "Mixed Evergreen Formation," *Pseudotsuga*-Sclerophyll in this case. These forests varied greatly with moisture regime but appeared to include: (1) a tree layer of *Pseudotsuga menziesii* (sometimes with *Pinus ponderosa* or *lambertiana* or both); (2) smaller trees of *Castanopsis chrysophylla*, *Lithocarpus densiflorus*, *Arbutus menziesii*, and *Quercus kelloggii*; and (3) understory shrubs such as *Quercus chrysolepis*, *Berberis nervosa*, *Rhus diversiloba*, and *Rosa gymnocarpa*. *Ceanothus integerrimus* and

Figure 84.—*Pinus ponderosa/Ceanothus cuneatus* **community which is one of the most xeric coniferous forest types found in the Rogue River valley (near Cave Junction, Oregon).**

118

Figure 85.—*Pinus ponderosa-Pseudotsuga menziesii / Arctostaphylos viscida* **community which is found on some of the Rogue River valley foothills (Ashland Research Natural Area, Rogue River National Forest, Oregon).**

Arctostaphylos viscida characterized the understory on driest sites. Gratkowski (personal communication) indicates the major conifers on the valley floor in this area are *Pinus ponderosa* and *Libocedrus decurrens* with *Arctostaphylos viscida* and *Ceanothus cuneatus* as understory dominants (fig. 84).

In the floristically poorer, eastern Siskiyou Mountains, Waring (1969) recognized a Ponderosa Pine Type as the most xeric of his coniferous types (fig. 85). *Pseudotsuga menziesii, Arbutus menziesii*, and, sometimes, *Abies concolor* were associated with the *Pinus ponderosa. Arctostaphylos patula, A. viscida, A. nevadensis, Achillea millefolium* var. *lanulosa, Solidago canadensis, Apocynum pumilum*, and *Lupinus* spp. typified the understory.

Grasslands

Grasslands occupied extensive areas of the interior valleys before they were settled (fig. 86) and continue to do so today. Many of the earlier prairies have been lost to forest and woodland (Johannessen et al. 1971), but new ones have been created by settlers and later farmers by clearing or burning or both (fig. 87). Other grasslands occupy sites that appear incapable of supporting tree growth, e.g., grass balds associated with self-mulching soils (Grumusols) or lithosolic, extremely xeric, southerly exposed slopes. Almost all grassland areas (and *Quercus* savanna) have been heavily grazed by domestic livestock—cattle, sheep, or Angora goats—and are extensively used as unimproved pastureland today (fig. 88).

The nature of the original grassland communities is strictly conjectural, since grazing and introduction of alien species have altered all stands to some degree. Turner (1969) has suggested these grasslands probably looked similar to parts of the "California annual-type grassland" with *Danthonia californica* and *Stipa* spp. typical dominant species. Habeck (1961) provided a list of grasses which may have been characteristic of dry and moist sites in the Willamette valley. Species on well-drained sites included *Agrostis hallii, Agropyron caninum, Bromus carinatus, B. vulgaris, Danthonia californica, Elymus glaucus, Festuca octoflora, F. californica, F. rubra, F. occidentalis, Melica subulata, Poa scabrella, Sitanion jubatum*, and *Stipa lemmonii*. Habeck (1961) suggests a large number of forbs were probably also present on the native prairies.

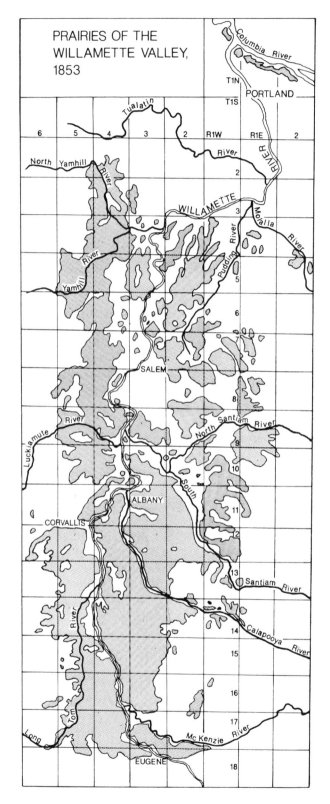

Figure 86.—Extent of prairie or grassland vegetation in the Willamette valley in 1853 (modified from Johannessen et al. (1971)).

Figure 87.—Many of the grasslands in the Interior Valley Zone were created by clearing or burning (or both) of forest lands; burning hill pasture in the Umpqua valley, Oregon.

Figure 88.—Grassland areas in the Interior Valley Zone are heavily grazed. Typically low-lying pasturelands are improved (foreground), and hill grasslands and oak woodland (background) are used as unimproved pasture (near Corvallis, Oregon; *photo courtesy Range Management, Oregon State University*).

There are very few descriptive data even for existing valley grasslands, and these are all confined to the vicinity of Corvallis in the Willamette valley (Livingston 1953, Turner 1969, Valassis 1955, Owen 1953), although the general patterns probably have much wider relevance. Typical constituent species are (introduced species are indicated by an asterisk):

Perennial Grasses—*Danthonia californica, Festuca rubra, F. arundinacea*, Agrostis hallii, A. idahoensis, Poa pratensis*, P. compressa*, Elymus glaucus, Danthonia intermedia, Holcus lanatus*, Stipa occidentalis* var. *minor, Sitanion hystrix, Carex* spp., *Lolium perenne*, Dactylis glomerata, Koeleria cristata,* and *Arrhenatherum elatius**.

Annual Grasses—*Bromus rigidus*, B. commutatus*, B. mollis*, Elymus caput-medusae*, Cynosurus echinatus*, Festuca bromoides*, F. myuros*, Avena fatua*, Aira caryophyllea*, Briza minor*,* and *Gastridium ventricosum**.

Forbs—*Torilis nodosa*, Daucus carota*, Ranunculus occidentalis, Lactuca serriola*, Sherardia arvensis*, Vicia americana, V. tetrasperma*, Erodium cicutarium*, Hypericum perforatum*, Taraxacum officinale*, Fragaria chiloensis, Plantago lanceolata*, Galium divaricatum*, Veronica peregrina, Lathyrus sphaericus*, Eriophyllum lanatum, Hypochaeris radicata*, Achillea millefolium* var. *lanulosa,* and *Sanicula bipinnatifida*.

There is obviously a very high proportion of introduced species in the existing communities; they include all the annual grass dominants. One of these, *Elymus caput-medusae*, is an extremely undesirable species which dominates many stands (Turner et al. 1963, Turner 1969).

121

Moir and Mika[2] have made the most comprehensive floristic analysis of near-natural prairie communities in the Willamette valley. Their study was confined, however, to a bottomland prairie located in the Willamette Floodplain Research Natural Area (Franklin et al. 1972); the tract had been grazed by domestic livestock but was never plowed. Three major prairie communities were identified: (1) *Deschampsia caespitosa*-dominated grassland, (2) *Rosa eglanteria*-dominated shrub thickets, and (3) a *Poa pratensis-Agrostis* spp. community which was typically ecotonal between the first two types. The *Deschampsia* community was a tall-grass type which occurred on the lower, flatter portions of the landscape (fig. 89). Important codominant species include *Holcus lanatus*, *Poa ampla*, *Juncus* spp., *Danthonia californica*, and *Bromus japonicus*. In slight depressions *Hordeum brachyantherum*, *Beckmannia syzigachne*, and *Alopecurus geniculatus* become more important. *Camassia quamash*, *Montia linearis*, and *Eleocharis acicularis* dominated the swales. The shrub thicket community occupied central portions of distinctive hummocks within the prairie proper (fig. 89). *Crataegus douglasii*, *Amelanchier alnifolia*, *Rhamnus purshiana*, *Rhus diversiloba*, and *Symphoricarpos albus* were major associates of the *Rosa*. Shrub cover was typically 60 to 90 percent. Certain species of tall forbs such as *Sidalcea campestris* were sometimes conspicuous in shrub thickets. The *Poa-Agrostis* community was of shorter stature than the *Deschampsia* type and, as mentioned, was often located in the ecotones between tall grass and shrub communities (fig. 89) or within openings in the shrub type. *Danthonia californica*, *Festuca pratensis*, *Carex* spp., *Geranium dissectum*, and *Aster chilensis* were other significant species in the *Poa-Agrostis* type. *Fraxinus latifolia* and other shrubs or trees appear to have invaded the *Poa-Agrostis* type much more aggressively than the *Deschampsia* type; *Fraxinus latifolia* did occur in each of the communities, however, suggesting a potentially forested climax in the absence of fire.

Turner (1969) was the only investigator who examined community composition on several sites. *Elymus caput-medusae* and *Danthonia californica* were generally the dominants on his study sites. At one location, Turner thought there might be three grassland types: *Festuca rubra*-dominated on the most mesic habitat, *Stipa occidentalis* var. *minor*-dominated on the most xeric, and *Danthonia californica*-dominated on the intermediate habitat. From his descriptions, we would judge that several of the native perennial grasses have considerable ability to resist grazing pressure and persist even though alien annuals are the most widespread dominants.

Livingston (1953) provided a short list of grasses for *Quercus* savanna used as unimproved pasture: *Melica geyeri*, *Dactylis glomerata*, *Poa compressa*, *Lolium perenne*, *Bromus mollis*, *Festuca myuros*, and *Cynosurus echinatus*.

The successional status of valley grasslands is little known. Johannessen et al. (1971) clearly feel that the vast majority, if not all, of the prairies are potentially forested if fire and other disturbances are eliminated. Moir and Mika (see footnote 2) found trees (*Fraxinus latifolia*) and tall shrubs in their prairie communities. Scattered bushes of *Rosa eglanteria* and *Rhus diversiloba* are found in many grasslands (fig. 90) and can become dominant over parts of pastures. Sprague and Hansen (1946) describe invasion of some grasslands by *Quercus* spp. It is our belief that most Willamette valley grasslands are seral communities created and maintained by fire or other human influences. Successional rates vary widely, however, and are probably much slower on some of the poorly drained, heavy floodplain soils than on better drained sites. Furthermore, other habitats appear to be climax grassland sites including some with relatively deep fine-textured, self-mulching soils (Grumusols) as well as xeric lithosolic sites. Both conditions are often noted as grass balds on valley hillsides.

---

[2] William Moir and Peter Mika. Prairie vegetation of the Willamette valley, Benton County, Oregon. Unpublished report on file at USDA Forest Service Forestry Sciences Laboratory, Corvallis, Oregon. 1972.

Figure 89.—Prairie communities in the floodplain of the Willamette valley (Willamette Floodplain Research Natural Area, William L. Finley National Wildlife Refuge, Corvallis, Oregon). Left: Shrub thickets dominated by *Rosa eglanteria* and *Poa pratensis-Agrostis* spp. community which is often ecotonal between the thickets on raised topography and *Deschampsia* community. Right: *Deschampsia caespitosa* community, typical of lower lying areas with seasonally wet soils.

Figure 90.—Grasslands are sometimes invaded by *Rosa eglanteria* or *Rhus diversiloba* in the Interior Valley Zone; *Rosa eglanteria* is common in this annual grassland dominated by *Bromus mollis*, *Cynosurus echinatus*, *Lolium perenne*, and *Daucus carota* (photo courtesy Range Management, Oregon State University).

123

Sclerophyllous Shrub Communities

Communities of sclerophyllous shrubs are conspicuous in southern interior valleys, especially the Rogue River valley (Gratkowski 1961a) (fig. 91). Very little is known about these communities. Gratkowski (1961a) lists *Ceanothus cuneatus* (especially on the most xeric sites) and *Arctostaphylos viscida* as major brushfield dominants in the valley bottoms. Other characteristic or abundant brushfield species listed by Gratkowski (1961a) are: *Ceanothus integerrimus, C. cordulatus, Rhus diversiloba, R. trilobata*, and *Lithocarpus densiflorus* with *Cornus glabrata* and *Quercus chrysolepis* on moister sites.

Detling (1961, 1968) views at least some of these shrub communities as southern Oregon chaparral, a northern extension of Californian chaparral types. He lists *Ceanothus cuneatus, Arctostaphylos viscida*, and *A. canescens*, as chief constituents of Oregon chaparral, *Ceanothus cuneatus* occupying the most xeric sites. Other chaparral species are *Cercocarpus betuloides, Eriodictyon californicum, Garrya fremontii, Rhamnus californica* var. *occidentalis, Rhus trilobata, Amelanchier pallida*, and *Chrysothamnus nauseosus* var. *albicaulis*. Detling (1961) indicates some chaparral communities are climax (e.g., *Ceanothus cuneatus* on the Rogue River valley floor), whereas others depend on fire for continuance. He found a *Pinus ponderosa/Arctostaphylos canescens* community in the Illinois valley in which *Pinus ponderosa* was believed climax, but indicates none of the chaparral communities (as he defined them) were normally associated with *Pseudotsuga menziesii*.

Figure 91.—Chaparral is conspicuous in southern interior valleys; *Ceanothus cuneatus* (left) and *Arctostaphylos viscida* (right), on the background slope, are typical in the Rogue River valley and are mixed with *Quercus* spp. in this area.

Riparian Communities

Hardwood forests are typical of riparian habitats and other poorly drained sites subject to annual flooding; however, these communities have not been studied. *Populus trichocarpa* (fig. 92) is one of the most characteristic dominants along the major rivers, with stands of this species attaining maximal development in the lower Willamette and Columbia Rivers (from the Columbia Gorge west). Nearly pure stands occur on many of the islands and line the shores in this area, often in association with communities of

Figure 92.—*Populus trichocarpa* is typical of riparian sites in the Interior Valley and *Tsuga heterophylla* Zones, often forming pure stands on islands in the lower Willamette and Columbia Rivers (near Wind River, Washington, Gifford Pinchot National Forest).

tall (up to 10 meters) *Salix* spp. (*Salix rigida, lasiandra, fluviatilis, sessilifolia,* and *scouleriana*). *Salix* spp. also often form a part of the nearly impenetrable understory found in many *Populus trichocarpa* stands. Interestingly, a *Populus trichocarpa* plantation was the first artificially established forest stand in the Pacific Northwest; this stand, planted on an island in the lower Willamette River, has since been harvested for pulpwood.

Fraxinus latifolia is a very characteristic species on seasonally flooded and swampy habitats in the interior valleys as well as in adjacent, higher elevation forest zones (fig. 93). Lichen-draped stands dominated by this species line many channels along sluggish streams and rivers in the Willamette and other interior valleys. Understories vary widely from almost nothing under dense stands, or in areas with recent silt deposits, to herbaceous (with *Carex* spp. typical constituents) or densely shrubby types.

Figure 93.—Forests of *Fraxinus latifolia* **are common on seasonally flooded habitats along the channels of sluggish streams and rivers in the Interior Valley Zone; shown here are external (left) and internal (right) views of such a forest (with some** *Quercus garryana*) **during a typical winter flood (note abundant cover of epiphytic lichens) (William L. Finley National Wildlife Refuge, Oregon).**

Several other tree species are characteristic of the floodplain, riparian, or gallery forests. *Acer macrophyllum* is common in some areas (fig. 94), although it is not particularly characteristic. In many low-lying areas, *Quercus garryana* is associated with the *Fraxinus*. *Alnus rubra* is strictly riparian in the Interior Valley Zone and is mixed with increasing proportions of *Alnus rhombifolia* from the Willamette valley south. As mentioned earlier, *Pinus ponderosa* is a flood plain forest constituent in some areas and *Umbellularia californica* is a riparian species in at least some portions of the Umpqua valley (see Myrtle Island Research Natural Area in Franklin et al. 1972) (fig. 95).

Successional Relationships in the Interior Valleys

Successional relationships of the communities within the Interior Valley Zone are essentially unknown except in the Willamette valley. Before the Willamette valley was settled, much of it was occupied by prairies and oak savannas (Smith 1949, Habeck 1961, Johannessen et al. 1971). Dense forests were confined primarily to the mountain foothills and floodplains (gallery forest). Indians were probably responsible for most of the fires which created and maintained these open conditions (Kirkwood 1902, Morris 1934, Johannessen et al. 1971).

The most notable change since the settlement of the Willamette valley (other than land conversion to farms and urban areas) has been the replacement of prairie and *Quercus* savanna by closed *Quercus* forest. Habeck (1961, 1962) has documented this change, using land survey records from the 1850's. Thilenius' (1964, 1968) detailed analyses of *Quercus garryana* stands confirm the fact that most closed canopy *Quercus* forests have originated

Figure 94.—*Acer macrophyllum* is a common hardwood on both riparian and upland sites in the Interior Valley Zone; this 100-year-old stand is located along the Santiam River near Jefferson, Oregon.

Figure 95.—*Umbellularia californica* sometimes occupies riparian habitats within the Interior Valley and adjacent forest zones in southwestern Oregon (Myrtle Island Research Natural Area in the North Umpqua River, Oregon).

127

since 1850. He found scattered large trees of open-grown form and averaging 237 years old, which were surrounded by smaller *Quercus* of forest-grown form, averaging 74 to 105 years of age (fig. 96).

Fire control activities instituted by the settlers are believed responsible for this major successional change. The hypothesis that open *Quercus* savannas were maintained by fire is strengthened by the fire resistance displayed by large, isolated trees during fall field burnings.

Figure 96.—Two-aged stands of *Quercus garryana* **are common in the Willamette valley consisting of scattered, large trees 200 to 300 years of age and of open-grown form (remnants of a former savanna stand) and a majority of smaller trees typically 75 to 100 years of age and of forest-grown form. Such forests resulted when burning of the** *Quercus* **savannas ceased with valley settlement, allowing establishment of closed forest (Pigeon Butte Research Natural Area, William L. Finley National Wildlife Refuge, Oregon).**

There are various opinions regarding the species which will succeed *Quercus garryana* and constitute the forest climax of the Willamette valley (fig. 97). *Quercus garryana* provides a favorable environment for establishment of *Pseudotsuga menziesii* seedlings (Sprague and Hansen 1946, Collins 1947, Owen 1953). Large *Quercus* often shelter abundant conifer reproduction. Old *Quercus* snags of open-grown form can be found in many foothill *Pseudotsuga menziesii* stands, victims of the conifer seedlings they sheltered (fig. 98). Consequently, Sprague and Hansen (1946) felt *Quercus garryana* stands would be replaced by *Pseudotsuga menziesii* and that these stands might, in turn, be replaced by climax forests of *Abies grandis* or *Abies grandis* and *Pseudotsuga menziesii*. Habeck (1962) agreed that *Pseudotsuga menziesii* was increasing in importance, and *Quercus garryana* was not reproducing under its own canopy. However, he felt that the significance of *Acer macrophyllum* had been overlooked and concluded *Quercus garryana* would be succeeded by *Pseudotsuga menziesii* or *Acer macrophyllum* or both. Johannessen et al. (1971) emphasize the fact that coniferous stands have increased in density since 1853 in the Willamette valley and have, in some cases, been grown, logged, and regenerated on sites which were grasslands or savannas prior to that time.

Thilenius (1964) concluded that several successional sequences are likely, depending on local conditions. These include replacement of *Quercus garryana* forest by: (1) *Pseudotsuga menziesii* and *Acer macrophyllum*, the former more abundant on drier and the latter on moister sites; (2) *Abies grandis*, either directly or with an interceding stand of *Pseudotsuga*; and (3) *Quercus garryana*, reproduction being sufficient in some communities (especially *Rhus diversiloba*) for it to qualify as climax. Thilenius (1964) also wondered whether the exotic *Prunus avium* might sometimes replace *Quercus* forest; the ability of *Pseudotsuga menziesii* seedlings to develop in understory *Prunus* thickets will be critical.

Successional sequences in the interior valleys to the south (e.g., Umpqua and Rogue River) have not been studied. Some which have been observed are replacement of (1) *Quercus* spp.-*Arbutus menziesii* stands by conifers, (2) *Pinus ponderosa* stands by *Pseudotsuga*, and (3) *Pseudotsuga menziesii* stands by *Abies grandis*. As to the climatic climax species in the Interior Valley Zone, there is evidence for *Abies grandis* in the Willamette valley and *Pseudotsuga menziesii* in the Umpqua valley. In various parts of the Rogue River valley, *Pseudotsuga*, *Pinus ponderosa-Quercus* spp., and chaparral may be the potential climatic climax.

We conclude (as did Thilenius 1964) that a wide variety of climax communities are possible in the Interior Valley Zone depending on the particular habitat. It seems most likely that the climatic or zonal climax vegetation is forest in all three valley systems and probably coniferous forest. This is certainly the case in the Willamette valley. The probable climatic climax species appear to be: (1) *Abies grandis* in the Willamette valley; (2) *Pseudotsuga menziesii* and *A. grandis* in the central (warmer and drier) and peripheral (cooler and wetter) portions of the Umpqua valley, respectively; and (3) *Pinus ponderosa*, *Pseudotsuga*, and *A. grandis* along a gradient from the central portion to the peripheries of the Rogue River valley.

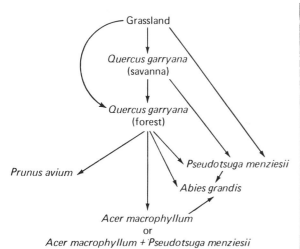

Figure 97.—Successional sequences suggested for upland sites in the central Willamette valley; any of the community types are possible climaxes on selected sites except, perhaps, for *Quercus garryana* savanna (interpreted from Sprague and Hansen 1946; Habeck 1961, 1962; Thilenius 1964).

Figure 98.—Old *Quercus* snags of open-grown form can be found in many Willamette valley *Pseudotsuga menziesii* stands, overtopped and killed by the *Pseudotsuga* reproduction they sheltered (*photo courtesy Range Management, Oregon State University*).

CHAPTER VI. FOREST ZONES OF SOUTHWESTERN OREGON

Southwestern Oregon is an extremely interesting and complex region environmentally, floristically, and synecologically. Climate ranges from cool and moist in the coastal regions to hot and dry in the interior valleys, which are the driest locales west of the Cascade Range. The geology and, consequently, soils are extremely varied. Floristically the region combines elements of the California, north coast, and eastern Oregon floras, with a large number of species indigenous only to the Klamath Mountains region (Whittaker 1960, 1961). This environmental and floristic diversity combines with a long history of pre-historic and historic disturbances, primarily by fire, to produce an extremely varied array of communities.

This chapter deals with portions of southwestern Oregon lying outside the *Tsuga heterophylla*, *Picea sitchensis* and Interior Valley Zones. It includes the main body of the Siskiyou Mountains[1] and western slopes of the Cascade Range. Most of the region is potentially occupied by forest communities of some kind.

The major southwestern Oregon tree species can be arrayed in relation to tolerance of moisture stress:

Species	Tolerance
Quercus garryana	High
Quercus kelloggii	
Pinus ponderosa	
Arbutus menziesii	
Libocedrus decurrens	
Pseudotsuga menziesii	
Pinus lambertiana	
Pinus monticola	
Abies concolor	
Chamaecyparis lawsoniana	
Abies magnifica **var.** *shastensis*	
Tsuga mertensiana	
Tsuga heterophylla	Very low

We might expect a systematic sequence of zones along the moisture and temperature gradients (which are broadly correlated with elevation) based on this array and the relative shade tolerance of the species:

Species	Tolerance
Tsuga heterophylla	High
Abies concolor	
Tsuga mertensiana	
Chamaecyparis lawsoniana	

---

[1] The Siskiyou Mountains are the northernmost range in the Klamath Mountains group (Irwin 1966). Since this is the only range in this group encountered in Oregon, we will use the term "Siskiyou Mountains."

130

Species	Tolerance
Pseudotsuga menziesii ⎱	
Libocedrus decurrens ⎰	
Abies magnifica var. *shastensis*	
Pinus lambertiana	
Pinus monticola	
Pinus ponderosa	
Arbutus menziesii	
Quercus garryana ⎱	
Quercus kelloggii ⎰	Low

In fact, history and environmental complexity often make it impossible to distinguish zones, particularly in the heavily disturbed valley regions. Furthermore, very few data on plant communities are available for the interior valleys and western slopes of the Cascade Range. Therefore, we have taken a more typological approach. The "zonal" outline is as follows (no climax implication is intended in naming the first three zones):

Interior Valley Zone (*Pinus-Quercus-Pseudotsuga*)
Mixed-Evergreen Zone (*Pseudotsuga*-sclerophyll)
Mixed-Conifer Zone (*Pseudotsuga-Pinus-Libocedrus-Abies*)
Abies concolor Zone
Abies magnifica shastensis Zone

The Interior Valley Zone has already been considered in Chapter V. The *Abies magnifica shastensis* Zone is bounded at its upper limits by the *Tsuga mertensiana* Zone (discussed earlier) wherever elevations are sufficiently great (Siskiyou Mountains and Cascade Range).

The arrangement of these zones in the Cascade Range and eastern and western Siskiyou Mountains is illustrated in figure 99. The major difference is the replacement of the Mixed-Conifer Zone by a Mixed-Evergreen Zone in the western Siskiyou Mountains. Waring (1969) has identified the boundary between these two zones (his eastern and western Siskiyou division) and related the boundary to environmental factors. It will be

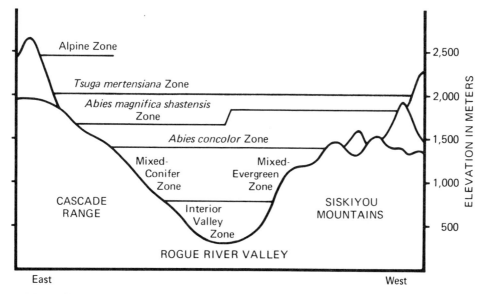

Figure 99.—Arrangement of vegetation zones in the Cascade Range and western Siskiyou Mountains of southwestern Oregon.

131

Figure 100.—Logging in southwestern Oregon's forested zones has typically been by clearcutting; because of greater environmental severity (hotter, drier) than in northwestern Oregon, forest regeneration is frequently poor as on this clearcut within the Mixed-Conifer Zone (*photo courtesy R. H. Waring*).

Figure 101.—Present logging activities on difficult-to-regenerate sites increasingly use shelterwood or selection systems; external (left) and internal (right) views of a partially cut, mixed-conifer forest in which logs were removed by helicopter (*photos courtesy R. H. Waring*).

obvious that most of the major formations found in southwestern Oregon are actually the northern extensions of formations which are considered typical of the California Coast Ranges and Sierra Nevada. As we will point out, separation of the Mixed-Conifer and *Abies concolor* Zones may not be justified on the basis of climax tree species, i.e., both may simply be aspects of a much broader *Abies concolor* Zone.

Fire has played an extremely important role in southwestern Oregon. Fire danger can reach very high levels during the long, hot, dry summers. Fires, both naturally caused and Indian-set, were common prior to the arrival of the white man. Early settlers were responsible for additional burning, but formal fire control activities have been in effect since the early 1900's. As will be seen, many of the communities are created and perpetuated by fires; e.g., some sclerophyllous shrub (chaparral) types and *Pinus attenuata* forests. Fire and fire control activities are having a profound influence on the shape of the future landscape.

Logging is now the major disturbance in the forests of southwestern Oregon. To a large extent, clearcutting continues. However, because environments are generally much more severe (temperatures higher and sites drier) than in northwestern Oregon, clearcutting has been much less successful in regenerating new forests (USDA Forest Service 1973) (fig. 100). Shelterwood cuttings, which provide some protection for the seedling environment, are being used more extensively to overcome this problem (fig. 101). Such systems are entirely appropriate to regeneration of *Pseudotsuga* and *Pinus* spp. in most of this region, and even selective logging may be necessary and appropriate on the most severe sites.

Mixed-Evergreen (*Pseudotsuga*-Sclerophyll) Zone

In the western Siskiyou Mountains, modal sites are generally occupied by a mixed forest of evergreen needle-leaved trees (upper strata) and sclerophyllous broad-leaved trees (lower strata) (fig. 102). Whittaker (1960, 1961), who has written the only significant publications on this zone, felt Mixed-Evergreen Forest should be recognized as a major community type of North America. Several studies of the vegetation in this zone are in progress and should greatly expand our knowledge of it during the next decade.

Figure 102.—Mixed-evergreen forest includes upper stratum of conifers and a lower stratum of sclerophyllous hardwoods; this stand is dominated by *Pseudotsuga menziesii* and *Arbutus menziesii* with an understory of *Lithocarpus densiflorus* (near Oregon Caves, Siskiyou National Forest, Oregon).

Environmental Features

Environmentally, this zone is almost unknown. It is relatively warm and wet during the winter months and hot and dry during the summer. Annual precipitation would appear to be between 600 and 1,700 millimeters or more, depending on elevation and distance from the coast. Less than 20 percent falls during the growing season (Gratkowski 1961a). Average annual temperatures of about 47° to 49° F., average July temperatures of 62° to 66° F., and average January temperatures of 32° to 36° F. are likely.

Soils have not been studied, but they are certainly diverse since there is a very wide variety of parent material. Differences in soils have important phytosociologic implications, as shown by Whittaker's (1960) contrasting of vegetation on diorite, gabbro, and serpentine in this area. Major soil great groups present in this zone probably include Haplohumults, Haploxeralfs, and Haplumbrepts (Reddish Brown Lateritic, Western Brown Forest, and Noncalcic Brown), as well as numerous shallow lithosolic soils.

Forest Composition

Without question, the most important tree species in the Mixed-Evergreen Zone are *Pseudotsuga menziesii* and *Lithocarpus densiflorus;* these are also judged to be the major climax tree species. A variety of other trees may be present, hardwoods being the more characteristic: *Arbutus menziesii*, *Castanopsis chrysophylla*, and *Quercus chrysolepis*. *Q. vaccinifolia* or *Q. sadleriana* or both are often shrubby components.[2] Conifers which may be present include *Pinus lambertiana*, *P. ponderosa*, and *Libocedrus decurrens*. *Pinus jeffreyi* is found on serpentine sites, and *Chamaecyparis lawsoniana* and *Acer macrophyllum* may be present on moister habitats. *Umbellularia californica* may occur as a shrub on ultrabasic soils. *Pinus attenuata* often regenerates after wildfires and will form extensive pure stands after repeated fires within this zone, particularly on the drier sites. It is capable of pioneering on either normal soils or soils derived from ultrabasic rock. *Pinus attenuata* is strictly dependent upon wildfires because of its totally serotinous cones. The clusters of unopened cones along the boles and branches give individuals and stands a distinctive appearance.

For published community descriptions within the Mixed-Evergeeen Zone, we are heavily dependent upon Whittaker (1960); we will use his description of vegetation growing on diorite, with average moisture conditions as modal for the zone. The upper canopy level is dominated by *Pseudotsuga menziesii*, although *Pinus lambertiana* is frequently present. The lower tree canopy of sclerophyllous trees is dominated by *Lithocarpus densiflorus* associated with *Quercus chrysolepis*, *Arbutus menziesii*, and *Castanopsis chrysophylla*. The shrub layer averages about 30-percent coverage and is typically composed of *Quercus chrysolepis*, *Berberis nervosa*, *Rubus ursinus*, *Rosa gymnocarpa*, and *Rhus diversiloba*. The herbaceous layer is not well developed (6.5-percent coverage) but includes *Whipplea modesta*, *Achlys triphylla*, *Trientalis latifolia*, *Goodyera oblongifolia*, *Pteridium aquilinum*, *Apocynum pumilum*, *Disporum hookeri*, *Lonicera hispidula*, *Festuca occidentalis*, and *Melica harfordii*.

Whittaker (1960) described a *Chamaecyparis lawsoniana-Pseudotsuga menziesii* forest on more mesic sites in which these species are dominants. Small sclerophyllous trees are present but not dominant. *Taxus brevifolia*, *Acer circinatum*, *Corylus cornuta* var. *californica*, and *Cornus nuttallii* are typical understory species along with *Gaultheria shallon*, *Berberis nervosa*, *Rubus ursinus*, *Linnaea borealis*, *Polystichum munitum*, and *Achlys triphylla*.

Whittaker (1960) refers to communities on more xeric dioritic sites as "Sclerophyll-*Pseudotsuga*" as they are characterized by an overstory (with less than 50-percent coverage) of *Pseudotsuga menziesii* and a closed canopy of sclerophylls. *Lithocarpus densiflorus* is characteristically the dominant sclerophyll, but *Arbutus menziesii* and *Quercus chrysolepis* are also abundant. Typical shrubs are *Rosa gymnocarpa*, *Rhus diversiloba*, and *Rubus ursinus; Pteridium aquilinum* is the only major herb. Whittaker (1960) indicates he found similar stands without *Pseudotsuga menziesii* or *Pinus lambertiana* on south slopes

---

[2] According to W. H. Emmingham, *Quercus vaccinifolia* in the Illinois River area is primarily, if not exclusively, a serpentine species and most understory *Quercus* on normal soils is a shrubby or sprout form of *Q. chrysolepis* or is *Q. sadleriana*.

(*Lithocarpus-Arbutus-Quercus chrysolepis* communities); he felt these were the result of more severe fires since *Pseudotsuga menziesii* seedlings were present. Gratkowski (1961a, 1961b) also mentions extensive tracts of *Lithocarpus-Arbutus* communities in the Siskiyou Uplands and indicates excellent conifer reproduction is often present where a seed source exists.

In Whittaker's (1960) study area, vegetation on gabbro was intermediate between that on diorite and on serpentine. The canopy levels were more open. *Umbellularia californica* and *Arctostaphylos cinerea* were added to the sclerophyllous species, and *Pinus ponderosa*, *Libocedrus decurrens*, and *Pinus lambertiana* were more important than on diorite.

Emmingham[3] recently completed a study of the vegetation along the Illinois River which is located largely within the Mixed-Evergreen Zone. In one area identified as typical of the zone (proposed Store Gulch Research Natural Area), *Pseudotsuga menziesii* is associated with *Lithocarpus densiflorus* and *Quercus chrysolepis* in the tree layer. *Pinus lambertiana*, *Arbutus menziesii*, and *Quercus kelloggii* are less abundant tree species. Except for shrubby *Lithocarpus* and *Quercus chrysolepis*, *Rhus diversiloba* is the major shrub with sporadic occurrences of *Berberis nervosa*, *Lonicera hispidula*, *Rosa* sp., *Holodiscus discolor*, *Corylus cornuta* var. *californica*, and *Philadelphus gordonianus*. Herb cover is sparse—most important are *Achlys triphylla*, *Apocynum* sp., *Campanula* sp., *Disporum hookeri*, *Goodyera oblongifolia*, *Polystichum munitum* var. *imbricans*, *Pteridium aquilinum*, *Hieracium albiflorum*, *Xerophyllum tenax*, and grasses. This community can perhaps best be called the *Pseudotsuga menziesii-Quercus chrysolepis-Lithocarpus densiflorus/Quercus chrysolepis-Lithocarpus densiflorus* type (note the dominance of two of the same species in both the overstory tree and shrub layers). It is often found on steep slopes and very stony soils sometimes referred to as rock mulch type. Toward the coast, a variant of this community with a nearly solid understory of *Vaccinium ovatum* is encountered.

Other communities in the Illinois River area are largely variants of the modal community. Several vary mainly in the proportions among the dominant *Quercus*, *Lithocarpus*, and *Pseudotsuga*. On shallow, stony soils or rock outcrops is a *Quercus chrysolepis-Lithocarpus densiflorus-Pseudotsuga menziesii/Rhus diversiloba/*moss community type. It often is found near canyon bottoms where erosion has prevented formation of deep soils. On slightly deeper soils than are found under the modal community or at higher elevations on rock mulch soils is a *Pseudotsuga menziesii-Lithocarpus densiflorus/Lithocarpus densiflorus/Polystichum munitum*-herb community type. *Pinus lambertiana* increases in importance in this type when found on hornblende diorite parent material as opposed to soils derived from the more typical metavolcanic rocks. Benches or ridgetops with deep soils are occupied by a *Pseudotsuga menziesii-Pinus* spp./*Lithocarpus densiflorus-Quercus chrysolepis-Castanopsis chrysophylla/Pteridium aquilinum* community. *Pseudotsuga*, *Pinus ponderosa*, and *P. lambertiana* are important trees. However, *Quercus kelloggii*, *Arbutus menziesii*, and *Castanopsis chrysophylla* are also significant tree components. *Lithocarpus* and *Quercus chrysolepis* are confined to the shrub layer. Finally, on relatively deep soils situated above 750-meter elevation is a *Pseudotsuga menziesii/Lithocarpus densiflorus-Quercus chrysolepis/Goodyera oblongifolia* community which resembles pure, nearly closed canopy *Pseudotsuga* types found elsewhere in southwestern Oregon. On north aspects, *Gaultheria shallon* may be an important shrub; elsewhere, the understory is poorly developed, and neither *Lithocarpus densiflorus* nor *Quercus chrysolepis* can maintain themselves as trees. Such sclerophyllous species may attain tree size after fires but are eventually overtopped and suppressed by *Pseudotsuga*.

---

[3] Personal communication, William H. Emmingham, Oregon State University, Corvallis.

Special Types

Except for the unique vegetation found on serpentine sites (discussed in Chapter XI), brushfields are the most conspicuous "special communities" found within the Mixed-Evergreen Zone (fig. 103). Many of the large and abundant brushfields found in the Siskiyou Uplands (Gratkowski 1961a) lie within this zone. Gratkowski (1961a) describes these communities as ". . . dense evergreen chaparral [which] is part of the Chaparral Association of the Broad Sclerophyll Formation. . . ." Typical constituent species are: *Arctostaphylos canescens, A. patula, Lithocarpus densiflorus, Quercus chrysolepis, Q. vaccinifolia, Q. sadleriana, Castanopsis chrysophylla* var. *minor, Garrya fremontii, G. buxifolia, Rhamnus californica, Ribes marshallii, Ceanothus cordulatus*, and *Berberis pumila*. Gratkowski (1961b) indicates that communities dominated by *Arctostaphylos patula, Ceanothus cordulatus*, and *Quercus chrysolepis* are particularly abundant.

The chaparral and forest communities appear to be successionally related in many cases; that is, the former are often fire-induced seral types, especially in moist areas (e.g., nearer the coast). Since most shrub dominants sprout after fire, successive burns can eliminate conifers and increase shrub density. On the other hand, chaparral is probably climax on many sites (Gratkowski 1961a): e.g., dry eastern slopes of the Siskiyou Mountains and sites with severe southerly exposures and shallow soils further west. Gratkowski (1961a) feels soil moisture during the dry summer season may be inadequate for tree growth on many of these sites.

Emmingham (personal communication) feels that the shrub fields which are climax, or only very slowly replaced by trees, can be identified by the dominance of hard-leaved shrubs, such as *Arctostaphylos canescens* and *A. patula*. Fields dominated by softer leaved shrubs or stands of *Lithocarpus densiflorus* and *Arbutus menziesii* are expected to succeed to conifers or conifer-*Lithocarpus* mixtures.

Grass balds of various compositions may also be encountered in this zone. At least some owe their origin to fire; the herbaceous vegetation provides an unfavorable habitat for tree regeneration, and heavy browsing of seedlings which do become established helps perpetuate herbaceous cover.

Figure 103.—Chaparral, forming fields of sclerophyllous shrubs, is very abundant in the Mixed-Evergreen Zone; this typical community is composed of *Arctostaphylos* spp., *Quercus chrysolepis*, and *Ceanothus* spp. (Siskiyou Mountains, Oregon).

Mixed-Conifer (*Pinus-Pseudotsuga-Libocedrus-Abies*) Zone

Mixed forests of *Pseudotsuga menziesii, Pinus lambertiana, P. ponderosa, Libocedrus decurrens*, and *Abies concolor* (or *A. grandis*) typify midelevations in the southwestern Oregon Cascade Range and eastern Siskiyou Mountains (Waring 1969). They are northern extensions of the well-known Sierran montane or mixed-conifer forest (Oosting 1956, Küchler 1964). The Mixed-Conifer Zone occurs from about 43° north latitude south along the western flanks of the Cascade Range at elevations of about 750 to 1,400 meters. It is also found in the eastern Siskiyou Mountains, but usually at slightly higher elevations (Waring 1969, Dennis 1959). The Mixed-Conifer Zone is generally bounded by the Interior Valley and the *Abies concolor* Zone at its lower and upper limits, respectively; to the north it grades into the *Tsuga heterophylla* Zone.

Environmental Features

Very few environmental data are available for the Mixed-Conifer Zone except for the eastern Siskiyou Mountains (Waring 1969). Precipitation varies from about 900 to 1,300 millimeters, with very little occurring during the summer months (table 18). Mean temperatures are about the same as in the *Tsuga heterophylla* Zone, but summers are distinctly warmer and drier. Waring (1969) has demonstrated that the moisture regime in mixed-conifer forests is more favorable than in the Ponderosa Pine and Black Oak types discussed earlier.

Forest soils in the Mixed-Conifer Zone of southwestern Oregon are extremely varied due to the complex geological history and topography. Soils typically belong to the Haplohumult, Haplumbrept, Xerumbrept, and Haploxerult soil great groups (Reddish Brown Lateritic, Gray-Brown Podzolic, Western Brown Forest, and Lithosolic great soil groups). Representative Haploxerults (Reddish Brown Lateritics) have thin organic layers (2 to 5 cm.), dark reddish-brown, slightly acid surface soils and red to dark-red strongly acid subsoils. Organic matter averages only 4 to 5 percent in the surface soil.

Table 18. — Climatic data from representative stations within the Mixed-Conifer Zone

Station	Eleva-tion	Lati-tude	Longi-tude	Temperature					Precipitation		
				Average annual	Average January	Average January minimum	Average July	Average July maximum	Average annual	June through August	Average annual snowfall
	Meters			- - - - - - - - - Degrees C. - - - - - - - - -					Millimeters		Centi-meters
Butte Falls, Oreg.	635	42°32'	122°33'	--	--	--	--	--	931	64	--
Prospect, Oreg.	630	42°44'	122°31'	9.9	1.9	−3.2	19.0	31.1	1,059	62	162
Siskiyou Station, Oreg.	1,295	42°03'	122°36'	9.1	1.3	−3.2	18.4	24.5	863	53	--
Toketee Falls, Oreg.	617	43°16'	122°26'	10.5	1.9	−1.4	20.4	30.4	1,219	85	114

Source: U. S. Weather Bureau (1965a); Johnsgard (1963).

Forest Composition

The major forest tree species in the Mixed-Conifer Zone are *Pseudotsuga menziesii*, *Pinus lambertiana*, *P. ponderosa*, *Libocedrus decurrens*, and *Abies concolor*.[4] These species occur in many combinations and degrees of mixture (Hayes 1959) (fig. 104). *Pseudotsuga menziesii* is probably the most abundant species, but it tends to decrease and *Pinus* spp.

---

[4] The *Abies* under discussion here is part of the *Abies grandis-A. concolor* species complex, widespread in southwestern and eastern Oregon. Some southwestern Oregon populations approach *Abies grandis* most closely morphologically; others approach *Abies concolor* morphologically and ecologically (Hamrick 1966, Hamrick and Libby 1972, Daniels 1969). For convenience, we will refer to all the morphologically variable populations in the Mixed-Conifer and *Abies concolor* Zones as *Abies concolor*.

Figure 104.—Major forest trees in the Mixed-Conifer Zone are *Pseudotsuga menziesii, Pinus lambertiana, Libocedrus decurrens, Pinus ponderosa,* **and** *Abies concolor;* **the first three of these are readily identifiable in this typical southwestern Oregon mixed-conifer stand.**

Figure 105.—*Abies concolor* **is often represented only by seedlings and saplings in existing mixed-conifer stands, such as this** *Pinus ponderosa* **forest (Rogue River National Forest, Oregon).**

tend to increase in importance from north to south within the zone. *Pinus lambertiana* and *P. ponderosa* usually occur as scattered individuals (Hayes and Hallin 1962) but give the forests much of their character. The proportion of *Libocedrus decurrens* tends to be greatest on relatively xeric sites in this zone. *Libocedrus* is clearly superior to most of its associates in drought tolerance, a probable consequence of extensive development of the seedling root system (Pharis 1967, Stein 1963). *Abies concolor* is often represented mainly by seedlings and saplings in existing mixed-conifer stands (fig. 105). Other typical tree species include *Acer macrophyllum*, *Arbutus menziesii*, and *Pinus monticola*. *Tsuga heterophylla* and *Thuja plicata* are frequently encountered on more mesic habitats in the northern parts of the Mixed-Conifer Zone. *Castanopsis chrysophylla* may occur as either a shrub or small tree.

The only detailed community analyses which have been reported for the Mixed-Conifer Zone are from the Abbott Creek Research Natural Area (Franklin et al. 1972) along the Rogue-Umpqua River divide (Mitchell 1972, Mitchell and Moir[5]). Three major forest

[5] Roderic J. Mitchell and William H. Moir. Vegetation of the Abbott Creek Research Natural Area. Unpublished manuscript on file at USDA Forest Service, Forestry Sciences Laboratory, Corvallis, Oregon (1973).

communities were identified along with a fourth type which is centered primarily in an *Abies concolor* Zone (*Abies concolor/Linnaea borealis* type): *Abies concolor-Tsuga heterophylla/Acer circinatum-Taxus brevifolia, Abies concolor/Whipplea modesta*, and *Pseudotsuga menziesii-Libocedrus decurrens/Arctostaphylos nevadensis*. These three communities reflect a moisture gradient. The *Abies-Tsuga/Acer-Taxus* community type is found on the moistest sites along stream terraces, on lower slopes, and in ravines. *Pseudotsuga menziesii, Pinus lambertiana*, and occasionally, *Pinus monticola* are seral species. *Abies concolor* and *Tsuga heterophylla* appear to share climax status. The relatively dense understory (fig. 106) contains a number of species characteristic of sites with low moisture stress (table 19) (Waring 1969, Waring et al. 1972, Minore 1972a). The shrub layer is particularly well developed. It is obvious that many species more characteristic of the *Tsuga heterophylla* Zone find their niche in this community—*Tsuga heterophylla, Taxus brevifolia, Acer circinatum, Linnaea borealis*, and even *Rhododendron macrophyllum*, for example.

The *Abies concolor-Pseudotsuga menziesii/Whipplea modesta* community is, in one form or another, the most extensive. *Pseudotsuga, Libocedrus decurrens*, and *Pinus lambertiana* characterize the overstory in mature stands. *Abies concolor* and *Pseudotsuga* dominate the reproductive size classes; hence, their climax designation. The understory is extremely sparse in this community (typically less than 20-percent total cover) (fig. 106) and few in species (table 19). The *Pseudotsuga menziesii-Libocedrus decurrens/Arctostaphylos nevadensis* community occupies the most xeric forested habitats. Both tree species characterize reproductive size classes as well as the overstory in these dry open forests (fig. 106), although *Pinus lambertiana* and even *P. ponderosa* may be present. *Abies concolor* is absent or a minor component. The understory is dominated by two mat-forming shrubs, *Arctostaphylos nevadensis* and *Ceanothus prostratus* (table 19). A scattering of larger shrubs (e.g., *Arctostaphylos patula* and *Garrya fremontii*) and herbs is also characteristic. Two crustose lichens, *Lecidea atrobrunnea* and *Rhizacarpon geographicum*, commonly coat rock surfaces.

The *Abies concolor/Linnaea borealis* community does occur on mesic sites within the Mixed-Conifer Zone (intermediate in environment between those occupied by the *Abies-Tsuga/Acer-Taxus* and *Abies/Whipplea* communities), but we will consider it in the discussion of the *Abies concolor* Zone.

Relevés are also available for stands in the Mixed-Conifer Zone in the central portion of the South Umpqua River drainage.[6] Major differences from the Abbott Creek area are occurrence of communities in which *Gaultheria shallon* is an understory dominant and a greater preponderance of stands in which *Tsuga heterophylla* is the projected climax dominant. Tentative community types include *Pseudotsuga menziesii/Bromus orcuttianus-Whipplea modesta, Pseudotsuga-Abies grandis/Holodiscus discolor-Gaultheria shallon, Pseudotsuga-A. grandis/Gaultheria*, and *Pseudotsuga-Tsuga heterophylla/Gaultheria/Linnaea borealis*. The *Pseudotsuga/Bromus-Whipplea* community has a very sparse shrub layer of *Berberis nervosa, B. aquifolium, Rosa gymnocarpa*, and *Rubus ursinus*, and a distinctive herb layer composed of *Bromus, Whipplea*, and a variety of associates, e.g., *Synthyris reniformis, Lathyrus polyphyllus, Adenocaulon bicolor, Satureja douglasii*, and *Festuca occidentalis*. *Pseudotsuga menziesii* is the tree species reproducing most effectively, but several other trees (*Libocedrus decurrens, Abies grandis*, and *Pinus lambertiana*) are usually also represented by seedlings and saplings. The *Pseudotsuga-Abies/Holodiscus-Gaultheria* and */Gaultheria* communities have well-developed shrub layers with the dominant *Gaultheria shallon* (average 60-percent cover) associated with *Holodiscus discolor* in the former type and *Berberis nervosa* and *Rubus ursinus* in both types. Herb layers are

---

[6] Data from South Umpqua Experimental Forest on file at USDA Forest Service, Forestry Sciences Laboratory, Corvallis, Oregon.

Figure 106.—Community types found within the Mixed-Conifer Zone of the Abbott Creek Research Natural Area (Rogue River National Forest, Oregon). Upper left—a dense shrubby understory of *Acer circinatum* and *Taxus brevifolia* characterizes the mesic *Abies concolor-Tsuga heterophylla/Acer-Taxus* community type; the major overstory species here is *Pseudotsuga menziesii*, but a large *Pinus lambertiana* and sapling *Tsuga heterophylla* are also visible. Upper right—the *Abies concolor/Linnaea borealis* community is also found on moist sites as well as in the higher elevation *Abies concolor* Zone; the understory is relatively rich but consists of low-growing species, primarily *Linnaea borealis* and *Rubus nivalis* in this nearly pure *Abies* stand. Lower left—an *Abies concolor-Pseudotsuga menziesii/Whipplea modesta* community showing typical overstory components (*Libocedrus decurrens* and *Pseudotsuga*) and abundant seedlings consisting of equal numbers of *Pseudotsuga* and *Libocedrus*. Lower right—dry site community of *Pseudotsuga-Libocedrus/Arctostaphylos nevadensis* including a *Pinus ponderosa* (behind stake); *Castanopsis chrysophylla* and *Arbutus menziesii* form a lower stratum of evergreen hardwoods in this stand but are not characteristic of the community.

141

Table 19. – Coverage of selected understory species in four forest community types found in the Mixed-Conifer and *Abies concolor* Zones of the Abbott Creek Research Natural Area[1] [2]

Layer and species	Community type[3]			
	Abco-Tshe/Acci-Tabr	Abco-Psme/Whmo	Psme-Lide/Arne	Abco/Libo
	- - - - - - - - - - - - - - - - - - Percent - - - - - - - - - - - - - - - - - -			
Tall shrubs:				
Castanopsis chrysophylla	tr	tr	1	4
Corylus cornuta californica	tr	tr	—	1
Taxus brevifolia	22	—	—	1
Acer circinatum	44	—	—	—
Low shrubs:				
Berberis nervosa	4	1	tr	3
Pachistima myrsinites	2	tr	tr	1
Rubus ursinus	3	tr	—	4
Symphoricarpos mollis	2	tr	tr	—
Vaccinium membranaceum	4	—	1	3
Gaultheria ovatifolia	tr	—	—	2
Rubus nivalis	—	—	—	5
Garrya fremontii	—	tr	1	—
Arctostaphylos nevadensis	—	—	14	—
Ceanothus prostratus	—	—	5	—
Arctostaphylos patula	—	—	tr	—
Herbs:				
Whipplea modesta	1	4	4	6
Disporum hookeri	tr	tr	tr	1
Goodyera oblongifolia	tr	tr	tr	tr
Hieracium albiflorum	tr	tr	tr	tr
Anemone deltoidea	1	tr	—	1
Iris chrysophylla	tr	1	—	tr
Achlys triphylla	4	tr	—	8
Bromus vulgaris	tr	tr	—	tr
Chimaphila umbellata	2	—	—	9
Viola sempervirens	tr	—	—	2
Linnaea borealis	6	—	—	18
Arenaria macrophylla	—	1	1	tr
Apocynum androsaemifolium	—	tr	—	—
Pyrola secunda	—	—	—	1

[1] Source: Roderic J. Mitchell and William H. Moir. Vegetation of the Abbott Creek Research Natural Area. Unpublished manuscript on file at USDA Forest Service Forestry Sciences Laboratory, Corvallis, Oregon.

[2] tr = coverage of less than 0.5 percent.

[3] Abco-Tshe/Acci-Tabr = *Abies concolor-Tsuga heterophylla/Acer circinatum-Taxus brevifolia*; Abco-Psme/Whmo = *Abies concolor-Pseudotsuga menziesii/Whipplea modesta*; Psme-Lide/Arne = *Pseudotsuga menziesii-Libocedrus decurrens/Arctostaphylos nevadensis*; Abco/Libo = *Abies concolor/Linnaea borealis*.

relatively sparse consisting mostly of *Whipplea modesta, Xerophyllum tenax, Linnaea borealis,* and *Polystichum munitum. Abies grandis* is clearly the major climax species, although *Pseudotsuga* and *Libocedrus* may be minor climax species in the /*Holodiscus-Gaultheria* community. The *Pseudotsuga-Tsuga*/*Gaultheria* community is the most mesic type. *Acer circinatum, Taxus brevifolia,* and *Rubus nivalis* are important additions to this community, with *Gaultheria shallon* and *Berberis nervosa* remaining the low shrub dominants. *Linnaea borealis, Polystichum munitum,* and *Whipplea modesta* are the most important herbs, although several other species have high constancies only in this community (*Trillium ovatum, Chimaphila umbellata, Viola sempervirens, Eurhynchium oreganum*). *Tsuga heterophylla* is the major climax tree species.

Minore (1972a) examined the distribution of forest species in the entire South Umpqua drainage and developed some environmental indices based upon plant indicators. His report gives no insight into community composition, but the original plot data have complete lists of vascular plants present.[7] Three environmental "cells" out of a 12-cell matrix of three moisture and four temperature classes include sufficient numbers of plots for examining shifts in species constancies—the warm dry, cool moist, and cold moist cells (table 20). The warm dry and cool moist environments correspond to the Mixed-Conifer Zone with the cold moist environment obviously tending to be more representative of the *Abies concolor* Zone which will be discussed next. Minore's data confirm the general pattern of shifts in species indicated by others as one moves from dry to moist sites within the Mixed-Conifer Zone.

The Mixed-Conifer forests have also been studied in the eastern Siskiyou Mountains (Waring 1969, Dennis 1959). The same tree species occur there, although *Libocedrus decurrens* appears to be less common than in the Cascade Range. Characteristic understory species are *Corylus cornuta* var. *californica, Holodiscus discolor, Castanopsis chrysophylla, Symphoricarpos mollis, Rubus ursinus, Rosa gymnocarpa, Adenocaulon bicolor, Hieracium albiflorum,* and *Senecio integerrimus.*

Waring (1969) and Waring et al. (1972) have identified the environmental niches (particularly temperature and moisture relations) for over 50 tree and understory species in the eastern Siskiyou Mountains, over half of which are found in the Mixed-Conifer Zone.

Successional Patterns

Successional relationships have not been studied in the Mixed-Conifer Zone. It is known that brushfields frequently develop on burned- or logged-over forest lands within the zone (Gratkowski 1961a, Hayes 1959) (fig. 107). Dominants in such communities include *Ceanothus velutinus, C. sanguineus, C. integerrimus, C. prostratus, C. cordulatus, Castanopsis chrysophylla, Quercus chrysolepis, Amelanchier alnifolia, Arctostaphylos canescens,* and *Lithocarpus densiflorus* (Gratkowski 1961a). These brushfield communities can significantly slow the rate of forest succession or, with repeated fire, become semipermanent communities. Under more normal conditions, they are overtopped and killed by conifers which often become established in their shade (fig. 108).

Ceanothus velutinus is important as a brushfield dominant or invader following logging or fire in the Mixed-Conifer Zone as well as many other zones: e.g., in parts of the *Tsuga heterophylla* (Morris 1958, Zavitkovski 1966), *Abies concolor,* and *A. magnifica shastensis* Zones, and in many of the forest types of eastern Oregon and Washington (Dyrness and Youngberg 1966, Mueggler 1965). In western Oregon, *Ceanothus velutinus* var. *laevigatus* and var. *velutinus* are generally found below and above 800 meters, respectively (Gratkowski

---

[7] Data on file at USDA Forest Service, Forestry Sciences Laboratory, Corvallis, Oregon.

Table 20. – Constancy values for important species in stands occupying different environments in the South Umpqua River drainage[1]

Life form and species	Environment[2]		
	Warm dry (n = 11)	Cool moist (n = 11)	Cold moist (n = 13)
	- - - - - - - - - - - - - - - - - - Percent - - - - - - - - - - - - - - - - - -		
Trees:			
Arbutus menziesii	91	9	0
Pinus ponderosa	73	9	0
Pinus lambertiana	73	45	23
Pseudotsuga menziesii	100	100	92
Libocedrus decurrens	64	64	69
Castanopsis chrysophylla	36	45	23
Tsuga heterophylla	0	64	15
Abies grandis	55	91	100
Abies magnifica shastensis	0	9	38
Shrubs:			
Rhus diversiloba	73	9	0
Berberis piperiana	73	45	0
Holodiscus discolor	91	64	23
Symphoricarpos albus	82	64	54
Rosa gymnocarpa	91	82	77
Rubus ursinus	91	91	54
Corylus cornuta californica	45	36	54
Gaultheria shallon	45	73	8
Berberis nervosa	45	91	69
Acer circinatum	9	36	54
Vaccinium membranaceum	0	0	54
Ribes binominatum	0	0	46
Herbs:			
Lonicera hispidula	91	9	0
Habenaria unalascensis	64	9	0
Fragaria vesca crinita	55	18	0
Carex concinnoides	64	27	0
Satureja douglasii	64	36	0
Cynoglossum grande	45	36	0
Madia madioides	82	36	8
Iris chrysophylla	73	27	15
Whipplea modesta	100	91	38
Pteridium aquilinum	100	55	38
Calypso bulbosa	45	73	0
Synthyris reniformis	73	64	46
Campanula prenanthoides	64	18	38

Table 20. — Constancy values for important species in stands occupying different environments in the South Umpqua River drainage[1] (Continued)

Life form and species	Environment[2]		
	Warm dry (n = 11)	Cool moist (n = 11)	Cold moist (n = 13)
	- - - - - - - - - - - - - - - - - - - Percent - - - - - - - - - - - - - - - - -		
Herbs (continued):			
Phlox adsurgens	64	36	46
Elymus glaucus	55	0	38
Festuca californica	55	9	31
Bromus orcuttianus	91	73	69
Disporum hookeri	91	82	77
Vicia americana	91	55	46
Goodyera oblongifolia	73	45	46
Hieracium albiflorum	82	91	92
Trientalis latifolia	82	82	85
Adenocaulon bicolor	73	82	77
Polystichum munitum	64	73	46
Festuca occidentalis	64	73	54
Linnaea borealis	55	82	38
Vancouveria hexandra	73	82	92
Osmorhiza chilensis	55	64	69
Achlys triphylla	45	73	85
Chimaphila umbellata	36	55	54
Viola sempervirens	36	64	69
Galium aparine	27	91	85
Trillium ovatum	27	64	85
Campanula scouleri	9	64	54
Pyrola picta	9	27	54
Bromus carinatus	18	9	62
Anemone deltoidea	0	64	46
Montia sibirica	0	36	62
Smilacina stellata	0	27	77
Clintonia uniflora	0	18	69
Galium oreganum	0	0	62
Phacelia mutabilis	0	0	54
Actaea rubra	0	0	46

[1]Data collected by Don Minore on file at USDA Forest Service Forestry Sciences Laboratory, Corvallis, Oregon.

[2]Environments are defined by temperature and moisture indices based on indicator species as outlined by Minore (1972a).

Figure 107.—Brushfields often develop on burned or logged-over lands within the Mixed-Conifer Zone; the Cat Hill brushfield, partially shown here, probably originated after a fire in the 1850's and has since been reburned (near Blue Rock, Rogue River National Forest, Oregon).

Figure 108.—Shrubs are major elements of early successional communities on many forest sites in the Mixed-Conifer Zone and are gradually overtopped and killed by coniferous trees; *Arctostaphylos* spp. giving way to a mixed stand of *Pseudotsuga menziesii* and *Pinus ponderosa* (Ashland Research Natural Area, Rogue River National Forest, Oregon).

146

1961a, Zavitkovski 1966). In this area, *Ceanothus velutinus* is generally absent from under-stories of older stands lacking recent disturbance, but it often appears in abundance follow-ing logging and slash burning (Morris 1958) (fig. 109). This reproduction is from seed stored in the forest floor (Gratkowski 1962); heat from fires and increased insolation breaks the seedcoat dormancy. The relationship of *Ceanothus velutinus* to establishment and growth of coniferous reproduction has been hotly debated for 50 years (Zavitkovski and Newton 1968). It can fix nitrogen (Wollum 1962, 1965; Wollum et al. 1968) and may provide a favorable microenvironment for establishment of conifer seedlings under some conditions (Gratkowski 1962, Zavitkovski and Newton 1968). On other sites, it may seriously hinder establishment of coniferous stands.

Figure 109.—After forest lands have been logged and burned, *Ceanothus velutinus* **reproduces from seed stored in the soil, despite its general absence from understories of older stands; it frequently hinders estab-lishment and growth of conifer seedlings in the Mixed-Conifer Zone.**

Successional relationships among the tree species are not completely known, but some general patterns are apparent. First, *Abies concolor* (or *Abies grandis*) is the major climax species over the entire Mixed-Conifer Zone; past fires and present logging activities have restrained *Abies* from attaining overstory dominance in much of the zone, but its domi-nance in reproductive size classes clearly shows the successional trend. On warm, dry forested habitats, *Pseudotsuga* or *Libocedrus* or both appear to be the major climax primarily because *Abies* is unable to tolerate these stressful environments; the successional relationship between *Pseudotsuga* and *Libocedrus* is not clear, however, and reproduction of either may dominate in a given stand. On the majority of mesic habitats, *Abies concolor* is again the major climax species since it is the most tolerant tree species present; how-ever, in northern portions of the Mixed-Conifer Zone, it is joined or successionally dis-placed by *Tsuga heterophylla* and, possibly, *Thuja plicata* on mesic forest habitats. *Pinus lambertiana* and *Pinus ponderosa* are considered to be seral species although *P. ponderosa* may achieve climax status on poorly drained or swampy sites (Stephens 1965).

The successional picture is incomplete without some consideration of the "gap-phase" phenomenon, however. Small openings are frequently scattered through mixed-conifer forests, and these provide favorable environments for regeneration of species which are theoretically seral and are considered relatively intolerant of shade (fig. 110). *Pinus lambertiana, Pseudotsuga menziesii, Libocedrus decurrens,* and even *Pinus ponderosa*

147

Figure 110.—Small openings are typical of mature mixed-conifer forests and provide opportunities for regeneration of less tolerant species; *Pseudotsuga menziesii, Pinus lambertiana, Abies concolor,* **and** *Lithocarpus densiflorus* **seedlings are present in this opening (Abbott Creek Research Natural Area, Rogue River National Forest, Oregon).**

perpetuate themselves in these forest "gaps" or openings. These openings are a pervasive feature of mixed-conifer forests; and, for that reason, it can be argued that the majority of the tree species are a part of the "climax" forest. Also, many of these so-called intolerant species show considerably more shade tolerance than commonly supposed (e.g., see Atzet and Waring 1970). In any case, it seems clear that, because of severe environmental stresses, forest gaps, and the long-lived nature of the trees, elimination of so-called seral species and attainment of dominance by the climax species are a very slow process in much of the Mixed-Conifer Zone.

Special Types

A major group of nonforested communities found within the Mixed-Conifer Zone is that found on very shallow soils. These communities occupy sites which are typically warm and extremely dry during the summer. One of the more common appears to be a "moss meadow" type in which *Rhacomitrium canescens* var. *ericoides* is the typical dominant (fig. 111); this community is also found in the Interior Valley Zone. Several spring annuals are also characteristic of this community. Other communities which have been described on dry, nonforested sites include: (1) an *Arctostaphylos nevadensis-Ceanothus prostratus* community dominated by these mat-forming shrubs but with a scattering of herbs such as *Senecio integerrimus exaltatus* and xerophytic ferns such as *Cryptogramma densa* (fig. 111); (2) grassy meadows on shallow, stony soils dominated in one area by *Stipa lemmonii, Stipa occidentalis,* and *Sitanion hystrix* (fig. 111) with many associated annual and perennial forbs—*Sanicula graveolens, Lomatium nudicaule, Gilia capitata, Navarretia divaricata, Perideridia bolanderi, Polygonum majus,* and *Collinsia parviflora*; and (3) "rock garden" types associated with rocky outcrops and characterized by *Sedum oregonense,* caespitose Polygonaceae (*Spraguea umbellata, Eriogonum compositum, E. nudum,* and *E. umbellatum*), and xeromorphic ferns (*Pellaea glabella, Cryptogramma densa,* and *Cheilanthes gracillima*) (fig. 111). These *Stipa* meadows and rock garden types are similar

Figure 111.—Xeric "meadow" communities are characteristic of sites with very shallow and often stony soils. Upper left—forest openings dominated by cryptogams, particularly *Rhacomitrium canescens* var. *ericoides,* are characteristic of sites with almost continuous bedrock at the surface. Upper right—*Arctostaphylos nevadensis-Ceanothus prostratus* form dense mats on some very stony slopes. Lower left—*Bromus carinatus-Stipa* spp.-*Sitanion hystrix-Eriogonum* spp.-*Senecio crassulus* community with scattered *Arctostaphylos patula.* Lower right—"rock garden" community associated with rock outcrops and dominated by *Sedum oregonense,* caespitose Polygonaceae, and xeromorphic ferns. (Photo at upper left from proposed Camas Swale Research Natural Area near Eugene, Oregon; remainder from Abbott Creek Research Natural Area, Rogue River National Forest, Oregon.)

in many respects to the nonforested communities described by Hickman (1968) in the Western Cascades Province of northern Oregon.

Among the special types of forest encountered in the Mixed-Conifer Zone, we again encounter swamps in which *Fraxinus latifolia* is a characteristic species. These communities occupy small poorly drained basins and often have a rich diversity of species including some found only here (e.g., *Fraxinus, Alnus* sp., several *Carex* spp., *Camassia quamash*), but also others widespread in the zone and often thought of as dry-site species—*Rhus diversiloba* and *Pinus ponderosa,* for example. As mentioned, *Pinus ponderosa* appears to be the climax tree on some of these poorly drained habitats (Stephens 1965).

Abies concolor Zone

The *Abies concolor* Zone, as discussed here and in the previous edition of this work, is a relatively narrow belt located at the upper margin of the Mixed-Conifer Zone. Its worthiness for zonal recognition is somewhat marginal, especially in southwestern Oregon; its status in California may be somewhat clearer. In light of the limited extent of the zone and its close relationship to the Mixed-Conifer Zone (in which we also identify *Abies concolor* or *grandis* as the major climax), the forests discussed here could be considered simply as upper altitudinal variants of the Mixed-Conifer Zone [Griffin's (1967) "White Fir Phase"] or even as stands ecotonal between the temperate Mixed-Conifer and subalpine *Abies magnifica shastensis* Zones. With these reservations in mind, we choose to continue to distinguish an *Abies concolor* Zone because: (1) in many areas, a fairly distinct belt of nearly pure *Abies concolor* stands occurs; (2) many species found in forests in the Mixed-Conifer Zone are drastically reduced in importance or absent in *Abies concolor* stands and, conversely, additional species not present in the mixed-conifer forests are conspicuous; and (3) there is clear evidence of a cooler (but not necessarily moister) environment in *Abies concolor* as compared with mixed-conifer stands (see table 2 in Waring et al. 1972).

Forests dominated by *Abies concolor* are the major feature of the *Abies concolor* Zone. This zone grades into the Mixed-Conifer and *Abies magnifica shastensis* Zones at its lower and upper limits, respectively. As mentioned, it occupies a relatively narrow elevational band, occurring at about 1,400 to 1,600 meters in the Cascade Range and 1,650 to 1,800 meters and 1,400 to 1,800 meters in the eastern and western Siskiyou Mountains, respectively (Waring 1969, Whittaker 1960). However, around Lake of the Woods and along the southwestern flank of the southern Oregon Cascade Range, there are extensive tracts at the appropriate elevations, and consequently, *Abies concolor* forests are widespread and well developed there (fig. 112). The *Abies concolor* Zone of southwestern Oregon extends around the southern end of the Cascade Range into southeastern Oregon, an area discussed later. Forests of the general type described here do occur at higher elevations along the western Cascades north to about the McKenzie River. The *Abies concolor* Zone correlates with the "White Fir Phase" of the Mixed-Conifer Forest in northern California (Griffin 1967) and is considered an element of Merriam's Canadian Life Zone (Dennis 1959).

Environmental Features

Climatic and edaphic data for the *Abies concolor* Zone are not available, but some evaluations of moisture, temperature, and nutrient regimes are, based upon monitoring of the plants themselves (Waring 1969, Waring et al. 1972, Griffin 1967). The zone does experience lower temperatures, less plant moisture stress, and less soil drought than the adjacent Mixed-Conifer Zone. It is the lowest zone where significant winter snow accumulations occur; Waring (1969) mentioned that heavy snowfalls are damaging to brittle-limbed species such as *Pinus ponderosa* and *Arbutus menziesii*. Haplorthods, Vitrandepts, and Haplumbrepts are typical soil great groups (Gray-Brown and Brown Podzols, Regosols, and Western Brown Forest soils).

Forest Composition

Abies concolor is the major tree species within the *Abies concolor* Zone, often forming pure or nearly pure stands (fig. 112). The most common associate is *Pseudotsuga menziesii*. *Pinus lambertiana*, *P. ponderosa*, and *P. monticola* may be present in small numbers. *Libocedrus decurrens* is often associated on mesic sites. *Abies magnifica shastensis* is

Figure 112.—*Abies concolor* forests are especially extensive and well developed along the southern and southwestern flanks of the Oregon Cascade Range; this pure *Abies concolor* forest is typical of those found near Mount McLoughlin (Rogue River National Forest, Oregon).

increasingly common toward the upper limits of the *Abies concolor* Zone. *Pinus contorta* is encountered as a pioneer species in the Cascade Range.

Abies concolor stands in the Western Cascades Province tend to be richer in species and have a higher coverage of herbs than those in the adjacent High Cascades Province. The only community descriptions we have for the Western Cascades Province are those of Mitchell (1972) and Mitchell and Moir (see footnote 5). They recognized an *Abies concolor/Linnaea borealis* community type in their study area which was often characterized by nearly pure stands of *Abies concolor* (fig. 106); as mentioned earlier, this community type also occurred on mesic habitats well within the Mixed-Conifer Zone. Data from the *Abies/Linnaea* community (table 19) indicate a relatively rich understory of herbs and low shrubs. *Linnaea borealis, Achlys triphylla,* and *Chimaphila umbellata* were the most important herbs, and *Rubus nivalis* was the shrub with highest cover.

Stand analyses from the High Cascades Province have not been published, but *Abies concolor/Castanopsis chrysophylla-Ceanothus velutinus* and *Abies concolor/Ceanothus velutinus* communities appear to be present. The latter (described in the section on the *Abies grandis* Zone in Chapter VII) is similar in many respects to the *Abies/Ceanothus*

151

association described by Dyrness and Youngberg (1966). In fact, many *Abies concolor* stands in the High Cascades Province have dense canopies with very depauperate under-stories primarily composed (around Lake of the Woods, for example) of a few ericads (e.g., *Chimaphila umbellata*) and scattered evergreen shrubs (e.g., *Castanopsis*).

Characteristic understory species in *Abies concolor* forests of the Siskiyou Mountains are (Whittaker 1960, Waring 1969, Dennis 1959):

Shrubs—*Holodiscus discolor, Rosa gymnocarpa, Berberis nervosa, Corylus cornuta* var. *californica, Acer glabrum* var. *douglasii, Rubus ursinus, R. nivalis, Amelanchier alnifolia,* and *Castanopsis chrysophylla.*

Herbs—*Campanula scouleri, Lathyrus polyphyllus, Anemone deltoidea, Achlys triphylla, Trientalis latifolia, Tiarella unifoliata, Galium triflorum, Adenocaulon bicolor, Vancouveria hexandra, Clintonia uniflora, Trillium ovatum, Hieracium albiflorum, Arenaria macrophylla, Phacelia heterophylla,* and *Fragaria vesca* var. *bracteata.*

Successional Patterns

Abies concolor appears to be the sole climax species on modal habitats (fig. 113). *Libocedrus decurrens* may be a climax associate on more mesic habitats and *Pseudotsuga menziesii* or *Libocedrus decurrens* or both on xeric habitats. *Abies concolor* probably replaces *Abies magnifica shastensis* on most sites where mixed stands of the two species occur.

Figure 113.—*Abies concolor* appears to be the sole climax species on modal sites in the *A. concolor* **Zone;** *Abies concolor* **regeneration completely dominates under this mixed stand of** *A. concolor, A. magnifica shastensis,* **and** *Libocedrus decurrens.*

Special Types

A variety of nonforested communities is found in the *Abies concolor* Zone. Brushfields similar to those found in the Mixed-Conifer Zone are encountered. *Ceanothus velutinus* remains a typical dominant, and *Ribes* spp. appear increasingly in the Cascade Range. *Arctostaphylos patula* is a major shrubby associate in the eastern Siskiyou Mountains (Dennis 1959).

Many different kinds of mountain meadows and barren openings are also found within the *Abies concolor* and adjacent *Abies magnifica shastensis* Zones. The compositions of these various meadow communities are not known with the exception of a species list from a series of mesic meadows in the Abbott Creek Research Natural Area (Dr. William H. Moir, personal communication). It includes as important components *Valeriana sitchensis*, *Veratrum californicum*, *Polygonum bistortoides*, *Mertensia ciliata*, *Hackelia jessicae*, *Rudbeckia occidentalis*, *Swertia umpquaensis*, *Agastache urticifolia*, *Ligusticum apiifolium*, *L. grayi*, *Lupinus* spp., *Orthocarpus imbricatus*, *Phleum alpinum*, *Bromus vulgaris*, *Elymus glaucus*, and *Melica spectabilis*. Moir found these particular meadows similar in many respects to the mesic tall forb communities described from the Wasatch Mountains of Utah.

There is extensive evidence of invasion of many of the meadows by tree species. *Libocedrus decurrens* is conspicuous as a pioneer tree in many areas (fig. 114). This is notable, for example, along the divide between the Rogue and Umpqua Rivers and on high ridges in the Umpqua River drainage and is occasionally observed as far north as 44° north latitude in the Western Cascades Province (H. J. Andrews Experimental Forest). The largest specimens of *Libocedrus* observed by the authors are found in and around high-elevation meadow communities. *Abies concolor* (or *grandis*) may also function as a pioneer in invading meadows along with *A. magnifica shastensis* at highest elevations.

Figure 114.—Invasion of meadows by *Libocedrus decurrens* **is common within the** *Abies concolor* **Zone (Abbott Creek Research Natural Area, Rogue River National Forest, Oregon).**

Abies magnifica shastensis Zone

Abies magnifica shastensis[8] dominates the forests between the subalpine *Tsuga mertensiana* Zone and the *Abies concolor* Zone (fig. 115). The *Abies magnifica shastensis* Zone is generally found at elevations between 1,600 and 2,000 meters in the Cascade Range and 1,800 and 2,200 meters in the Siskiyou Mountains (Dennis 1959, Whittaker 1960). It is well developed on the western slopes of the Cascade Range (e.g., in the vicinity of Crater Lake National Park), but it also occurs on the east side of the Cascade divide to nearly 43° north latitude. The forests of this zone are closely allied with *Abies magnifica* or Red Fir forests of the California Cascade Range and Sierra Nevada (Oosting and Billings 1943, Küchler 1964, Griffin 1967). They are generally considered a part of Merriam's Canadian Life Zone (Bailey 1936, Dennis 1959).

Figure 115.—Pure, even-aged stands of *Abies magnifica shastensis* are common within the *A. magnifica shastensis* Zone of southwestern Oregon (Rogue River National Forest, Oregon).

---

[8] Taxonomic controversy surrounds this southern Oregon *Abies* (Franklin 1964). It has been referred to as both *Abies procera* and *A. magnifica shastensis*. Populations in southwestern Oregon appear to be part of a species complex involving *Abies procera* and *A. magnifica*; it has been suggested these constitute hybrid populations between these species or a midpoint on an ecocline involving the whole complex. For convenience, we will refer to these populations as *Abies magnifica shastensis* which they resemble ecologically. Although Whittaker (1960) referred to his populations as *Abies nobilis* (synonym for *A. procera*), he stated (personal communication) that elements of *Abies magnifica*, *A. magnifica shastensis*, and *A. procera* were all present.

154

Environmental Features

Environmental data are lacking for the *Abies magnifica shastensis* Zone. The Crater Lake climatic station (table 13) is near its boundary with the *Tsuga mertensiana* Zone. Two major climatic features are known: (1) much of the annual precipitation falls as snow, which accumulates in winter snowpacks with maximum depths of 2 meters or more; (2) critical plant moisture stresses do not occur during the short summers (Waring 1969, Griffin 1967). Soils tend toward Haplorthods (Brown Podzolic) with well-developed mor humus layers; podzolic A2 horizons are not encountered.

Forest Composition

Abies magnifica shastensis is the major tree species within the *Abies magnifica shastensis* Zone (fig. 115). *Abies concolor, Pinus monticola, P. contorta,* and *Tsuga mertensiana* are the most common associates. Many other species are not common but may be present, especially on specialized habitats, including *Pseudotsuga menziesii, Pinus ponderosa, Libocedrus decurrens, Picea engelmannii, Abies amabilis,* and *A. lasiocarpa.*

The composition and density of understory in *Abies magnifica shastensis* stands vary widely. Volland[9] describes several stand types along the crest of the High Cascades, and Mitchell (1972) and Mitchell and Moir (see footnote 5) describe several stands located along the Rogue-Umpqua River divide (Abbott Creek Research Natural Area in Franklin et al. 1972). The High Cascade communities (fig. 116) include the following:

Name	Major associated trees	Characteristic understory plants
Abies magnifica shastensis/ Rubus ursinus/Bromus vulgaris	*Abies concolor, Pseudotsuga menziesii, Pinus monticola*	*Rubus ursinus, Ribes lobbii, Symphoricarpos* sp., *Berberis nervosa, Bromus vulgaris*
Abies magnifica shastensis/ Chimaphila umbellata		*Chimaphila umbellata, Carex pensylvanica*
Abies magnifica shastensis/ Vaccinium ovalifolium/herb	*Tsuga mertensiana, Pinus monticola, Abies concolor*	*Vaccinium ovalifolium, Chimaphila umbellata, Smilacina stellata, Rubus ursinus, Osmorhiza* sp., *Pyrola* spp., *Bromus vulgaris*
Abies magnifica shastensis- Tsuga mertensiana/ Vaccinium scoparium	*Pinus contorta*	*Arctostaphylos nevadensis, Chimaphila umbellata, Carex pensylvanica, Vaccinium scoparium*
Abies magnifica shastensis/ Castanopsis chrysophylla- Arctostaphylos nevadensis		*Arctostaphylos nevadensis, A. patula*

---

[9] Personal communication, Leonard A. Volland, January 21, 1973.

Figure 116.—Several major *Abies magnifica shastensis* communities have been recognized in the High Cascades Province of southern Oregon (all photos from proposed Wickiup Springs Research Natural Area, Rogue River National Forest, Oregon). **Upper left:** *A. magnifica shastensis/Chimaphila umbellata* community with *Carex pensylvanica* **and** *Bromus vulgaris* **as major associates. Upper right: Depauperate variant of preceding community with only very scattered colonies of the aforementioned herbs. Lower left:** *A. magnifica shastensis-Tsuga mertensiana/Vaccinium scoparium* **community found at higher elevations in the zone. Lower right:** *A. magnifica shastensis/Vaccinium ovalifolium/*herb community found on moister sites in the zone.

The *Abies magnifica shastensis/Rubus/Bromus* type is typically a mixed-conifer forest found at the lower elevational limits of *A. magnifica shastensis* which is clearly seral to *Abies concolor* on this habitat. The *A. magnifica shastensis/Chimaphila* type is a depauperate community which characterizes a large proportion of the zonal forests. The *A. magnifica shastensis/Vaccinium ovalifolium*/herb type is a richer community found on more mesic habitats. A wide variety of succulent herbs with substantial coverage are associated with the *Vaccinium*. The *A. magnifica shastensis-Tsuga mertensiana/ Vaccinium scoparium* community is particularly common near the upper limits of *A. magnifica shastensis* forests. The *A. magnifica shastensis/Castanopsis-Arctostaphylos* community is found on the eastern slopes of the High Cascades. A depauperate community, it is typically associated with very stony soils, usually having exposed igneous rock talus. At higher elevations, rock stripes are frequently associated features.

Abies magnifica shastensis communities in the Abbott Creek Research Natural Area are a complex of varied herb-rich types (fig. 117). *A. magnifica shastensis* is considered to be climax with *Tsuga mertensiana* and, possibly, *Abies concolor* as climax associates and *Libocedrus decurrens* as a pioneer associate. Major understory species are *Ribes binominatum, R. viscosissimum, Rubus ursinus, Asarum caudatum, Adenocaulon bicolor, Achlys triphylla, Rubus parviflorus, Vicia americana, Smilacina stellata, Osmorhiza chilensis, Circaea alpina, Vancouveria hexandra, Galium oreganum, Hieracium albiflorum, Galium triflorum,* and *Trientalis latifolia.* The *Abies magnifica shastensis* forests along the Rogue-Umpqua and Umpqua-Willamette River divides appear to be, in general, more luxuriant than the *A. magnifica shastensis* forests in the adjacent High Cascades Province and Siskiyou Mountains because many northern elements (e.g., *Abies amabilis*) find their southern limits there.

Figure 117.—*Abies magnifica shastensis* **forests on moist sites are typified by lush herbaceous understories and mesic shrubs such as** *Ribes binominatum* **or** *Vaccinium ovalifolium*; **note the dense cover of herbs and deciduous shrubs on the right photo and scattering of** *Tsuga mertensiana* **saplings and poles on the left photo (Abbott Creek Research Natural Area, Rogue River National Forest, Oregon).**

Community analyses are not available from the Siskiyou Mountains; but typical species listed for *Abies magnifica shastensis* forests are (Waring 1969, Whittaker 1960, Dennis 1959):

Shrubs—*Vaccinium membranaceum, Ribes marshallii, R. viscosissimum, Arctostaphylos patula, A. nevadensis,* and *Castanopsis chrysophylla.*

Herbs—*Anemone deltoidea, A. oregana, Valeriana sitchensis, Arenaria macrophylla, Campanula scouleri, Achlys triphylla, Arnica latifolia, A. cordifolia, Osmorhiza chilensis, Hieracium albiflorum, Viola glabella, Polemonium californicum,* and *Stellaria crispa.*

From these data and our own observations, some general community patterns appear in *Abies magnifica shastensis* forests. Modal habitats have relatively depauperate understories (*Chimaphila umbellata* or *Carex pensylvanica* types). Moister habitats are occupied by communities with rich herbaceous understories usually with shrub associates (*Vaccinium ovalifolium*- or *Ribes* sp.-herb types). On colder, more rigorous habitats, *Tsuga mertensiana* becomes a major associate and *Vaccinium scoparium* typifies the understory. Under dense forest stands on modal habitats, the understory may be nearly absent and consist only of small ericads and orchids, many of which are nearly or completely aclorophyllous [the "*Pirola-Corallorrhiza* union" of Oosting and Billings (1943) and elements of Furman and Trappe's (1971) "mycotrophic aclorophyllous angiosperms"]: *Chimaphila umbellata, C. menziesii, Pyrola secunda, P. picta, P. aphylla, P. dentata, Corallorhiza maculata, C. mertensiana, Pterospora andromedea, Allotropa virgata,* and *Sarcodes sanguinea.*

Successional Patterns

Successional relationships in the *Abies magnifica shastensis* Zone are not completely understood. Any of the tree species may invade an area directly following fire or logging, but *Pinus contorta* and *P. monticola* are confined strictly to a seral role on normal forest sites. Interesting two-storied stands of scattered old *Abies magnifica shastensis* over a *Pinus contorta* lower canopy are sometimes encountered; reproduction of *Abies magnifica shastensis* is typically present within the understory.

Abies magnifica shastensis exhibits varied behavior. In many stands it appears to be climax; i.e., it is reproducing in sufficient numbers to maintain the population. In other areas, *Abies magnifica shastensis* stands are apparently succeeded by *Abies concolor* or *Tsuga mertensiana* (particularly at the lower and upper limits of the zone, respectively), or even by *Abies amabilis* (on some protected sites toward the north end of the zone). This variable behavior is interesting since *Abies magnifica* is the major climax species in the Sierran *Abietum magnificae* (Oosting and Billings 1943), whereas *Abies procera* is never climax in the northern Cascade Range (Franklin 1966, Thornburgh 1969).

Special Types

Brushfields are encountered with the *Abies magnifica shastensis* Zone. *Ribes* spp., *Vaccinium membranaceum,* and *Ceanothus velutinus* are typical dominants. Dennis (1959) described two shrub communities within this zone in the eastern Siskiyou Mountains—*Arctostaphylos patula/Ceanothus velutinus* and *Artemisia tridentata/Lonicera conjugialis.* Both included scattered *Abies magnifica shastensis* and *Pinus monticola.* Shrub communities dominated by *Cercocarpus ledifolius* are also encountered in the Siskiyou Mountains.

A variety of wet and dry meadow communities is associated with the *Abies magnifica shastensis* forest, as mentioned earlier. Volland (personal communication) describes small nonforested openings in *Abies magnifica shastensis* forest dominated by *Lupinus latifolius, Carex pensylvanica, Sitanion hystrix,* and *Stipa occidentalis* with occasional *Bromus carinatus* and *Haplopappus bloomeri* (fig. 118). *Abies* appears to be slowly invading these meadows.

Figure 118.—A variety of meadows and nonforested openings are found in the *Abies magnifica shastensis* **Zone; this opening is dominated by** *Carex pensylvanica, Sitanion hystrix,* **and** *Stipa occidentalis* **and has abundant gopher activity as well as some invading tree seedlings and saplings (near Freye Lake, Winema National Forest, Oregon).**

CHAPTER VII. FOREST ZONES OF EASTERN OREGON AND WASHINGTON

Conifer forests are conspicuous in the dry interior regions. They clothe the eastern slopes of the Cascade Range and extend, with only minor interruptions, around the northern edge of Washington to the northern Rocky Mountains. Forests also dominate in the Blue, Ochoco, and Wallowa Mountains of Oregon.

In this region, elements of the continental Rocky Mountain forests meld with some of those from coastal areas. In addition, forest species mix with species from steppe and shrub-steppe communities, especially near lower timberline. It is a country typified by the "western yellow pine" (*Pinus ponderosa*) forests.

Forests of the interior region have been studied by foresters and ecologists for many years. Shantz and Zon (1924) and Hansen (1947) provided generalized accounts. Notable synecological studies include those of Daubenmire (1952, 1953, 1956, 1966, 1969b), Daubenmire and Daubenmire (1968), Driscoll (1962, 1964a, 1964b), Hall (1967, 1968, 1971), Berry (1963), Dealy (1971), Dyrness and Youngberg (1958, 1966), McLean (1970), Swedberg (1961), Trappe and Harris (1958), Volland (1963), West (1964, 1969a, 1969b), Tisdale (1968), and McMinn (1952). Weaver (1943, 1955, 1959, 1961, 1964, 1968) has written extensively about the role of fire in these forests. The intricate relationship between *Pinus ponderosa* and *Pinus contorta* in the central Oregon pumice region has received special consideration by Kerr (1913), Munger (1914), Tarrant (1953), Youngberg and Dyrness (1959), Cochran (1972), and Berntsen (1967).

Through these studies, many forest associations have been identified, characterized by different climax tree species: *Juniperus occidentalis*, *Pinus ponderosa*, *Pseudotsuga menziesii*, *Abies grandis*, *Thuja plicata*, *Tsuga heterophylla*, *Abies lasiocarpa*, *Pinus contorta*, and *Libocedrus decurrens*. *Abies concolor*[1] and *Abies magnifica shastensis*[2] associations, so widespread in California and southwestern Oregon, are also encountered on eastern slopes of the southern Oregon Cascade Range. At higher elevations, near the crest of the Cascade Range, occur *Tsuga mertensiana* and sometimes *Abies amabilis* associations. These have already been discussed.

This abundance of climax forest types and the complex array of seral communities result from an abundance of coniferous species and environmental diversity. Climax associations are generally arrayed elevationally as zones, the consequence of differing responses of tree species to temperature and moisture gradients interacting with differing degrees of tolerance. Daubenmire (1966) has illustrated this phenomenon for forests of eastern Washington and northern Idaho (fig. 119). At various points along the gradient, increasingly tolerant species enter the forest communities producing a sequence of steps based on the climax tree species.

---

[1] *Abies grandis* and *Abies concolor* form a continuously varying biological complex in eastern Oregon. Most of the *Abies grandis* found in northeastern Oregon (Blue Mountains and Cascade Range) intergrades with *Abies concolor*. In this area we will, for convenience, refer to the populations as *Abies grandis* since the complex resembles this species morphologically. In south-central Oregon, we will refer to the complex as *Abies concolor*. These two species and the zones they typify occupy analogous positions synecologically and environmentally in their respective "areas."

[2] See footnote 8 in Chapter VI.

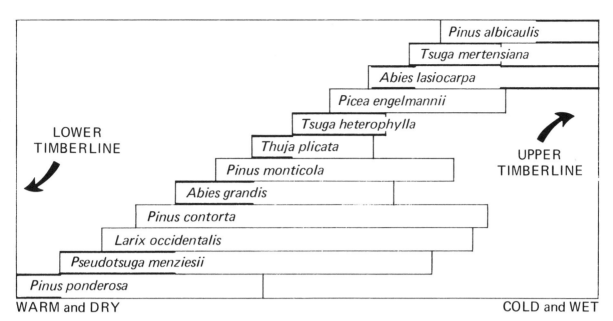

Figure 119.—Coniferous trees in eastern Washington and northern Idaho showing the usual order in which the species are encountered with increasing altitude; horizontal bars designate upper and lower limits of species relative to the climatic gradient, portions of a species range where it is climax in the face of intense competition indicated by heavy lines (from Daubenmire 1966).

For discussion of interior forests, we will group the series of climax associations into seven zones:

> *Juniperus occidentalis*
> *Pinus ponderosa*
> *Pinus contorta*
> *Pseudotsuga menziesii* (plus *Libocedrus decurrens*)
> *Abies grandis* (plus *Abies concolor*)
> *Tsuga heterophylla* (plus *Thuja plicata*)
> *Abies lasiocarpa*

All these zones delineate important phytogeographic units, but they do not all occur on a single mountain slope nor do they necessarily occur as sequential belts. Furthermore, the *Pinus contorta* Zone does not really qualify as part of a zonal series since *P. contorta* does not occur as a climatic climax. Since climax *P. contorta* forests are so conspicuous in parts of eastern Oregon, they are treated at the zonal level, however.

Typical zonal sequences at various locations in eastern Oregon and Washington are indicated in table 21. Obviously, a great variation is possible. Lack of a zone in a particular area can result from species absence, as a consequence of macroclimate or history, or from localized edaphic conditions. For example, in northern Idaho, *Tsuga heterophylla*, *Thuja plicata*, and *Abies grandis* Zones are all common (Daubenmire and Daubenmire 1968). To the south, along the Rocky Mountains within Idaho, *Tsuga*, *Thuja*, and finally *Abies grandis* drop out in turn, gradually altering the zonal sequence. Similarly, *Tsuga heterophylla* and *Thuja plicata* are absent from the eastern slopes of the southern Oregon Cascade Range and Blue Mountains, whereas *Juniperus occidentalis* is generally absent from eastern Washington. Examples of the influence of localized edaphic and climatic conditions include the *Pinus ponderosa-Pinus contorta* complex on Mazama pumice in south-central Oregon and the tendency on finer textured soils for elimination of a

Pinus ponderosa Zone between steppe and the *Pseudotsuga menziesii* Zone (as illustrated by Brayshaw 1965).

A most important regional variation in zonal sequences involves the relative importance of the *Abies grandis* and *Pseudotsuga menziesii* Zones. The *Abies grandis* Zone (including *Abies concolor*) is probably the most extensive forested zone in eastern Oregon; the *Pseudotsuga menziesii* Zone is poorly represented or absent. Indeed, *Pseudotsuga* is essentially totally absent from the forests on the eastern slopes of the southern Oregon Cascade Range and the mountains of the Basin and Range Province to the east! Conversely, on eastern slopes of the northern Washington Cascade Range and in the Okanogan Highlands Province, the *Pseudotsuga menziesii* Zone becomes relatively more important than the *Abies grandis* Zone until, in adjacent British Columbia, the latter is absent (McLean 1970, Brayshaw 1965).

Table 22 provides an overview of forest composition and successional relationships in the forested zones of eastern Oregon and Washington.

Table 21. — Typical zonal forest sequences at locations in eastern Washington and Oregon adapted from the indicated sources[1]

Northwestern Washington and northern Idaho (Daubenmire 1952, 1966)	Eastern slopes, Washington Cascade Range (Franklin and Trappe 1963)	Eastern slopes, central Oregon Cascade Range (Swedberg 1961; West 1964)	Eastern slopes, southern Oregon Cascade Range (Dyrness and Youngberg 1958, 1966)	Ochoco and Blue Mountains, Oregon (Hall 1967)
		Juniperus occidentalis	*Juniperus occidentalis*	(*Juniperus occidentalis*)
Pinus ponderosa	(*Pinus ponderosa*)	*Pinus ponderosa*	*Pinus contorta*	*Pinus ponderosa*
Pseudotsuga menziesii	*Pseudotsuga menziesii*	(*Pseudotsuga menziesii*)	*Pinus ponderosa*	
Abies grandis	*Abies grandis*	*Abies grandis*	*Abies concolor*	*Abies grandis*
Tsuga heterophylla	(*Tsuga heterophylla* and/or *Abies amabilis*)	(*Abies amabilis*)	(*Abies magnifica shastensis*)	
Abies lasiocarpa or *Tsuga mertensiana*	*Abies lasiocarpa* or *Tsuga mertensiana*	*Tsuga mertensiana*	(*Tsuga mertensiana*)	(*Abies lasiocarpa*)

[1] Species names in parentheses indicate the enclosed zone may be absent or only fragmentally represented.

Table 22. — Importance and ecological role of major tree species in representative forest zones and locales in eastern Oregon and Washington[1]

Species	Eastern Washington Cascade Range					Central Oregon Cascade Range					Ochoco and Blue Mountains		Southern Oregon Cascade Range		
	Pinus ponderosa Zone	Pseudotsuga menziesii Zone	Abies grandis Zone	Tsuga heterophylla Zone	Abies lasiocarpa Zone	Juniperus occidentalis Zone	Pinus ponderosa Zone	Pseudotsuga menziesii Zone	Abies grandis[2] Zone	Tsuga mertensiana Zone	Pinus ponderosa Zone	Abies grandis[2] Zone	Pinus contorta Zone	Pinus ponderosa Zone	Abies concolor Zone
Abies amabilis	–	–	–	–	c	–	–	–	–	c	–	–	–	–	–
Abies grandis or concolor	–	–	C	S	s	–	–	–	C	s	–	C	–	–	C
Abies lasiocarpa	–	–	–	s	C	–	–	–	s	C	–	–	–	–	–
Abies magnifica shastensis	–	–	–	–	–	–	–	–	–	–	–	–	–	–	s
Libocedrus decurrens	–	S	S	s	s	–	–	c	s	–	–	–	–	–	s
Larix occidentalis	–	S	S	s	s	–	–	s	s	–	–	S	–	–	–
Larix lyallii	–	–	–	–	s	–	–	–	–	–	–	–	–	–	–
Picea engelmannii	–	–	s	s	Sc	–	–	–	s	s	–	s	–	–	–
Pinus contorta	–	S	S	s	s	–	–	s	S	S	–	S	C	S	s
Pinus lambertiana	–	–	–	–	–	–	–	–	s	–	–	–	–	–	s
Pinus monticola	–	–	S	S	s	–	–	–	s	s	–	s	–	–	s
Pinus ponderosa	C	S	S	s	–	–	C	S	S	s	C	s	c	C	S
Juniperus occidentalis	–	–	–	–	–	C	s	s	–	–	s	–	–	s	–
Populus tremuloides	s	s	s	–	–	–	s	s	s	–	–	–	s	s	s
Pseudotsuga menziesii	–	C	S	S	s	–	–	C	S	s	–	Sc	–	–	–
Thuja plicata	–	–	–	sc	–	–	–	–	–	–	–	–	–	–	–
Tsuga heterophylla	–	–	–	C	s	–	–	–	–	–	–	–	–	–	–
Tsuga mertensiana	–	–	–	–	c	–	–	–	s	C	–	–	–	–	s

[1] C = major climax species, c = minor climax species, S = major seral species, and s = minor seral species.
[2] Abies grandis-A. concolor complex.

Juniperus occidentalis Zone

The *Juniperus occidentalis* Zone is the northwestern representation of the Pinyon-Juniper Zone so conspicuous in the Great Basin region (Billings 1951). It is generally a savanna zone (fig. 120), occupying habitats intermediate in moisture between *Pinus ponderosa* forest and steppe or shrub-steppe (Driscoll 1964b). The *Juniperus occidentalis* Zone is found only in eastern Oregon, reaching maximum development in central Oregon around the Deschutes, Crooked, and John Day Rivers. Similar communities occur at various localities throughout southeastern Oregon. For example, Faegri (1966) has described an area of *Juniperus occidentalis-Artemisia* vegetation situated on the slopes of Steens Mountain just below a subalpine *Populus tremuloides* belt. Driscoll (1964b) has suggested a physiographic subdivision of the zone into three units based on soil parent materials. Elevational range of the zone appears to be between about 760 and 1,400 meters, although most stands Driscoll (1964b) sampled were between 1,200 and 1,400 meters.

Figure 120.—The *Juniperus occidentalis* **Zone is primarily a savanna region, ecotonal between** *Pinus ponderosa* **forest and** *Artemisia* **shrub-steppe (near Bend, Oregon).**

Environmental Features

The *Juniperus occidentalis* Zone is the most xeric of the tree-dominated zones in the Pacific Northwest. Annual precipitation is low. At Bend, in the *Pinus ponderosa-Juniperus occidentalis* transition, it averages 312 millimeters, but in the center of the zone 200 to 250 millimeters is typical (table 23). Most precipitation falls during the winter, and the hot summer months are often completely dry.

The zone includes Camborthids (Sierozems), Haplargids (Sierozem and Brown soils), and Haploxerolls (Chestnut) soil great groups. Haplargids (Brown soils and associated Regosols) are most common. Surface soils are typically light colored, coarse textured (sandy loams), low in organic matter (e.g., 1 to 4 percent), and slightly acid (pH 6.0) to neutral. Soils average around 76 centimeters in depth, although roots may penetrate underlying cracked bedrock. Subsoils typically have white calcareous or siliceous deposits on peds or rocks.

Table 23. — Climatic data from representative stations within the *Juniperus occidentalis* zone

Station	Eleva-tion	Lati-tude	Longi-tude	Temperature					Precipitation		
				Average annual	Average January	Average January minimum	Average July	Average July maximum	Average annual	June through August	Average annual snowfall
	Meters			- - - - - - - - - Degrees C. - - - - - - - - -					Millimeters		Centi-meters
Bend, Oreg.	1,097	44°04′	121°19′	8.0	−0.9	−6.5	17.9	28.7	318	50	92
Redmond, Oreg.	913	44°17′	121°10′	9.1	−.1	−5.9	18.8	29.5	216	42	61
Prineville, Oreg.	866	44°21′	120°54′	8.0	−.56	−6.6	18.1	29.8	238	41	35

Source: U. S. Weather Bureau (1965a).

Community Composition

Driscoll (1962, 1964a, 1964b) has provided the only comprehensive description of communities in the *Juniperus occidentalis* Zone. His general description of the vegetation of the zone is as follows (Driscoll 1964b):

Juniperus occidentalis is the dominant tree species of the area. An occasional *Pinus ponderosa* may be found in canyon bottoms or on north slopes where soil moisture is more effective. Natural wide spacing of individual junipers provides the aspect of a savanna. . . . *Artemisia tridentata* is most often the dominant shrub in the understory. Occasionally it is displaced wholly or to codominance by *Purshia tridentata*. Other shrubs characteristic of the area are *Chrysothamnus nauseosus*, *C. viscidiflorus*, *Tetradymia canescens*, *Leptodactylon pungens*, and *Artemisia arbuscula*. *Ribes cereum*, *Grossularia velutina*, [*Ribes velutinum*], and *Grayia spinosa* occur infrequently. Suffrutescents are represented by various species of *Eriogonum*.

Agropyron spicatum and *Festuca idahoensis* are the characteristic grasses of relatively undisturbed communities. *Poa secunda* and *Stipa thurberiana* are common. Other grasses include *Sitanion hystrix*, *Stipa comata*, *Bromus tectorum*, *Festuca octoflora*, and *Koeleria cristata*.

Forbs commonly do not constitute major components of relatively undisturbed communities. Some of the more common perennial forbs are *Agoseris* sp., *Achillea millefolium*, *Eriophyllum lanatum*, *Astragalus* spp., *Erigeron linearis*, and *Lupinus* spp.

From the 11 associations and variants described by Driscoll (1964a, 1964b), we have chosen five which typify major variations in *Juniperus occidentalis* communities (table 24). The *Juniperus/Artemisia/Agropyron* association was found on well-drained loamy soils and, therefore, is designated as the climatic climax. *Artemisia tridentata* (average maximum height of 0.6 m.) and *Agropyron spicatum* typified the shrub and herb components, both attaining maximum status here. *Chrysothamnus nauseosus* was the only other shrub. *Stipa thurberiana*, *Poa sandbergii*, *Lomatium triternatum*, *Bromus tectorum*, and *Festuca octoflora* were other notable species (table 24).

The remaining four associations shown in table 24 are all interpreted as constituting topoedaphic climax associations. The *Juniperus/Festuca* association is characterized by

165

Table 24. – Constancy and average cover (constancy/cover) of important species in five associations found within the *Juniperus occidentalis* Zone of central Oregon (from Driscoll 1964b)

Life-form and species	*Juniperus/ Artemisia/ Agropyron*	*Juniperus/ Festuca*	*Juniperus/ Agropyron*	*Juniperus/ Artemisia- Purshia*	*Juniperus/ Artemisia/ Agropyron- Astragalus*
	Association				
	———————————— Percent ————————————				
Trees:					
Juniperus occidentalis	100/10	100/77	100/43	100/7	100/28
Shrubs:					
Artemisia tridentata	100/8	100/1	100/1	100/8	100/7
Chrysothamnus nauseosus	100/1	20/tr	34/tr	–	34/tr
Chrysothamnus viscidiflorus	–	100/tr	84/1	100/2	50/1
Purshia tridentata	–	100/1	100/tr	100/6	34/tr
Graminoids:					
Festuca idahoensis	60/1	100/11	100/2	100/2	–
Agropyron spicatum	100/9	100/2	100/5	100/1	100/7
Poa sandbergii	100/1	100/1	100/1	60/tr	100/1
Koeleria cristata	–	100/tr	100/tr	100/tr	100/1
Bromus tectorum	100/2	100/tr	100/1	100/3	100/1
Stipa thurberiana	80/2	80/1	100/tr	60/tr	50/tr
Festuca octoflora	100/1	–	34/tr	–	34/tr
Sitanion hystrix	40/tr	–	–	100/tr	17/tr
Herbs:					
Astragalus sp.	100/tr	60/tr	50/tr	–	–
Astragalus lectulus	–	60/tr	17/tr	–	100/tr
Achillea millefolium	100/tr	60/tr	100/tr	100/tr	100/tr
Lomatium triternatum	100/1	100/tr	34/tr	20/tr	–
Collinsia parviflora	100/tr	100/tr	100/tr	100/tr	84/tr
Gayophytum nuttallii	–	80/tr	84/tr	100/1	50/tr
Phlox douglasii	20/tr	100/tr	100/tr	20/tr	100/1
Eriophyllum lanatum	–	100/tr	68/tr	40/tr	34/tr
Cryptantha ambigua	20/tr	80/tr	100/tr	100/1	100/tr
Erigeron linearis	20/tr	80/tr	100/tr	100/tr	100/tr
Collomia grandiflora	–	80/tr	68/tr	100/tr	17/tr
Mentzelia albicaulis	–	–	–	100/1	–
Montia perfoliata	–	20/tr	–	100/tr	–
Eriogonum vimineum var. *baileyi*	–	60/tr	68/tr	20/tr	84/tr
Eriogonum umbellatum	20/tr	80/tr	34/tr	20/tr	17/tr
Penstemon humilis	–	80/tr	17/tr	–	–

tr = trace.

high cover of *Juniperus occidentalis* and *Festuca idahoensis* and a notable lack of shrub cover. The *Juniperus/Agropyron* association also has low amounts of shrub cover and is typified mainly by the occurrence of such grass species as *Agropyron spicatum*, *Festuca idahoensis*, *Poa sandbergii*, and *Bromus tectorum*. The *Juniperus/Artemisia-Purshia* association is lowest in *Juniperus occidentalis* and highest in shrub coverage (*Artemisia tridentata* and *Purshia tridentata*). The *Juniperus/Artemisia/Agropyron-Astragalus* association (fig. 121) was found to occupy the most xeric sites. *Festuca idahoensis* is generally found only in *Juniperus* shade in stands typical of this association. Of the few characteristic perennial herbs, *Astragalus lectulus* was most abundant within this association.

A *Juniperus occidentalis/Artemisia tridentata/Carex filifolia* community has been described in the Horse Ridge Research Natural Area east of Bend, Oregon (Franklin et al. 1972). This community is similar to the *Juniperus/Festuca* community of Driscoll (1946b) but occupies sites that are slightly more moist. The understory in these stands is dominated by *Artemisia tridentata*, *Carex filifolia*, and *Festuca idahoensis* (fig. 122). Other commonly occurring species are *Agropyron spicatum*, *Sitanion hystrix*, *Koeleria cristata*, *Stipa thurberiana*, *Eriogonum umbellatum*, *Senecio canus*, *Collinsia parviflora*, *Linanthus septentrionalis*, *Astragalus curvicarpus*, *Tetradymia glabrata*, *Erigeron filifolius*, and *Chrysothamnus viscidiflorus*.

Figure 121.—The *Juniperus occidentalis/Artemisia tridentata/Agropyron spicatum-Astragalus lectulus* **association was the most xeric described by Driscoll (1964b).**

Figure 122.—**A stand typical of the** *Juniperus occidentalis/Artemisia tridentata/Carex filifolia* **community; graminoids visible are** *Carex filifolia, Festuca occidentalis,* **and** *Agropyron spicatum* **(Horse Ridge Research Natural Area).**

Succession

Very little is known concerning successional relationships in the *Juniperus occidentalis* Zone. Burning can kill *Juniperus occidentalis* (fig. 123) and temporarily produce an herb- or shrub-dominated community which is gradually reinvaded by trees. Sparsity of *Juniperus occidentalis* in some associations was explained in this way (Driscoll 1964b). The fire-sensitive *Purshia tridentata* is similarly affected. Grazing by cattle can result in a reduction of the preferred *Agropyron spicatum* and *Festuca idahoensis*; deer browsing affects primarily *Purshia tridentata* and *Juniperus*.

Figure 123.—Fire-killed *Juniperus occidentalis* **clump, probably caused by a lightning strike (Horse Ridge Research Natural Area).**

Pinus ponderosa Zone

Pinus ponderosa forests are widely distributed in eastern Oregon and Washington. They occupy: (1) a narrow band (15 to 30 km. wide) on the eastern flanks of the entire Cascade Range (generally); (2) much of the high pumice plateau extending east from the High Cascades Province; (3) large areas in the Blue Mountains Province (Ochoco, Blue, and Wallowa Mountains) of northeastern Oregon and extreme southeastern Washington; and (4) extensive tracts in the Okanogan Highlands Province of northeastern Washington. The band of *Pinus ponderosa* forests generally increases in elevation from north to south. Throughout much of Washington they are at approximately 600 to 1,200 meters in elevation. In the Blue Mountains Province and northeastern Oregon generally, the elevational range is about 900 to 1,500 meters. Elevations in the south-central Oregon pumice area are considerably higher—about 1,450 to 2,000 meters.

At their upper limits, *Pinus ponderosa* forests may grade into forests of *Pseudotsuga menziesii*, *Abies grandis*, or *A. concolor* depending on the locale (table 21). Throughout much of Oregon, they abut *Artemisia tridentata* steppe or open *Juniperus occidentalis-Artemisia tridentata* woodland at their lower limits. At lower elevational limits in Washington, *Pinus ponderosa* forests grade into either grassland or *Artemisia* steppe. At some locations in northern Oregon and southern Washington, there is an ecotonal belt of *Quercus garryana* between *Pinus ponderosa* forest and steppe.

The discussion of the *Pinus ponderosa* Zone which follows considers seral *Pinus ponderosa* forests as well as forests in which the *Pinus* is climax. The *Pinus ponderosa* Zone, in the strict sense, includes only the latter; in this narrower definition, the *Pinus ponderosa* Zone correlates with the Ponderosa Pine-Bunchgrass Zone of Krajina (1965) and Brayshaw (1965). It is important to realize that in many locations there is no belt of climax *Pinus ponderosa* forests between steppe and areas of *Pseudotsuga menziesii* (Brayshaw 1965, Johnson 1959) or *Abies grandis* (Hall 1967) climax. The *Pinus ponderosa* Zone, more broadly defined, correlates roughly with Merriam's Arid Transition Zone (Barrett 1962, Bailey 1936) and includes representation of Küchler's (1964) Ponderosa Shrub and Western Ponderosa Forests.

Environmental Features

The climate of the *Pinus ponderosa* Zone is characterized by a short growing season and minimal summer precipitation (table 25). Average annual precipitation ranges from about 355 to 760 millimeters, much of it falling as winter snow. Diurnal summer temperatures fluctuate widely, with hot days and cold nights. In many areas, frost may occur any night of the year. The months of July, August, and September are very dry, with rainfall averaging less than 25 millimeters. Much of this summer rain is ineffective, as it usually comes during brief, high-intensity convection storms. Winter temperatures are generally low; as a result, snow often accumulates to considerable depths.

Since *Pinus ponderosa* occupies drier sites than any other forest type (except *Juniperus occidentalis*), its distribution is closely correlated with supplies of available soil moisture. This is often reflected by a distinct relationship between occurrence of *Pinus ponderosa* and soil texture, especially at the dry end of its range. Many studies have shown better survival and growth of *Pinus ponderosa* on coarse-textured, sandy soils than on fine-textured, clayey soils (Pearson 1923, Howell 1932, Stone and Fowells 1955, Fowells and Kirk 1945). These effects can be largely attributed to more extensive root proliferation on coarse-textured soils. On xeric sites, a mosaic of *Pinus ponderosa* communities (on coarse-textured soils) and steppe or shrub-steppe communities (on finer textured soils) is common (Hall 1967). An extreme example is a disjunct stand of *Pinus ponderosa* in central Oregon, located 64 kilometers from the nearest other *Pinus ponderosa* forest (fig. 124). Despite less than 250 millimeters annual precipitation, abundant regeneration attests that the stand is maintaining itself. Berry (1963) attributed existence of this anachronistic forest primarily to the uniformly sandy soils it occupies.

Table 25. — Climatic data from representative stations in the vicinity of the *Pinus ponderosa* Zone

Station	Eleva-tion	Lati-tude	Longi-tude	Temperature					Precipitation		
				Average annual	Average January	Average January minimum	Average July	Average July maximum	Average annual	June through August	Average annual snowfall
	Meters			- - - - - - - - - - Degrees C. - - - - - - - - - -					Millimeters		Centi-meters
Chiloquin, Oreg.	1,280	42°35′	121°47′	6.3	−3.1	−9.4	16.3	28.4	441	43	--
Joseph, Oreg.	1,341	45°21′	117°15′	6.4	−5.3	−10.6	17.5	26.6	411	94	--
Starkey, Oreg.	1,036	45°14′	118°23′	6.3	−3.6	−10.0	16.2	28.0	457	89	--
Cle Elum, Wash.	579	47°11′	120°55′	7.4	−3.2	−7.7	18.1	27.8	537	41	219
Leavenworth, Wash.	344	47°34′	120°40′	8.4	−5.1	−9.9	20.4	30.8	590	44	252
Northport, Wash.	411	48°55′	117°47′	8.3	−4.5	−8.2	20.6	31.4	477	96	141

Source: U. S. Weather Bureau (1956, 1965a, 1965b) and Johnsgard (1963).

Figure 124.—Coarse-textured soils favor development of *Pinus ponderosa* in forest-steppe ecotonal areas; this disjunct stand, the Lost Forest, is an extreme example, occupying sandy soils located 64 kilometers within the central Oregon shrub-steppe.

Intensity of soil profile development varies with elevation and parent material. At lower elevations within the zone, soils tend to be coarse textured and generally have weakly differentiated A, B, and C horizons. Haplumbrepts (Western Brown Forest soils) are the most common soils showing substantial profile development. They have moderately dark-colored and thick A horizons grading into B horizons distinguished by color, and sometimes by structure, since clay eluviation is not characteristic. Surface soils are generally slightly acid, and reaction often becomes more neutral with depth. Soils on moister, cooler sites generally show some evidence of podzolization—e.g., they may have moderately thick accumulations of duff and litter and thin A1 horizons underlain by distinct, light-colored A2 (bleicherde) horizons. Soil reaction ranges from slight to medium acidity. These soils are generally classified as Haplorthods (Gray Wooded soils).

The *Pinus ponderosa* Zone includes large areas of immature regosolic soils, particularly the 5,000,000-hectare pumice plateau of south-central Oregon. These Vitrandepts are developed in deposits of dacitic and rhyolitic pumice erupted from Mount Mazama (Crater Lake) and Newberry Crater, respectively. Thin A horizons have moderate to low organic matter content and grade into relatively unweathered pumice sand and gravel. A finer textured buried soil is generally encountered at ½ to 3 meters. These coarse-textured pumice soils apparently enable *Pinus ponderosa* to extend its range east into areas where the vegetation would otherwise be *Artemisia* steppe. The frequent coincidence of eastern boundaries of *Pinus ponderosa* forest and pumice soil areas provides evidence for this.

170

Forest Composition

Pinus ponderosa is associated with a rich variety of tree species. Only four of these—*Juniperus occidentalis*, *Populus tremuloides*, *Pinus contorta*, and *Quercus garryana*—are generally associates in climax *Pinus ponderosa* stands (the narrowly defined *Pinus ponderosa* Zone). Even these associates are restricted to specific habitats in the zone or geographically. *Juniperus occidentalis* occurs as a minor component of xeric *Pinus ponderosa* stands in much of southeastern Oregon (Dealy 1971; Swedberg 1961, 1973; West 1964). Groves of *Populus tremuloides* occur on riparian and poorly drained wet areas throughout the *Pinus ponderosa* Zone and adjacent forest and steppe zones as well (fig. 125) (Johnson 1961, Dealy 1971, Daubenmire 1952). *Quercus garryana* is associated only on the east slopes of the Cascades in northern Oregon and southern and central Washington. *Pinus contorta* and *P. ponderosa*, frequent seral associates in more mesic zones (e.g., *Pseudotsuga menziesii* Zone), are the sole constituents of extensive forests in the pumice region of south-central Oregon (Dyrness and Youngberg 1966).

The remaining tree species found in *Pinus ponderosa* stands are generally present only on sites where *Pinus ponderosa* is seral; e.g., more mesic zones such as the *Abies grandis* or *Pseudotsuga menziesii*. Any of these may occur as accidentals within climax *Pinus*

Figure 125.—Groves of *Populus tremuloides* **are common on riparian and poorly drained habitats within the** *Pinus ponderosa* **Zone (Colville Indian Reservation, Washington).**

ponderosa stands, however. *Abies grandis, Pseudotsuga menziesii, Larix occidentalis,* and *Pinus monticola* are associated with *Pinus ponderosa* essentially throughout northeastern Oregon and eastern Washington (including eastern slopes of the Cascade Range) Swedberg 1961, 1973; Hall 1967; Daubenmire and Daubenmire 1968). *Libocedrus decurrens* occurs only along the eastern slopes of the central and northern Oregon Cascades (Swedberg 1961, 1973; West 1969a; Sherman 1969). *Abies concolor* is a major constituent of some seral *Pinus ponderosa* stands of the southern and central Oregon Cascade Range and pumice region (Volland 1963, Dyrness and Youngberg 1966).

Community composition in *Pinus ponderosa* stands varies widely with geographic location, soils, elevation and aspect, and successional status. The history of stand disturbances, such as by fire and logging, influence overstory density which, in turn, can have profound effects on understory composition and density (Moir 1966, Robinson 1967, Sherman 1966). The open nature of typical mature *Pinus ponderosa* stands (fig. 126)

Figure 126.—The open nature of many mature *Pinus ponderosa* stands provides niches for heliophytic species, including many typical of steppe and shrub-steppe communities; *Pinus ponderosa-Juniperus occidentalis/Artemisia tridentata* savanna (Lost Forest Research Natural Area, northern Lake County, Oregon; photo courtesy C. Johnson).

provides abundant niches for heliophytic species, including many typical of steppe and shrub-steppe communities.

Pinus ponderosa community types or associations have been identified in many localities: (1) south-central Oregon (Volland 1963, Dyrness 1960, Dyrness and Youngberg 1966, Dealy 1971), (2) the Blue Mountains of northeastern Oregon,[3] and (3) eastern Washington (Daubenmire 1952, Daubenmire and Daubenmire 1968). West (1964, 1969a) and Swedberg (1961) found a continuum viewpoint useful in interpreting *Pinus ponderosa* communities along environmental gradients on the eastern slopes of Oregon's Cascade Range. All these studies show that some characteristic understory dominants, such as *Festuca idahoensis* and *Purshia tridentata*, occur throughout much of the zone—from northeastern Washington to south-central Oregon. However, many others have more restricted distributions, so that the character of the understory tends to vary with locale. Consequently, communities will be considered by geographic location.

In eastern Washington, six *Pinus ponderosa* associations have been recognized (Daubenmire and Daubenmire 1968) (table 26):

> *Pinus ponderosa/Symphoricarpos albus,*
> *Pinus ponderosa/Physocarpus malvaceus,*
> *Pinus ponderosa/Festuca idahoensis,*
> *Pinus ponderosa/Agropyron spicatum,*
> *Pinus ponderosa/Stipa comata,* and
> *Pinus ponderosa/Purshia tridentata.*

The first two associations comprise a shrubby group found on deep, fine-textured, fertile soils. *Pinus* reproduction is sparse but sufficient to produce an all-aged forest. The *Pinus ponderosa/Symphoricarpos albus* association is considered a climatic climax (Daubenmire 1952), since it occurs on loamy soils and undulating topography. The understory is dominated by a nearly continuous, 0.5- to 1-meter-tall cover of low, deciduous shrubs, mainly *Symphoricarpos albus, Spiraea betulifolia* var. *lucida, Rosa woodsii,* and *R. nutkana.* There is a rich variety of herbaceous, mainly perennial, associates, of which several grasses (e.g., *Calamagrostis rubescens,* rhizomatous *Agropyron spicatum*) have the highest constancies. The *Pinus ponderosa/Physocarpus malvaceus* association occupies slightly more mesic sites and adds a taller shrub layer (2 m.) of *Physocarpus malvaceus, Holodiscus discolor,* and *Ceanothus sanguineus* to the *Symphoricarpos* understory discussed above.

The other four *Pinus ponderosa* associations—*Festuca idahoensis, Agropyron spicatum, Stipa comata,* and *Purshia tridentata*—Daubenmire and Daubenmire (1968) refer to as a grassy group (fig. 127). Understories are dominated by xerophytic grasses; soils are stony, coarse textured, or shallow; and reproduction of *Pinus* is episodic. The first three associations were grouped in a *Pinus ponderosa/Agropyron spicatum* association in an earlier study (Daubenmire 1952). Each has an understory dominated almost exclusively by a single large perennial bunchgrass—*Festuca idahoensis,* caespitose *Agropyron spicatum,* or *Stipa comata.* The *Pinus ponderosa/Purshia tridentata* association has a *Purshia*-dominated shrub layer superimposed on a variety of perennial grasses including *Festuca idahoensis,* caespitose *Agropyron spicatum, Stipa comata,* and *Aristida longiseta.* In some stands, forbs such as *Balsamorhiza sagittata* and *Erigeron compositus* are abundant. Two of these associations—the *Pinus/Agropyron* and *Pinus/Festuca*—extend into British Columbia (McLean 1970).

Cooke (1955) has provided data on occurrence of fungi, mosses, and lichens in *Pinus ponderosa/Agropyron spicatum* and *Pinus ponderosa/Symphoricarpos albus* stands in eastern Washington and adjacent Idaho.

Studies have indicated that soil moisture regime is the most important single factor influencing the distribution of these climax vegetation types (McMinn 1952, Daubenmire 1968b). Supplies of available soil moisture are exhausted early in the growing season in

---

[3] F. C. Hall (1967) and personal communication.

Table 26. – Constancy and average cover (constancy/coverage) of selected understory species in *Pinus ponderosa* associations in eastern Washington and northern Idaho (calculated from Daubenmire and Daubenmire 1968)

Life-form and species	Association[1]					
	Pipo/Syal	Pipo/Phma	Pipo/Feid	Pipo/Agsp	Pipo/Stipa	Pipo/Putr
	———————————————— Percent ————————————————					
Tall shrubs:						
Amelanchier alnifolia	88/2	100/3	11/tr[2]	–	29/tr	83/NA[3]
Prunus virginiana melanocarpa	50/1	57/5	–	–	–	100/NA
Crataegus douglasii	43/1	86/1	–	–	–	17/NA
Holodiscus discolor	–	86/16	–	–	14/tr	17/NA
Medium shrubs:						
Symphoricarpos albus	100/46	100/32	11/tr	–	–	83/NA
Rosa woodsii and nutkana	100/6	100/3	–	20/tr	–	67/NA
Physocarpus malvaceus	13/tr	100/62	–	–	–	–
Spiraea betulifolia lucida	62/9	100/7	–	–	–	–
Low shrubs:						
Berberis repens	13/tr	57/1	–	–	–	–
Eriogonum heracleoides	13/tr	–	–	20/tr	–	50/NA
Eriogonum niveum	–	–	11/tr	20/tr	71/1	17/NA
Arceuthobium campylopodum	–	–	66/tr	40/tr	43/tr	50/NA
Perennial graminoids:						
Agropyron spicatum	75/5	29/2	56/10	100/61	29/1	100/NA
Festuca occidentalis	50/1	29/2	11/tr	–	–	–
Calamagrostis rubescens	62/10	86/5	–	–	–	17/NA
Bromus vulgaris	13/tr	71/2	–	–	–	–
Carex geyeri	25/20	57/12	–	–	–	–
Elymus glaucus	88/5	43/1	–	–	–	–
Poa compressa	50/6	29/2	–	–	–	–
Festuca idahoensis	13/tr	–	100/76	–	14/3	67/NA
Koeleria cristata	–	–	78/4	20/3	–	67/NA
Poa sandbergii	–	–	56/1	100/11	86/14	67/NA
Stipa occidentalis	–	–	11/tr	–	57/39	17/NA
Perennial forbs:						
Achillea millefolium lanulosa	88/tr	71/tr	89/tr	60/2	71/tr	100/NA
Lithospermum ruderale	62/tr	14/tr	33/1	60/tr	14/1	67/NA
Tragopogon dubius	50/tr	14/tr	44/tr	60/tr	29/tr	67/NA
Brodiaea douglasii	62/tr	57/tr	11/tr	–	–	50/NA
Erythronium grandiflorum	38/5	86/4	–	–	–	–
Fragaria sp.	38/tr	86/1	22/tr	–	29/tr	50/NA
Galium boreale	25/1	100/1	–	–	–	–
Osmorhiza chilensis	13/1	71/2	–	–	–	–
Potentilla gracilis	62/tr	43/tr	11/tr	–	14/4	–
Sisyrinchium inflatum	50/tr	43/tr	89/1	20/2	29/tr	–
Vicia americana	38/2	57/2	–	–	–	–
Apocynum androsaemifolium	25/1	–	11/7	–	–	50/NA
Balsamorhiza sagittata	25/6	–	89/4	80/2	57/1	83/NA
Ranunculus glaberrimus	25/tr	–	78/3	80/6	57/tr	67/NA
Antennaria dimorpha	–	–	56/tr	40/tr	86/1	17/NA
Erigeron compositus	–	–	11/tr	20/tr	71/1	–

174

Table 26. (Continued)

| Life-form and species | Association[1] | | | | | |
	Pipo/Syal	Pipo/Phma	Pipo/Feid	Pipo/Agsp	Pipo/Stipa	Pipo/Putr
	------------------------------Percent-----------------------------					
Perennial forbs: (cont'd.)						
Lithophragma bulbifera	–	–	89/2	80/3	43/1	33/NA
Lotus nevadensis	–	–	56/6	60/tr	–	–
Frasera albicaulis	–	–	11/tr	–	29/4	50/NA
Arabis holboellii pendulocarpa	–	–	–	–	–	67/NA
Melilotus alba	–	–	–	–	–	50/NA
Annuals:						
Collinsia parviflora	62/tr	43/tr	89/1	100/2	100/1	100/NA
Montia perfoliata	88/3	86/2	11/tr	–	–	33/NA
Galium aparine	75/6	71/4	–	–	–	17/NA
Bromus japonicus	13/3	–	22/tr	40/tr	14/tr	50/NA
Bromus tectorum	50/tr	–	78/tr	80/1	86/2	83/NA
Montia linearis	25/tr	–	89/2	100/1	86/2	–
Myosotis micrantha	13/tr	–	100/3	60/2	86/2	–
Draba verna	–	14/tr	100/1	100/3	86/3	50/NA
Epilobium paniculatum	–	–	89/tr	100/tr	71/1	83/NA
Festuca microstachys	–	–	33/tr	60/tr	29/2	67/NA
Madia exigua	–	–	33/1	20/1	71/1	50/NA
Microsteris gracilis	–	–	11/tr	20/2	71/1	50/NA
Stellaria nitens	–	–	78/tr	60/1	71/1	50/NA
Holosteum umbellatum	–	–	22/tr	60/1	29/2	–
Bromus commutatus	–	–	11/tr	–	–	67/NA

[1] Pipo/Syal = *Pinus ponderosa/Symphoricarpos albus* association; Pipo/Phma = *Pinus ponderosa/Physocarpus malvaceus* association; Pipo/Feid = *Pinus ponderosa/Festuca idahoensis* association; Pipo/Agsp = *Pinus ponderosa/Agropyron spicatum*; Pipo/Stipa = *Pinus ponderosa/Stipa* sp. association; Pipo/Putr = *Pinus ponderosa/Purshia tridentata* association.

[2] tr = trace.

[3] NA = data were not collected.

areas of the more xeric habitat types; e.g., *Pinus ponderosa/Agropyron spicatum*. Mesic associations are characterized by a delayed onset of soil drought.

Six climax *Pinus ponderosa* associations have been identified in the Blue Mountains Province (see footnote 3). The *Pinus ponderosa/Agropyron spicatum* and *Pinus ponderosa/Purshia tridentata/Agropyron spicatum* associations are often found in areas transitional between steppe or shrub-steppe and forest. In the extreme southern Blue Mountains, a *Pinus ponderosa/Purshia tridentata/Carex rossii* association is found on some coarse-textured soils. *Sitanion hystrix* and *Stipa occidentalis* are common associates in the herb-poor understory. *Pinus ponderosa/Elymus glaucus* communities are limited to areas adjacent to dry meadows. The *Pinus ponderosa/Festuca idahoensis* associations are characterized by an abundance of other grasses and sedges in the understory, including *Agropyron spicatum*, *Sitanion hystrix*, *Calamagrostis rubescens*, *Carex rossii*, and *C. geyeri*. At higher elevations, adjacent to *Abies grandis* forests, the *Pinus ponderosa/Carex geyeri* association occurs. Constituent species include *Cercocarpus ledifolius* and *Poa nervosa*.

Many *Pinus ponderosa* stands on the eastern slopes of Washington's Cascade Range (Rummell 1951, Weaver 1961) and in the Blue Mountains (Hall 1967) have an understory

Figure 127.—Grassy *Pinus ponderosa* **communities often consist of a mosaic of arborescent and herbaceous patches, the latter developing well only in treeless openings; a** *Pinus ponderosa/Agropyron spicatum* **community in eastern Oregon.**

dominated by *Calamagrostis rubescens* or mixed *Calamagrostis* and *Carex geyeri*. Almost without exception, *Pinus ponderosa* is seral to *Pseudotsuga menziesii* or *Abies grandis* in these stands where a principal understory dominant is *Calamagrostis*. F. C. Hall (personal communication) identified two communities of this type in the Blue Mountains, one with herbaceous associates (e.g., *Arnica cordifolia*, *Hieracium albiflorum*, and *Carex concinnoides*), the other with herbs and a shrubby layer of *Symphoricarpos albus*, *Spiraea*, and *Rosa*.

Understory vegetation found in *Pinus ponderosa* stands on pumice soils in south-central Oregon differs considerably from that found on nearby residual soils. Total plant cover tends to be lower, especially in the more xeric communities, and the herbaceous flora is more depauperate on pumice soils. Sclerophyllous shrubs such as *Arctostaphylos patula*[4] and *Ceanothus velutinus* assume much more importance in these areas (fig. 128).

Communities occurring on pumice soils, in order of increasing effective moisture, are (Dyrness and Youngberg 1966) (table 27):

Pinus ponderosa/Purshia tridentata,
Pinus ponderosa/Purshia tridentata/Festuca idahoensis,
Pinus ponderosa/Purshia tridentata-Arctostaphylos patula,
Pinus ponderosa/Ceanothus velutinus-Purshia tridentata, and
Pinus ponderosa/Ceanothus velutinus.

With the exception of the seral *Pinus/Ceanothus*, these communities are considered to be edaphic climaxes because of the immaturity of the pumice soils. The *Pinus/Purshia* community is situated at lowest elevations and is characterized by open stands having

---

[4] This taxon has been variously identified as *Arctostaphylos patula* Greene, *A. parryana* var. *pinetorum* (Rollins) Wiesl. & Schr., *A. patula* ssp. *platyphylla* (Gray) P. V. Wells, and *A. obtusifolia* Piper. These revisions were made to distinguish the nonsprouting types prevalent in eastern Oregon from the crown-sprouting *Arctostaphylos patula*. For convenience, we will refer to this entire group as *Arctostaphylos patula* Greene.

176

Figure 128.—Sclerophyllous shrubs such as *Ceanothus velutinus* **are important understory species in** *Pinus ponderosa* **stands on pumice soils in south-central Oregon; a** *Pinus ponderosa/Ceanothus velutinus* **community.**

little advance pine regeneration (fig. 129). Grass and herbaceous cover is sparse; characteristic species are *Stipa occidentalis, Sitanion hystrix, Gayophytum nuttallii,* and *Cryptantha affinis.* The *Pinus/Purshia/Festuca* community occurs on finer textured soils derived from water-lain pumice deposits. Numerous dense stands of *Pinus ponderosa* seedlings and saplings are present in the understory. Characteristic species include *Stipa occidentalis, Carex rossii, Achillea millefolium* var. *lanulosa, Paeonia brownii,* and *Eriophyllum lanatum.*

The *Pinus/Purshia-Arctostaphylos* community is situated at slightly higher elevations than the *Pinus/Purshia* and is accompanied by greater amounts of tree reproduction and seral *Pinus contorta.* This community shares many species with the *Pinus/Purshia* but has additional herbs such as *Phacelia heterophylla, Fragaria chiloensis* and *Epilobium angustifolium.* In the *Pinus/Ceanothus-Purshia* community, *Ceanothus velutinus* replaces *Arctostaphylos patula* in the shrub layer. More abundant tree reproduction and occasional patches of *Salix* sp. indicate more mesic conditions. Characteristic grasses and herbs are *Stipa occidentalis, Carex rossii, Sitanion hystrix, Apocynum androsaemifolium,* and *Hieracium cynoglossoides.* In the *Pinus/Ceanothus* community, *Abies concolor* generally dominates the tree reproduction (fig. 128) and is, therefore, assumed to be climax (Volland 1963, Dyrness and Youngberg 1966). The *Pinus/Ceanothus* type is restricted to higher elevations, and characteristic species include *Chimaphila umbellata* and *Pyrola picta.*

Recently, three additional *Pinus ponderosa* communities occupying pumice soils on the east slopes of the Oregon Cascades have been described.[5] These communities are largely differentiated from those discussed above by the occurrence of substantial quantities of *Carex pensylvanica* and are designated as the *Pinus ponderosa/Purshia tridentata/Carex pensylvanica, Pinus ponderosa/Purshia tridentata-Ceanothus velutinus/Carex pensylvanica,* and *Pinus ponderosa/Purshia tridentata-Arctostaphylos patula/Carex pensylvanica* communities.

---

[5] USDA Forest Service. Vegetation-site key and descriptions of least-disturbed plant communities within the pumice deposition zone, Winema National Forest. Report on file at the Forestry Sciences Laboratory, Corvallis, Oregon, 1972.

Table 27. – Constancy and average cover (constancy/coverage) of selected understory species in *Pinus ponderosa* communities on pumice soils in south-central Oregon (from Dyrness and Youngberg 1966)

Life-form and species	Association				
	Pinus/ Purshia	Pinus/ Purshia/ Festuca	Pinus/ Purshia-Arctostaphylos	Pinus/ Ceanothus-Purshia	Pinus/ Ceanothus
	---------------------- Percent ----------------------				
Tree regeneration:					
Pinus ponderosa	83/2	100/12	100/4	100/5	100/2
Pinus contorta	33/tr	20/tr	50/1	33/tr	100/1
Abies concolor	—	—	—	—	83/2
Shrubs:					
Purshia tridentata	100/19	100/14	100/15	100/2	50/tr
Arctostaphylos patula	33/tr	60/tr	100/14	100/1	100/6
Ceanothus velutinus	—	—	50/1	100/25	100/33
Graminoids:					
Stipa occidentalis	100/5	100/3	100/4	100/2	100/2
Carex rossii	100/1	100/1	100/1	100/2	83/1
Sitanion hystrix	100/1	100/1	100/1	100/1	83/tr
Festuca idahoensis	17/tr	100/14	17/tr	—	—
Herbs:					
Gayophytum nuttallii	100/1	80/tr	100/1	100/1	100/tr
Cryptantha affinis	83/tr	100/tr	100/tr	100/tr	83/tr
Collinsia parviflora	100/tr	100/tr	67/tr	83/tr	50/tr
Viola purpurea	100/1	100/tr	100/tr	100/tr	67/tr
Senecio integerrimus	67/1	60/tr	—	—	17/tr
Lomatium triternatum	33/tr	100/tr	—	17/tr	—
Arabis rectissima	67/tr	20/tr	17/tr	83/tr	33/tr
Madia minima	67/tr	80/tr	—	—	—
Achillea millefolium lanulosa	—	100/1	—	—	—
Antennaria corymbosa	—	40/tr	—	17/tr	—
Paeonia brownii	—	20/tr	—	—	—
Eriophyllum lanatum	—	40/tr	17/tr	—	—
Phacelia heterophylla	33/tr	20/tr	100/tr	—	33/tr
Lupinus caudatus	—	100/tr	83/tr	—	17/tr
Phlox gracilis	—	80/tr	33/tr	—	—
Fragaria chiloensis	—	100/tr	100/2	17/tr	100/1
Antennaria geyeri	—	80/tr	83/tr	67/tr	33/tr
Epilobium angustifolium	—	—	100/tr	33/tr	100/1
Hieracium cynoglossoides	—	—	17/tr	50/tr	33/tr
Pyrola picta	—	—	17/tr	—	67/tr
Chimaphila umbellata	—	—	—	67/tr	50/tr
Apocynum androsaemifolium	—	—	—	100/1	50/tr

tr = trace.

178

Figure 129.—*Pinus ponderosa/Purshia tridentata* **communities in the central Oregon pumice region are characterized by open stands with little** *Pinus* **regeneration.**

Height growth of *Pinus ponderosa* saplings on pumice soils is often very slow for the first 50 to 80 years, after which the growth rate greatly accelerates. Studies by Hermann and Petersen (1969) indicate that the accelerated growth rate is due to roots reaching the buried soil beneath the pumice. They attributed increased rates of growth to improvement of moisture relations and, possibly, nutrition within the buried soil.

On residual soils in south-central Oregon, *Festuca idahoensis* is much more widespread, generally dominating the herbaceous layer under *Pinus ponderosa* stands (Dealy 1971). Apparently, the climatic climax in this area is represented by the *Pinus ponderosa/ Purshia tridentata/Festuca idahoensis* association. On residual soils, this association includes such species as *Chrysothamnus viscidiflorus, Berberis repens, Poa nervosa,* and *Balsamorhiza sagittata,* which are absent in pumice soil areas. In addition to the modal *Pinus/Purshia/Festuca* association, Dealy (1971) has recognized two phases—*Festuca idahoensis* and *Carex rossii.* The *Festuca* phase is marked by a substantial decrease in abundance of *Purshia tridentata* (2-percent cover versus 7-percent for modal *Pinus/ Purshia/Festuca* stands). The *Carex rossii* phase is characterized by less *Pinus ponderosa* cover, greater amounts of *Purshia tridentata,* and a decided dominance of *Carex rossii* in the herb layer.

Other forest communities occupying residual soils in the *Pinus ponderosa* Zone of south-central Oregon include *Pinus ponderosa/Arctostaphylos patula/Festuca idahoensis, Pinus ponderosa/Cercocarpus ledifolius/Festuca idahoensis,* and *Pinus ponderosa/ Artemisia tridentata/Bromus carinatus* (Dealy 1971). The presence of *Abies concolor* regeneration within the *Pinus/Arctostaphylos/Festuca* community indicates that it is a seral grouping within the *Abies* Zone. The *Pinus/Cercocarpus/Festuca* community is found on drier sites than those characteristic of the *Pinus/Purshia/Festuca* association. Species of secondary importance include *Juniperus occidentalis, Purshia tridentata,*

179

Artemisia tridentata, Chrysothamnus viscidiflorus, Carex rossii, Sitanion hystrix, Agropyron spicatum, and *Stipa occidentalis*. Important additional species in the *Pinus ponderosa/Artemisia tridentata/Bromus carinatus* association include *Symphoricarpos albus, Eriogonum microthecum, Poa nervosa, Sitanion hystrix, Stipa occidentalis, Achillea millefolium*, and *Lathyrus* sp.

Successional Patterns

The importance of fire in shaping the vegetation within the *Pinus ponderosa* Zone is stressed by virtually every ecologist who has worked there. Fire scars at the base of almost every large tree offer abundant evidence of repeated fires. Before fire control was initiated about 1900, fires burned through *Pinus ponderosa* stands at intervals variously reported as 8 to 20 years (Weaver 1955, 1959; Soeriaatmadja 1966; Hall 1967). Generally, these were ground fires which consumed only surface organic debris, including branches and down trees, a portion of the understory vegetation, and many of the young tree seedlings.

Because *Pinus ponderosa* is more fire resistant than most associated tree species, past fires have had a profound effect on its distribution. Although young *Pinus ponderosa* seedlings are readily killed by fire, older trees possess thick bark which offers effective protection from fire damage. Competing tree species, such as *Abies grandis* and *Pseudotsuga menziesii*, are considerably less fire tolerant, especially in the sapling and pole size classes. As a result, periodic fires in the past served to maintain *Pinus ponderosa* in ecotonal areas where, without fire disturbance, the climax tree species would have attained dominance (Weaver 1955, 1959, 1961). Fire control activities during the past 60 to 70 years have, on the moister sites, resulted in gradual replacement of *Pinus ponderosa* by such species as *Abies concolor, A. grandis, Libocedrus decurrens*, and *Pseudotsuga menziesii* (Swedberg 1961; Johnson 1961; West 1969a, 1969b).

Fire has also influenced the understory vegetation. Several workers reported that burning substantially reduces shrub cover and increases grass cover, especially on more xeric sites (Brayshaw 1965, Hall 1967, Daubenmire and Daubenmire 1968). *Purshia tridentata* is most readily eliminated of the common shrubs, although on some sites fire reduction and consequent competition from increases in canopy density may have the same effect (Sherman 1966). On the other hand, the shrubs *Arctostaphylos patula* and *Ceanothus velutinus* may increase in importance following severe fires (fig. 130), since heat generated by burning aids in germination of seeds of these species (Gratkowski 1962, Johnson 1961). In the absence of fire, these species apparently perpetuate themselves, at least in pumice soil areas, by vegetative regeneration (Dyrness 1960).

Dense, stagnated stands of *Pinus ponderosa* saplings are common throughout the zone (fig. 131), especially on shallow, stony soils of low productivity. These stands have been attributed to fire exclusion during the past half century (Weaver 1955, 1959, 1961). It is claimed that, previously, periodic fires regulated amounts of advance regeneration and resulted in the open, grassy, parklike stands that early settlers described. However, Daubenmire and Daubenmire (1968) feel that the dense, patchy, episodic reproduction in their grassy *Pinus ponderosa* associations cannot be completely attributed to fire control since a patchy structure is absent in *Pinus/Symphoricarpos* and *Pinus/Physocarpus* stands. Furthermore, it is possible that many *Pinus ponderosa* thickets actually originated with fires. Brayshaw (1965) observed that burning immediately preceding a heavy seed year assisted establishment of unusually numerous pine seedlings by greatly reducing competition from other vegetation.

In many areas, there is a strong tendency for climax *Pinus ponderosa* forests to be even aged by small groups rather than to be truly uneven aged. Daubenmire and Daubenmire

(1968) commented that grassy *Pinus ponderosa* stands consist of ". . . a mosaic of dense patches of trees, each tending to be distinctive in height and age." West (1969b) recognized four structural patterns in central Oregon *Pinus ponderosa* forests: (1) differences in amount of *Pinus ponderosa* regeneration along a moisture gradient, (2) a mosaic of virtually even-aged regeneration groups as a result of past fires, (3) variation in stand density within even-aged groups primarily as a result of chance factors during establishment, and (4) a tendency toward regular dispersal of individual trees within a regeneration group due to competition.

Figure 130.—Fires in *Pinus ponderosa* forests on the east slope of Oregon's Cascade Range may give rise to communities dominated by sclerophyllous shrubs such as *Arctostaphylos patula* and *Ceanothus velutinus*.

Figure 131.—Dense, stagnated thickets of *Pinus ponderosa* saplings are common on shallow, stony soils in the *Pinus ponderosa* Zone; these are frequently attributed to the exclusion of periodic, natural wildfires during the last 50 years.

West (1968) reported that at least 15 percent of *Pinus ponderosa* seedlings develop from unrecovered rodent caches in central Oregon (fig. 132). Rodent caches are also significant in the establishment of *Purshia tridentata* (Sherman and Chilcote 1972). Cache placement pattern is strongly correlated with sites which are free of pine litter. As a result of fire exclusion, litter-free sites are becoming less abundant, and a decrease in *Purshia* populations may be anticipated. Heavy needle litter associated with dense stands of *Pinus ponderosa* also has a negative effect on the establishment and growth of understory herbs and grasses (Pase 1958, Moir 1966). McConnell and Smith (1971) found that heavy needle litter significantly decreased initial survival of *Festuca ovina* and concluded that "thick accumulations of needle litter (and its byproducts) under dense ponderosa pine canopies may exert a strong selective influence on the eventual composition and cover of understory vegetation."

Heavy livestock grazing has, in some respects, effects on understory vegetation opposite to those of burning. That is, heavy grazing pressures often favor shrubs at the expense of grass cover. For example, Brayshaw (1965) reports that heavy grazing may extend the *Symphoricarpos* type into areas formerly dominated by *Agropyron*. On the other hand, grazing in *Pinus/Symphoricarpos* stands may eliminate all native shrubs and herbs, resulting in a sward of *Poa pratensis* and *P. compressa* (Daubenmire and Daubenmire 1968). Changes in grass species may also occur as a result of grazing pressure. Under-

Figure 132.—*Pinus ponderosa* **seedlings often develop from unrecovered seed caches of rodents in the Oregon Cascade Range.**

Figure 133.—**Logging in the** *Pinus ponderosa* **Zone is usually carried out on a selective basis to eliminate overmature, low-vigor trees (of which several are visible here) susceptible to attack by the western pine beetle (Pringle Falls Research Natural Area, Deschutes National Forest, Oregon).**

stories in the *Pinus ponderosa/Festuca idahoensis* association in northeastern Washington can be shifted, apparently irreversibly, to domination by *Bromus tectorum, Linaria dalmatica*, and *Hypericum perforatum* (Daubenmire and Daubenmire 1968).

Logging in the *Pinus ponderosa* Zone is usually carried out on a single-tree selection basis. Most often, older trees which are more susceptible to attack by the western pine beetle (*Dendroctonus brevicomis*) are removed first, leaving the younger, more vigorous trees in the stand as growing stock (fig. 133). Selective logging of *Pinus ponderosa* in areas transitional to *Abies grandis* or *A. concolor* substantially accelerates the successional trend toward dominance by these climax tree species. Some forest managers attempt to reserve these sites for the growth of *Pinus ponderosa* through the removal of *Abies* seed sources wherever practical (Volland 1963).

Pinus ponderosa seedlings are commonly hand planted following logging or wildfire, often with little success. Frequently, the principal cause of seedling mortality is a lack of available soil moisture during the dry summer months. However, it is becoming increasingly apparent that another threat to *Pinus ponderosa* plantations is widespread mortality caused by pocket gophers (*Thomomys* spp.) (Crouch 1971).

Logging also affects the understory vegetation and reproduction of *Pinus ponderosa*. In one study (Garrison and Rummell 1951; Garrison 1961, 1965), selective logging by tractors reduced herb and shrub coverage by 33 percent, denuding some areas and burying others in slash. The understory required about 14 years to recover to nearly its original condition. Deforestation generally results in dominance by those understory species present before logging, with invading species playing only a very minor role in the postdisturbance period (Daubenmire 1952); a notable exception is *Crataegus douglasii* on *Pinus/Symphoricarpos* sites (Daubenmire and Daubenmire 1968).

Nonforested Communities

Pinus ponderosa forests form a mosaic with shrub-steppe or steppe communities in many locales (fig. 134). For example, in eastern Washington, *Pinus ponderosa* stands first appear within a matrix of steppe, and increase in extent in more mesic areas until steppe or shrub-steppe communities constitute islands within a matrix of *Pinus ponderosa* forest (Daubenmire and Daubenmire 1968). Many of the nonforest communities in the *Pinus ponderosa* Zone are identical with those found within the steppe zones, e.g., the *Purshia tridentata/Festuca idahoensis* association (Daubenmire 1970); others occur only in forest zones. McLean (1970) describes an *Artemisia tridentata/Festuca idahoensis* community with scattered trees in the *Pinus ponderosa* Zone of southern British Columbia.

One of the best described *Pinus ponderosa*-steppe mosaics is that encountered in the Ochoco Mountains of eastern Oregon (Hall 1967). Hall described three associations which occur in natural forest openings:

Artemisia rigida/Poa sandbergii with *Sitanion hystrix* and *Trifolium macrocephalum* as major associates;

Artemisia arbuscula/Agropyron spicatum with *Poa sandbergii* and *Purshia tridentata* as major associates; and

Artemisia arbuscula/Festuca idahoensis with *Phlox douglasii, Balsamorhiza serrata*, and *Poa sandbergii* as associates.

Hall (1967) attributed the presence of these nonforested communities to soil conditions, specifically to periodic moisture saturation, heavy soils which impede tree root penetration, and soil drought during the summer. He also mentioned a *Purshia tridentata-Cercocarpus ledifolius*-dominated community, characteristic of rock outcrops in the lower part of the *Pinus ponderosa* Zone.

Figure 134.—*Pinus ponderosa* **forest and steppe or shrub-steppe often form an intricate mosaic in ecotonal areas with the communities on coarser and finer textured soils, respectively (near Tonasket, Okanogan National Forest, Washington).**

Daubenmire (1970) mentioned two communities, encountered as edaphic climaxes within the forested zones of eastern Washington. The *Artemisia rigida/Poa sandbergii* association is widely distributed on lithosolic soils from steppe through the *Pseudotsuga menziesii* Zone. The *Festuca idahoensis-Eriogonum heracleoides* association is found only in forest "parks" and includes *Antennaria rosea* and *Castilleja miniata* as characteristic species.

In south-central Oregon, many nonforested openings on stream terraces and flood plains are occupied by a *Purshia tridentata/Stipa occidentalis-Carex pensylvanica* community (see footnote 5). Soils in these areas are generally derived from deep alluvial deposits of pumice or basic scoria. Other characteristic species include *Haplopappus bloomeri, Eriogonum umbellatum, Ribes cereum, Bromus carinatus, Senecio canus, Aster canescens, Spraguea umbellata,* and *Phacelia* sp. Dealy (1971) describes an *Artemisia tridentata/Stipa occidentalis-Lathyrus* community occurring on soils derived from basalt. Species composition of this community closely resembles the understory characteristic of the *Pinus ponderosa/Artemisia tridentata/Bromus carinatus* association. Accordingly, Dealy (1971) recognizes the *Artemisia/Stipa-Lathyrus* community as a fire-caused seral grouping within the *Pinus/Artemisia/Bromus* habitat type.

Severe fires or destructive logging practices can eliminate *Pinus ponderosa* and give rise to shrub-dominated communities, as mentioned earlier. Reestablishment of trees in such communities may be very slow.

184

Pinus contorta Zone

Pure, or nearly pure, stands of *Pinus contorta* are widely distributed throughout forested areas of eastern Oregon and Washington. The majority are seral, having developed after fire or logging. However, *Pinus contorta* is considered an edaphic or topoedaphic climax on many sites in the pumice plateau of south-central Oregon. Since other areas where *Pinus contorta* may be a climax tree species are small and very scattered, our *Pinus contorta* Zone will be limited to the pumice plateau; note that "zone" is not used here in a strict synecological sense but rather to identify a large area in which, because of soils and topography, *Pinus contorta* is the major climax type. However, it is interesting that Moir (1969) has recently provided experimental evidence for a *Pinus contorta* Zone in the Rocky Mountains.

Pinus contorta is well known for rapid invasion of severely disturbed sites, justifying its designation as a pioneer species. Pollen analyses indicate it was one of the first trees to occupy the infertile materials deposited at the close of the last glacial period in the southern Oregon Cascades (Hansen 1946). After gradual replacement by *Pinus ponderosa* during the succeeding 25 centuries, *Pinus contorta* quickly regained dominance with the widespread deposition of Mount Mazama pumice 6,600 years ago. It has retained dominance on pumice and ash soils at several other locations in the Oregon and central Washington Cascades (Roach 1952, Herring 1968). *Pinus contorta* forests are extensive on volcanic ash and pumice in the Blue Mountains of northeastern Oregon (Trappe and Harris 1958) but are rarely found on south slopes where this material is thin or absent. These forests are apparently fire created and generally seral to *Abies grandis*.

Pinus contorta has unusually wide ecologic amplitude—thriving on wet, poorly drained sites, as well as coarse-textured, droughty soils. After fires, prolific seeding allows it to quickly invade and establish dominance, especially on more extreme sites where competitors are absent. For this reason, pure stands of *Pinus contorta* are often found only on very wet or dry soils, and the more productive, medium-textured soils support other tree species (Stephens 1966).

Environmental Features

The climate of the *Pinus contorta* Zone is characterized by (1) low summer rainfall, (2) wide diurnal temperature fluctuations, especially in the summer, and (3) a relatively short growing season (table 28). Since elevations within the zone range from approximately 1,200 to 1,525 meters, growing-season frosts are not uncommon, and much of the precipitation occurs as snow. Average annual precipitation ranges from about 350 to 700 millimeters.

Topographically, the pumice plateau includes broad, level areas in enclosed depressions, gently rolling terrain, and numerous small volcanic cones. *Pinus contorta* stands are generally on nearly level terrain in depressions locally known as lodgepole flats. Temperature measurements show that cold-air drainage from surrounding slopes often produces substantially lower nighttime temperatures in these low-lying areas. For example, Berntsen (1967) found minimum spring temperatures were as much as 4.5° C. higher on a slope supporting *Pinus ponderosa* than those registered in an adjacent *Pinus contorta* stand at the base of the slope.

Climax stands of *Pinus contorta* are found on both well and poorly drained pumice soils. All are Vitrandepts (Regosols) developed in aerially deposited pumice, mostly from Mount Mazama (Crater Lake). Poorly drained soils occupy low topography in basinlike depressions or areas adjacent to intermittent streams. These soils have a dark-gray sandy loam or loam A1 horizon underlain by a light-gray to white C horizon composed of

Table 28. — Climatic data from representative stations within the *Pinus contorta* Zone

Station	Elevation	Latitude	Longitude	Temperature					Precipitation		
				Average annual	Average January	Average January minimum	Average July	Average July maximum	Average annual	June through August	Average annual snowf...
	Meters			- - - - - - - - - Degrees C. - - - - - - - - -					Millimeters		Centimeters
Chemult, Oreg.	1,449	43°14'	121°47'	5.3	−3.9	−11.1	15.3	28.2	676	66	417
Crescent, Oreg.	1,357	43°27'	121°42'	5.1	−3.1	−11.2	14.3	26.8	485	56	---
Lapine, Oreg.	1,336	43°40'	121°29'	5.6	−3.3	−10.8	15.5	29.8	361	64	---

Source: U. S. Weather Bureau (1965a) and Johnsgard (1963).

Figure 135.—Better drained pumice soils in south-central Oregon typically have dark-colored A1 and AC horizons overlying the lighter colored gravelly C; note the abrupt boundary between Mazama pumice and buried soil.

pumice gravel and sand. Depth to water table is generally 60 centimeters or less throughout the growing season. On better drained sites, the pumice soil exhibits (1) a thin, dark grayish-brown, loamy coarse sand A1 horizon underlain by (2) a yellowish-brown, gravelly, loamy coarse sand AC horizon which grades, in turn, into (3) relatively fresh, unweathered pumice gravel and sand (fig. 135).

The pumice soils of the area under consideration have some distinctive chemical and physical properties which are partially responsible for the vegetational features (e.g., see Cochran 1969, 1971; Youngberg and Dyrness 1965).

Forest Composition

Seventeen *Pinus contorta*-dominated communities have been identified in the pumice soil region of central Oregon (Youngberg and Dahms 1970, Dealy 1971),[6] but it is the climax species in only eight of these:

Community	*Author*
Pinus contorta/Purshia tridentata	Youngberg and Dahms 1970 ([6])
Pinus contorta/Purshia tridentata-Ribes cereum	Youngberg and Dahms 1970 ([6])
Pinus contorta/Purshia tridentata/Festuca idahoensis	Youngberg and Dahms 1970 ([6])
Pinus contorta/Arctostaphylos uva-ursi	Youngberg and Dahms 1970 ([6])
Pinus contorta/Danthonia californica	Dealy 1971
*Pinus contorta/Vaccinium uligonosum/*herb	([6])
*Pinus contorta/Carex-Elymus-*herb	([6])
Pinus contorta/Stipa occidentalis	([6])

The *P. contorta/Purshia* and *P. contorta/Purshia-Ribes* communities are found on well-drained soils with the former the most widespread and characteristic community of the entire zone (fig. 136). The sparse understory in the *P. contorta/Purshia* type is similar in composition to that occurring in nearby *Pinus ponderosa/Purshia tridentata* stands. *Purshia tridentata, Stipa thurberiana, Sitanion hystrix,* and *Carex rossii* are characteristic. *Ribes cereum* and *Fragaria chiloensis* are added to *Purshia* as diagnostic species in the *P. contorta/Purshia-Ribes* type in which the *Stipa, Sitanion,* and *Carex* are also characteristic.

The climax *Pinus contorta/Purshia tridentata/Festuca idahoensis* community is found on seasonally wet soils. *Purshia* and *Festuca* are diagnostic and *Ribes viscosissimum, Penstemon* sp., *Sitanion hystrix,* and *Carex rossii* are characteristic understory species. *Pinus contorta* is found in this community on better drained soils but is not climax there (Youngberg and Dahms 1970), and a similar *Pinus contorta/Festuca idahoensis* community is found on well-drained soils in the Silver Lake area in which *P. contorta* is successional to *Abies concolor* (Dealy 1971).

The *Pinus contorta/Arctostaphylos uva-ursi* community is found on seasonally ponded soils which have a water table at approximately 70 centimeters during the summer (Youngberg and Dahms 1970). Diagnostic species are *Arctostaphylos* and *Trifolium longipes. Fragaria chiloensis, Ribes viscosissimum, Festuca idahoensis, Stipa thurberiana,* and *Carex rossii* are also characteristic. Dealy (1971) describes the *P. contorta/Danthonia californica* community as "one step above a meadow." It occupies a moist site and has a rich herbaceous understory in which *Danthonia, Carex* sp., *Deschampsia elongata, Juncus* sp., *Koeleria cristata, Muhlenbergia filiformis, Antennaria corymbosa, Achillea millefolium, Aster* sp., *Fragaria virginiana,* and *Trifolium* sp. are conspicuous.

---

[6] See footnote 5.

Figure 136.—The *Pinus contorta/Purshia tridentata* **community is widespread in the central Oregon pumice region;** *Pinus contorta* **achieves climax status on poorly drained sites and in frosty depressions in this area (Pringle Falls Research Natural Area, Deschutes National Forest, Oregon).**

The *Pinus contorta/Vaccinium uliginosum*/herb and *Pinus contorta/Carex-Elymus*-herb communities also occupy imperfectly drained pumice soils which are seasonally ponded. Species which distinguish the *Pinus/Vaccinium*/herb community are *Vaccinium caespitosum, V. uliginosum, Lonicera* sp., *Spiraea douglasii, Danthonia intermedia, Elymus glaucus, Calamagrostis inexpansa,* and *Fragaria chiloensis.* The *Pinus/Carex-Elymus*-herb community is characterized by *Carex lasiocarpa, C. nebrascensis, Elymus glaucus, Arnica chamissonis,* and *Fragaria chiloensis.*

The *Pinus contorta/Stipa occidentalis* community generally occupies lower slopes and basinlike depressions which are subject to frequent low temperatures as a result of cold air drainage. Because of the scattered distribution of *Pinus contorta* and the exposure of large expanses of bare soil, these areas are sometimes locally termed "pumice deserts." Cover of *Pinus contorta* seldom exceeds 30 percent; and *Purshia tridentata,* where present, is generally restricted to tree clumps. Although very scattered in occurrence, common herbaceous species include *Stipa occidentalis, Carex rossii, Eriogonum umbellatum, Lupinus lepidus, Viola nuttallii* var. *vallicola,* and *Spraguea umbellata.*

Successional Patterns

Kerr (1913) first reported the interesting forest patterns in this area. Pure stands of *Pinus contorta* occupy depressions and broad, level areas at lower elevations, but on every hill and slight rise in topography it is replaced by *Pinus ponderosa.* Munger (1914) concluded that *Pinus contorta* owed its widespread distribution to frequent fires and

188

contended that it was continually encroaching on sites previously occupied by *Pinus ponderosa*. Later, however, the effects of fire were discounted when soil investigations disclosed that many sites supporting pure stands of *Pinus contorta* were subject to permanent or fluctuating high water tables (Tarrant 1953). Further study of the area revealed many sites where climax stands of *Pinus contorta* were growing on well-drained soils. Since these were generally located in enclosed depressions suggestive of frost pockets, it was postulated that *Pinus contorta* occupied these sites because of its greater resistance to low temperatures, especially during the early growing season (Youngberg and Dyrness 1959). Berntsen (1967) confirmed this hypothesis and indicated selection for *Pinus contorta* occurs during seed germination. Emerging *Pinus contorta* seedlings were appreciably more resistant to low temperatures than *Pinus ponderosa* seedlings.

Consequently, two types of climax *Pinus contorta* stands have been recognized: (1) an edaphic climax in poorly drained soil areas and (2) a topoedaphic climax in frost pocket depressions where the soil is often well drained. Questions still remain as to factors responsible for favoring *P. contorta* over *P. ponderosa* on poorly drained sites, as seedlings of both species appear almost equally tolerant of very wet soils (Cochran 1972); hence, differential tolerance of species to high soil water content during early seedling stage is not the factor favoring *P. contorta*'s occupancy of wet sites. The transition between nearly pure stands of *Pinus contorta* and those of *Pinus ponderosa* on adjacent slopes is often strikingly abrupt, although there frequently is a narrow band where the two species intermingle. Climax *Pinus contorta* stands are generally even aged and nearly pure, except for occasional clumps of *Populus tremuloides* in poorly drained areas. As stands reach old age and begin to break up, they are quickly replaced by abundant *Pinus contorta* seedlings (fig. 137). Cones in these populations are nonserotinous; therefore, seed fall occurs virtually every year (Mowat 1960).

Figure 137.—*Pinus contorta* **regeneration is often excessive on disturbed sites or in decadent stands, resulting in dense, stagnated patches of saplings.**

Nonforested Communities

Numerous wet meadows are interspersed with *Pinus contorta* stands at lowest elevations within the zone (fig. 138). These meadows are on very poorly drained pumice soils and exhibit a mesic flora as a result of the shallow water table. Principal grasses are *Poa* spp., *Deschampsia caespitosa*, and *Sitanion hystrix*. In addition, *Juncus* spp. and sedges such as *Carex praegracilis* and *Carex nebrascensis* are common. Herbaceous species vary appreciably depending on grazing pressure; however, two of the more common are *Trifolium longipes* and *Ranunculus occidentalis* (Dyrness 1960).

Figure 138.—Numerous wet meadows dominated by grasses and sedges are interspersed with *Pinus contorta* stands at lower elevations in the *Pinus contorta* Zone (Bear Flat between Chemult and Silver Lake, Oregon).

Pseudotsuga menziesii Zone

Associations in which *Pseudotsuga menziesii* is climax are the next zonal step in some areas. In most of eastern Washington and adjacent British Columbia, a *Pseudotsuga menziesii* Zone is apparently well developed (Daubenmire 1952, 1966; Daubenmire and Daubenmire 1968; Brayshaw 1965; McLean 1970). In eastern Oregon, it is conjectural (West 1964, 1969a; Swedberg 1961) or absent (Dyrness and Youngberg 1966, Hall 1967), except in parts of the Wallowa Mountains (Johnson 1959, Head 1959). A typical elevational range for the zone is 600 to 1,300 meters in northeastern Washington. The *Pseudotsuga menziesii* Zone correlates with the Interior Douglas-fir Zone in British Columbia (Krajina 1965) and the lower part of the Canadian Life Zone (Barrett 1962).

The *Pseudotsuga menziesii* Zone normally abuts the *Abies grandis* and *Pinus ponderosa* Zones at its upper and lower limits, respectively (table 21). In places, climax *Pinus ponderosa* forests may be absent and the *Pseudotsuga menziesii* Zone borders shrub-steppe; this is particularly common on fine-textured soils in north-central Washington. In other locations, *Abies grandis* forests are absent and *Abies lasiocarpa* abuts the upper limits of the *Pseudotsuga menziesii* Zone; e.g., see McLean (1970).

190

Environmental Features

The *Pseudotsuga menziesii* Zone is more mesic than the *Pinus ponderosa* Zone. Temperatures undoubtedly average cooler and annual precipitation higher (Krajina 1965), although Daubenmire (1956) could not identify climatic criteria distinguishing the zones at their ecotone. It is known that soil moisture conditions are more favorable in the *Pseudotsuga menziesii* Zone (Daubenmire 1968b, McMinn 1952, Brayshaw 1965).

Forest Composition

Pseudotsuga menziesii, Pinus ponderosa, P. contorta, and *Larix occidentalis* are the major tree species in the *Pseudotsuga menziesii* Zone (fig. 139). *Quercus garryana* may be present along the flanks of the Cascade Range from about 45° to 47° north latitude; in some locations, there even appears to be a belt or zone of *Quercus* woodland between coniferous forest (*Pseudotsuga* or *Pinus ponderosa*) and the steppe. *Libocedrus decurrens* also occurs from Mount Hood south. *Populus tremuloides* and *Juniperus occidentalis* are minor associates.

Pseudotsuga menziesii associations have been identified only in eastern Washington and the Wallowa Mountains (Daubenmire 1952, Daubenmire and Daubenmire 1968, Johnson 1959, Head 1959). Daubenmire recognized two associations with shrubby understories in his study area. The *Pseudotsuga menziesii/Symphoricarpos albus* association

Figure 139.—Mixed stands of *Pseudotsuga menziesii* **and** *Larix occidentalis* **are common in parts of the** *Pseudotsuga menziesii* **Zone; note reproduction of the climax** *Pseudotsuga* **(Colville National Forest, Washington).**

has an understory dominated by low shrubs: *Symphoricarpos albus, Spiraea betulifolia,* var. *lucida, Rosa woodsii,* and *R. nutkana.* Taller shrubs belonging to the *Physocarpus* union (Daubenmire 1952) are absent. Similar communities were noted in the Wallowa Mountains (Johnson 1959) and on the eastern slopes of the Cascade Range (see Wolf Creek Research Natural Area in Franklin et al. 1972). The *Pseudotsuga menziesii/Physocarpus malvaceus* association includes the aforementioned shrubs as understory constituents and *Physocarpus malvaceus* and *Holodiscus discolor.*

The *Pseudotsuga menziesii/Calamagrostis rubescens* association is probably the most widespread, existing in the Wallowa Mountains (Head 1959) as well as in eastern Washington (Daubenmire and Daubenmire 1968). It has a nearly shrub-free understory, dominated by *Calamagrostis rubescens* along with other herbs such as *Carex concinnoides, C. geyeri,* and *Arnica latifolia.* Daubenmire and Daubenmire (1968) have recognized a phase of this type in which *Arctostaphylos uva-ursi* is an understory dominant. Similar forest communities occur in adjacent British Columbia (Brayshaw 1965, McLean 1970). McLean (1970) has also recognized *Pseudotsuga menziesii/Agropyron spicatum* and *Pseudotsuga menziesii/Festuca idahoensis* communities in southern British Columbia, however.

Quercus garryana communities, which are probably within this zone, have been noted on south-facing slopes within the Mill Creek Research Natural Area (Franklin et al. 1972). The scrubbier stands of *Quercus* have understories dominated by *Elymus glaucus* in association with *Symphoricarpos albus, Carex geyeri,* and various forbs. Stands of larger *Quercus* include scattered *Pinus ponderosa* and a ground vegetation characterized by *Carex geyeri, Purshia tridentata, Amelanchier alnifolia, Stipa* spp., and *Agropyron spicatum* (rhizomatous habit).

Successional Patterns

Any of the four major tree species—*Pseudotsuga menziesii, Pinus ponderosa, P. contorta,* and *Larix occidentalis*—may dominate forest stands in the *Pseudotsuga menziesii* Zone. *Pseudotsuga menziesii* is climax (fig. 140), but the other three are better adapted to fires, which were common prior to 1900. Hence, any of these species may form pure or nearly pure stands, depending upon fire history and available seed source. This is one of the zones in which fire protection and selective logging of *Pinus ponderosa* are accelerating natural trends toward elimination of the pine.

As mentioned, a zone of climax *Pseudotsuga menziesii* forest is largely conjectural on the eastern slopes of the central Oregon Cascade Range. There are *Pinus ponderosa* forests with *Pseudotsuga menziesii* and *Libocedrus decurrens* reproduction (West 1964, 1969a; Swedberg 1961; Sherman 1969) suggesting either one or both are climax. However, at least small numbers of *Abies grandis* seedlings are normally present whenever significant numbers of *Pseudotsuga* are encountered. The question thus remains whether habitats with potential climax communities of *Pseudotsuga menziesii* exist in this area or whether the zonal sequence is *Pinus ponderosa-Abies grandis.* A better case can be made for climax *Libocedrus decurrens* habitats in at least some areas [see West's (1964, 1969a) Warm Springs transect and Sherman (1969)]. *Libocedrus* appears tolerant of more xeric habitats than *Pseudotsuga menziesii* and is more shade tolerant than *Pinus ponderosa.* There is more evidence for a *Pseudotsuga* Zone on the eastern slopes of the northern Oregon Cascade Range (see Mill Creek and Persia M. Robinson Research Natural Areas in Franklin et al. 1972).

The *Pseudotsuga menziesii* Zone definitely seems to be absent in south-central Oregon and the Ochoco and Blue Mountains (Hall 1967). Hall recognized no association in which *Pseudotsuga menziesii* was a major climax dominant.

Figure 140.—*Pseudotsuga menziesii* **replacing** *Pinus ponderosa* **(Meeks Table Research Natural Area, Snoqualmie National Forest, Washington).**

Special Types

As mentioned earlier the *Artemisia rigida/Poa sandbergii* and *Festuca idahoensis-Eriogonum heracleoides* communities can occur in openings in the *Pseudotsuga menziesii* Zone of eastern Washington. In southern British Columbia, McLean (1970) found the *Festuca-Eriogonum* community at 800- to 900-meter elevation between the *Pinus ponderosa* and *Pseudotsuga* Zones and, in one area, extending to almost 2,000 meters where it merged with alpine grassland. A second nonforest type in McLean's (1970) *Pseudotsuga* Zone was the *Artemisia tripartita/Agropyron spicatum* community.

Abies grandis Zone

The *Abies grandis* Zone is the most extensive midslope forest zone in the Oregon and southern Washington Cascade Range and Blue Mountains of eastern Oregon. The *Abies grandis* Zone typically occurs at 1,100 to 1,500 meters in the central Oregon Cascade Range, 1,500 to 2,000 meters in the Ochoco and Blue Mountains (Hall 1967), and 1,650 to 2,000 meters (*Abies concolor* Zone) in south-central Oregon (Dyrness and Youngberg 1966). The *Abies grandis* Zone has no counterpart in British Columbia, and climax *Abies grandis* forests are included in the *Tsuga heterophylla* series in the northern Rocky Mountains (Daubenmire and Daubenmire 1968). This zone is analogous to the upper part of Merriam's Canadian Life Zone (Barrett 1962) and the forests within Küchler's (1964) Grand Fir-Douglas-Fir type.

The *Abies grandis* Zone is typically bounded by the *Abies lasiocarpa* Zone at its upper limits and the *Pseudotsuga menziesii* or *Pinus ponderosa* Zones at its lower limits. However, locally at higher elevations or on more mesic sites in the Cascade Range, it may abut forests in which *Abies amabilis*, *Tsuga heterophylla*, *Thuja plicata*, or *Tsuga mertensiana* are climax. At its lower limits, it may be adjacent to *Artemisia* steppe without an intervening belt of climax *Pseudotsuga menziesii* or *Pinus ponderosa* (Hall 1967). In northern Idaho and western Montana, an *Abies grandis* Zone regularly intervenes between a *Pseudotsuga* and *Tsuga* or *Thuja* Zone.

Environmental Features

The *Abies grandis* Zone provides the most moderate environmental regime of any of the east-side forest zones (except for the areas where *Tsuga* or *Thuja* are present) (table 29). Neither moisture nor temperature conditions are extreme. Precipitation is generally higher and temperatures lower than in lower forested zones. Daubenmire (1956, 1968b) concluded that differences in summer dryness and soil drought differentiate the *Abies grandis* and *Tsuga heterophylla* from the *Pseudotsuga menziesii* associations. Soil drought is of minor ecologic significance in the *Abies grandis* Zone (Daubenmire 1968b, McMinn 1952). The *Abies grandis* Zone is distinguished climatically from subalpine forests by its higher temperatures (Daubenmire 1956) and lesser accumulations of snow.

Table 29. — Climatic data from representative stations in the vicinity of the *Abies grandis* Zone

Station	Eleva-tion	Lati-tude	Longi-tude	Temperature					Precipitation		
				Average annual	Average January	Average January minimum	Average July	Average July maximum	Average annual	June through August	Average annual snowfall
	Meters			- - - - - - - - - Degrees C. - - - - - - - - - -					Millimeters		Centi-meters
Parkdale, Oreg.	530	45°31′	121°36′	8.4	−1.3	5.1	17.6	26.9	1,064	43	231
Meacham, Oreg.	1,234	45°30′	118°24′	6.4	−3.6	−7.1	17.6	25.4	835	115	396
Minam, Oreg.	1,092	45°41′	117°36′	--	--	--	--	--	632	82	--
Mount Adams Ranger Station, Wash.	597	46°00′	121°32′	8.1	−2.1	−6.1	18.4	28.6	1,157	44	309
Lake Wenatchee, Wash.	600	47°50′	120°48′	--	--	--	--	--	988	51	450
Plain, Wash.	549	47°46′	120°40′	7.2	−4.6	−9.7	18.3	28.2	632	43	--

Source: U. S. Weather Bureau (1965a, b).

Soils generally exhibit minimal development in the *Abies grandis* Zone but are relatively deep, largely due to accumulations of volcanic ash throughout much of the zone's range. The dominant soil processes are podzolic, and typical soil great groups are Haplorthods and Haplumbrepts (Gray Wooded, Brown Podzolic, and Western Brown Forest). Vitrandepts (regosolic soils) with A-C profile sequences in pumice or ash parent materials are also common. Thin (2 to 5 cm.), mull-type humus layers are typical along with relatively thin, brown or dark-brown, weakly to moderately acid A1 horizons. Continuous bleached A2 horizons (bleicherde) are not common. Clay eluviation is minimal in B horizons which are distinguished primarily by color. *Abies concolor* sites in the pumice region are distinguished by A1-AC-C sequences. A horizons average only 30 centimeters thick; they are generally slightly acid (pH 6.2) with 8 to 9 percent of organic matter.

Forest Composition

Major tree species in the *Abies grandis* Zone are *Abies grandis* (or *A. concolor*), *Pinus ponderosa*, *P. contorta*, *Larix occidentalis*, and *Pseudotsuga menziesii*. Any of the four associates listed here may dominate seral forest stands. Many other species are present in limited numbers or in localized areas such as *Picea engelmannii*, *Abies lasiocarpa*, *Libocedrus decurrens*, *Pinus lambertiana*, *P. monticola*, *Tsuga mertensiana*, and *Abies magnifica* var. *shastensis*.

Composition of *Abies grandis* (or *A. concolor*) associations has been studied in eastern Washington (Daubenmire 1952, Daubenmire and Daubenmire 1968), the central Oregon Cascade Range (Swedberg 1961, Sherman 1969, West 1964), the Ochoco and Blue Mountains (Hall 1967, 1971), the Wallowa Mountains (Johnson 1959, Head 1959), and south-central Oregon (Dyrness and Youngberg 1966). Most *Abies grandis* associations can be related to two major groups: (1) *Abies grandis/Pachistima myrsinites* and *Abies grandis/Vaccinium membranaceum*[7] (Daubenmire 1952, Daubenmire and Daubenmire 1968, Head 1959, Swedberg 1961, Hall 1967), and (2) the *Abies grandis/Calamagrostis rubescens* (Hall 1967, Johnson 1959).

The *Abies grandis/Pachistima myrsinites* and /*Vaccinium membranaceum* associations have well-developed herbaceous understories. Characteristic species include:

Shrubs—*Rosa gymnocarpa*, *Pachistima myrsinites*, *Ribes lacustre*, and *Vaccinium membranaceum*; and

Herbs—*Bromus vulgaris*, *Galium triflorum*, *Smilacina stellata*, *Thalictrum occidentale*, *Arnica cordifolia*, *Mitella stauropetala*, *Arenaria macrophylla*, *Hieracium albiflorum*, *Linnaea borealis*, *Adenocaulon bicolor*, *Anemone piperi*, *A. lyallii*, *Viola glabella*, *Trillium ovatum*, *Clintonia uniflora*, *Asarum caudatum*, *Lupinus latifolius*, and *Rubus lasiococcus*.

Under very dense canopies, a group of heliophytes forms a sparse understory (fig. 141) [the *Pirola-Corallorrhiza* union of Oosting and Billings (1943) or "mycotrophic achlorophyllous angiosperms" of Furman and Trappe (1971) again]; typical species are *Chimaphila umbellata*, *C. menziesii*, *Corallorhiza maculata*, *Pyrola asarifolia*, *P. secunda*, *P. picta*, *Monotropa hypopitys*, and *Listera convallarioides*.

The *Abies grandis/Calamagrostis rubescens* association is typically found on volcanic ash soils in the Ochoco and Blue Mountains[8] and apparently occurs at least sporadically along the eastern slopes of the Cascade Range as well (see Meeks Table Research Natural Area in Franklin et al. 1972). *Pinus ponderosa/Calamagrostis rubescens* or *Pseudotsuga menziesii/Calamagrostis rubescens* communities are often seral stages of this association in the Blue and Ochoco Mountains (fig. 142) (e.g., see Canyon Creek and Ochoco Divide Research Natural Areas in Franklin et al. 1972); however, as mentioned earlier, a climax *Pseudotsuga/Calamagrostis* community does occur in northeastern Washington where an *Abies/Calamagrostis* community is absent (Daubenmire and Daubenmire 1968).

---

[7] F. C. Hall (personal communication) has described three related *Abies grandis* associations in the Blue Mountains Province to which we have referred by the collective term, "the *Abies grandis/Vaccinium membranaceum* type." These are: *Abies grandis/Vaccinium membranaceum* (by far the most widespread), *Abies grandis/Bromus vulgaris* (southern Blue Mountains), and *Abies grandis/Linnaea borealis* (northern Blue Mountains).

[8] F. C. Hall (1967, 1971), and personal communication.

Figure 141.—Understory vegetation is sparse under dense stands in the *Abies grandis* **Zone;** *Abies grandis-Pseudotsuga menziesii* **forest in the Blue Mountains of Oregon (range pole is marked in 1-foot segments).**

The understory of the *Abies/Calamagrostis* association is rich in herbs: *Calamagrostis rubescens, Carex geyeri, C. concinnoides, Arnica cordifolia, Lupinus caudatus* (or *laxiflorus*), and *Hieracium albiflorum*. The shrub cover is normally sparse, but Hall (1967, 1971) and Johnson (1959) have recognized variants in which understory shrubs such as *Spiraea betulifolia* var. *lucida, Rosa* sp., *Salix* sp., and *Symphoricarpos albus* are typical [conifer/*Spiraea*/*Calamagrostis* community type of Hall (1971), for example]. Hall (1971) also found it useful to separate *Calamagrostis* communities growing on residual soils from those on volcanic ash soils because of differences in response to grazing.

On upper slopes in the Ochoco and Blue Mountains, the *Abies/Calamagrostis* and *Abies/Vaccinium* associations generally occupy the south and north slopes, respectively.

The *Abies concolor/Ceanothus velutinus* association of south-central Oregon has much higher shrub and lower herbaceous coverage than the *Abies grandis* associations (Dyrness and Youngberg 1966). *Ceanothus velutinus, Arctostaphylos nevadensis*, and *A. patula* are the major shrubs (26.5-, 2.8-, and 0.5-percent cover, respectively). *Stipa occidentalis, Carex rossii, Sitanion hystrix, Gayophytum nuttallii, Cryptantha affinis, Fragaria chiloensis, Epilobium angustifolium, Chimaphila umbellata*, and *Apocynum androsaemifolium* make up most of the sparse herbaceous layer which totals only 3.1-percent cover.

Abies concolor communities with herb rather than shrub-dominated understories have been identified in south-central Oregon: *Pinus ponderosa-Abies concolor/Festuca idahoensis* (Dealy 1971) and *Pinus-Abies/Carex rossii* (Hall on Goodlow Mountain Research Natural Area in Franklin et al. 1972). *Abies concolor* appears to be the climax species in both communities based on reproductive success. The *Pinus-Abies/Festuca* type has a grass-dominated understory in which *Festuca idahoensis, Carex rossii, Stipa occidentalis, Poa nervosa, Sitanion hystrix, Fragaria virginiana*, and *Arnica cordifolia* are conspicuous (Dealy 1971). *Carex rossii* is the major understory species in the *Pinus-Abies/Carex* type (fig. 143).

196

Figure 142.—This *Pinus ponderosa/Calamagrostis rubescens* community is being invaded by reproduction of *Abies grandis* which will eventually form the climax stand; *Pseudotsuga/Calamagrostis* or *Pinus/Calamagrostis* communities seral to *Abies grandis* are common in the Blue and Ochoco Mountains and in at least a few localities on the eastern slope of the Cascade Range (Ochoco Divide Research Natural Area, Ochoco National Forest, Oregon; *photo courtesy Dr. Fred Hall*).

Figure 143.—*Pinus ponderosa-Abies concolor/Carex rossii* community (Goodlow Mountain Research Natural Area, Fremont National Forest, Oregon; *photo courtesy Dr. Fred Hall*).

Successional Patterns

Data on early successional stages following logging or burning on *Abies grandis* habitats are scarce. *Cirsium vulgare* was a major species on a clearcut and burned area in the Wallowa Mountains (Pettit 1968); *Ceanothus sanguineus, Physocarpus malvaceus, Spiraea betulifolia* var. *lucida, Astragalus* sp., *Fragaris* spp., *Carex rossii, Epilobium paniculatum, Bromus tectorum, and Rumex acetosella* were also noted. Later in preforest succession, shrub communities are encountered in the *Abies grandis* Zone, at least in eastern Washington and northern Idaho (Daubenmire 1952, Mueggler 1965). Typical shrub invaders (Daubenmire and Daubenmire 1968) are *Salix scouleriana, Spiraea betulifolia* var. *lucida, Ceanothus velutinus, C. sanguineus, Amelanchier alnifolia, Sambucus cerulea,* and *S. racemosa* var. *arborescens.* Mueggler (1965) related the composition of these shrub communities to type of originating disturbance and age.

Throughout much of the *Abies grandis* and *Abies concolor* Zones in the Cascade Range, *Ceanothus velutinus* is again a major brushfield dominant, often behaving ecologically as described for southwestern Oregon. *Ceanothus velutinus* and species such as *Castanopsis chrysophylla* and *Arctostaphylos patula* may form a sclerophyll scrub following wildfires in the central and southern Oregon *Abies grandis-A. concolor* Zones.

As mentioned, *Pinus contorta, P. ponderosa,* and *Pseudotsuga menziesii* often dominate seral stands in the *Abies grandis* Zone. Fires were most important in instigating these stands, although they were generally less frequent than in the more xeric *Pinus ponderosa*

Zone. In the Blue Mountains, *Pinus contorta* is a common seral species on *Abies grandis/Vaccinium membranaceum* habitats, and, as reviewed earlier, *Pinus ponderosa* is typically successional to *Abies grandis/Calamagrostis rubescens*. Most of the *Pinus contorta* forest described by Trappe and Harris (1958) is seral to *Abies grandis* (fig. 144). Sites occupied by *Pinus contorta* forests may go through a period of dominance by *Pseudotsuga menziesii* or *Picea engelmannii* prior to development of climax *Abies grandis* forests. *Pinus ponderosa* is the typical seral species on *Abies concolor/Ceanothus velutinus* habitats (Dyrness and Youngberg 1966), and *Pseudotsuga menziesii* is typical on *Abies grandis/Pachistima myrsinites* habitats in eastern Washington (Daubenmire 1952).

Larix occidentalis is a major seral dominant in parts of the Blue Mountains (*Abies grandis/Vaccinium membranaceum* habitats) and eastern Washington and can form a nearly pure type following repeated fires (fig. 145). *Picea engelmannii* is not ubiquitous in the *Abies grandis* Zone but may be common in some areas; it is replaced only gradually since reproductive vitality is often high. *Pinus monticola* is an important species in localized areas. *Libocedrus decurrens* occurs sporadically as a seral species within the zone in the Cascade Range south of Mount Hood.

Most of the seral species attain optimum growth in the *Abies grandis* Zone. Daubenmire (1961) found that *Pinus ponderosa* grew better on *Abies grandis/Pachistima myrsinites* habitats than on sites where *Pseudotsuga menziesii* or *Pinus ponderosa* were climax. In Montana, *Larix occidentalis* grew better in the *Abies grandis* than in the cooler *Abies lasiocarpa* or drier *Pseudotsuga menziesii* Zones (Roe 1967). Selective logging of *Pinus ponderosa* within the *Abies grandis* Zone has produced strong successional pressures toward its elimination since reproduction is composed of more tolerant species. Forestry practices are shifting toward clearcutting of old-growth forest, partially to eliminate rotted *Abies* and partially to favor the *Pinus*.

Abies grandis or *A. concolor* are the major climax species. In some areas, *Pseudotsuga menziesii* may play a minor climax role. There are locations where *Abies grandis*-dominated forests occur which are themselves subject to replacement by the more tolerant *Abies amabilis*, *Tsuga heterophylla*, or *Thuja plicata;* these are considered elsewhere in this chapter.

Figure 144.—*Pinus contorta* **is a common seral species in the** *Abies grandis* **zone of eastern Oregon;** *Abies grandis* **and** *Picea engelmannii* **succeeding** *Pinus contorta* **in the Blue Mountains of Oregon (range pole is marked in 1-foot segments).**

Figure 145.—*Larix occidentalis* forms nearly pure, seral stands in parts of the *Abies grandis* and *Pseudotsuga menziesii* Zones after fires; *Larix* reproduction with scattered *Pinus contorta* (Katherine Creek, Wallowa-Whitman National Forest, Oregon).

Special Types

Nonforest communities are common in many parts of the *Abies grandis* Zone. Some of these (discussed earlier) are successional to forest, and others are climax shrub or grassland communities which are covered in the discussions of the *Pinus ponderosa* and *Abies lasiocarpa* Zones. Mountain meadows are the most important single group of permanent nonforest communities in the *Abies grandis* Zone.

Mountain meadow communities are conspicuous, essentially permanent, herbaceous communities, typically found on relatively gentle topography along and near the heads of stream courses (fig. 146). These are not subalpine communities; and though associated with zones from the *Abies lasiocarpa* to *Pinus ponderosa*, they are probably most typical of the *Abies grandis* Zone.

Deschampsia caespitosa, a perennial grass, typifies the dense herbacoues cover of climax mountain meadow communities (Reid and Pickford 1946) (fig. 147). Other major species are *Festuca rubra*, *Carex* spp., *Juncus balticus*, *Aster occidentalis*, *Polygonum bistortoides*, *Trifolium* spp., and *Senecio* spp.

Most *Deschampsia* meadows have been overgrazed by domestic livestock (fig. 148) and have deteriorated into other kinds of communities. Reid and Pickford (1946) recognized four major steps in deterioration: (1) perennial grass or climax, (2) mixed grass and weed (fig. 148), (3) perennial weed, and (4) annual weed. Serious erosional problems are associated with these changes in community composition. Major dominants in the perennial weed stage are *Senecio* spp., *Achillea millefolium* var. *lanulosum*, *Wyethia* spp., *Potentilla* spp., *Aster occidentalis*, *Taraxacum officinale*, and *Poa pratensis*. The annual weed stage is characterized by *Bromus mollis*, *Muhlenbergia filiformis*, *Polygonum douglasii*, and *Madia* spp. Development of climax vegetation from these deteriorated

Figure 146.—Mountain meadows are common on gentle topography along and near the heads of stream courses in the *Abies grandis* Zone; they constitute an important grazing resource.

Figure 147.—*Deschampsia caespitosa,* a perennial grass, dominates mountain meadow communities in good condition.

Figure 148.—Most mountain meadows have been severely overgrazed by domestic livestock, resulting in major compositional changes.

Figure 149.—Parkland community of *Stipa occidentalis* var. *minor, Phlox diffusa,* **and** *Artemisia rigida* **with forest of** *Pinus ponderosa* **in the background (Meeks Table Research Natural Area, Snoqualmie National Forest, Washington;** *photo courtesy Dr. Fred Hall*).

communities is usually extremely slow, even if grazing is completely eliminated.

Juniperus occidentalis or sclerophyllous shrub communities are other nonforested types found on dry, steep slopes or shallow soils in the *Abies grandis* Zone. In the Canyon Creek Research Natural Area, *J. occidentalis* characterizes shallow soils with *Cercocarpus ledifolius, Poa sandbergii, Bromus tectorum, Carex geyeri, Crepis acuminata,* and *Achillea millefolium* as major associates (Hall in Franklin et al. 1972). *Ceanothus velutinus* communities characterize rock outcrops in association with *Prunus emarginata, Salix scouleriana, Carex geyeri,* and *Poa pratensis.* In the Ochoco Mountains, a *J. occidentalis-Prunus emarginata/Festuca idahoensis* community occupies shallow-soiled steep slopes in some parts of the *Abies grandis* Zone (Hall on Ochoco Divide Research Natural Area in Franklin et al. 1972).

Tiedemann and others (Franklin et al. 1972) have described an interesting vegetational mosaic on Meeks Table Research Natural Area which is apparently located within the *Abies grandis* Zone. Here, a grassland community dominated by *Stipa occidentalis* var. *minor, Poa sandbergii, Phlox diffusa,* and *Artemisia rigida* is most common (fig. 149). On rocky outcrops with little soil, there is a lithosolic community characterized by *Eriogonum douglasii, Poa sandbergii, Artemisia rigida,* and *Purshia tridentata;* it shows a surprising similarity to some of the lithosolic communities found well within the steppe (Daubenmire 1970). Forests of *Pinus ponderosa* and *Pseudotsuga menziesii* with reproduction of *Pseudotsuga* and *Abies grandis* characterize the remainder of the landscape. These forests occupy soils which are generally much deeper and have substantially lower bulk densities than soils under the *Stipa occidentalis* var. *minor* community.

Tsuga heterophylla Zone

Habitats where *Tsuga heterophylla* or *Thuja plicata* are climax are encountered on the eastern slopes of the Cascade Range in Washington and northern Oregon. These constitute the *Tsuga heterophylla* Zone. It is essentially an eastern extension of the widespread coastal *Tsuga heterophylla* Zone, although the composition of seral forests and understory is somewhat altered. This inland *Tsuga heterophylla* Zone correlates with the Interior Western Hemlock Zone in British Columbia (Krajina 1965, Bell 1965, Smith 1965) and the *Tsuga heterophylla* series of Daubenmire and Daubenmire (1968).

Interior *Tsuga heterophylla* and *Thuja plicata* forests are most common in the eastern Washington Cascades at elevations between 800 and 1,200 meters. In Oregon, they are increasingly rare to their southern limit at about 44°30' north latitude. Both species are absent from the Blue and Wallowa Mountains. Habitats sufficiently mesic to support *Tsuga heterophylla* or *Thuja plicata* are very often disjunct, occurring within areas of *Abies grandis* Zone.

Environmental Features

The *Tsuga heterophylla* Zone occurs under what appears to be the most equitable climatic regime of all the interior forest zones (Daubenmire 1956). Krajina (1965) indicated a precipitation range of 560 to 1,700 millimeters and mean annual temperatures of 2.5° to 7.5° C. for the zone in British Columbia. Major soil great groups are Haplorthods, Cryorthods, and Xerochrepts (Brown Podzolic, Podzolic, and Gray-Brown Podzolic) including some soils with gleyed subsoils. Alban (1967) has described some contrasting influences of *Tsuga* and *Thuja* on soil properties in this zone.

Forest Composition

Forest tree species found within the *Tsuga heterophylla* Zone include most of those found in the adjacent *Abies grandis* and *Abies lasiocarpa* (or *Tsuga mertensiana*) Zones. *Pseudotsuga menziesii*, *Abies grandis*, and *Pinus monticola* are the most abundant seral species (table 22). It is in this zone in northern Idaho that the well-known western white pine (*Pinus monticola*) forests are best developed (fig. 150). *Thuja plicata* and *Tsuga heterophylla* are typically present together, although *Thuja plicata* is sometimes found on sites too xeric for *Tsuga heterophylla*.

Daubenmire and Daubenmire (1968) have described four *Tsuga* or *Thuja* associations in eastern Washington and northern Idaho: *Tsuga heterophylla/Pachistima myrsinites*, *Thuja plicata/Oplopanax horridum*, *Thuja plicata/Pachistima myrsinites*, and *Thuja plicata/Athyrium filix-femina*. The *Tsuga heterophylla/Pachistima myrsinites* is the most important. *Thuja plicata* is usually present, but it does not reproduce well. The understory is composed of the floristically rich *Pachistima* union (Daubenmire 1952) discussed earlier, the most constant species being *Pachistima myrsinites*, *Tiarella unifoliata*, *Vaccinium membranaceum*, *Clintonia uniflora*, and *Gymnocarpium dryopteris*. The *Thuja plicata/ Oplopanax horridum* association is found in low-lying situations in association with the *Tsuga/Pachistima* association. Stands belonging to the *Thuja/Oplopanax* have a dense understory including a shrub layer dominated by *Oplopanax horridum*, many members of the *Pachistima* union, *Athyrium filix-femina*, and *Dryopteris austriaca*. Successional relations in stands are unclear; reproductive vigor variously indicates *Tsuga heterophylla* and/or *Thuja plicata* may be climax. Associations comparable to both of these were recognized by Bell (1965).

The *Thuja plicata/Pachistima myrsinites* and *Thuja plicata/Athyrium filix-femina* associations constitute a comparable pair outside the range of *Tsuga heterophylla*. Daubenmire and Daubenmire (1968) have noted that the *Thuja* associations tend to have a southerly distribution and *Tsuga* associations a more northerly distribution in their study area; this is consistent with our observations that *Thuja* extends its range into drier climatic regions than *Tsuga*, provided adequate soil moisture is available. The *Pachistima*

Figure 150.—Many seral species, such as the *Pinus monticola* **shown here, attain their best development in the mesic** *Tsuga heterophylla* **and** *Abies grandis* **Zones.**

union typifies the understory in the *Thuja/Pachistima* association and a variety of mesic-site herbs and *Alnus sinuata*, in the *Thuja/Athyrium* association. *Abies grandis* is a major long-lingering seral species in the *Thuja plicata/Pachistima myrsinites* association; *Thuja plicata* is the major climax species in both.

Our observations suggest that the *Tsuga heterophylla* and *Thuja plicata* communities found in the valleys on the eastern slopes of the Washington Cascade Range (north of about 46°15′ north latitude) are very similar to those described by Daubenmire and Daubenmire (1968). One difference is increasing admixtures of western slope species (not found in extreme eastern Washington and northern Idaho) as the crest of the range is approached.

Successional Patterns

Early stages in forest succession on disturbed sites have been discussed by Daubenmire (1952), Daubenmire and Daubenmire (1968), and Mueggler (1965). Shrub communities are often conspicuous prior to development of a tree overstory. Most of the *Tsuga heterophylla* Zone on the eastern slopes of the Cascade Range is occupied by relatively youthful stands in which the seral *Abies grandis*, *Pseudotsuga menziesii*, *Pinus monticola*, *Larix occidentalis*, *Pinus contorta*, and *Picea engelmannii* are dominant. *Tsuga heterophylla* and *Thuja plicata* are often represented only by reproduction in these stands.

The successional relationships between *Tsuga heterophylla*, *Thuja plicata*, and *Abies grandis* are complex. Daubenmire and Daubenmire (1968) state:

> Three species of trees in the *Tsuga* series [*Abies grandis*, *Tsuga heterophylla*, *Thuja plicata*] show . . . an ability to continue reproduction . . . but all have distinctive aute-cologies so that for the most part only one is the climax dominant in any one habitat type.

It is on this basis they recognize each as the major climax in one or more of the *Tsuga heterophylla* series of associations.

Abies lasiocarpa Zone

Climax forests of *Abies lasiocarpa* characterize the subalpine forest zone at many locations in eastern Washington and Oregon. The *Abies lasiocarpa* Zone is well represented on the high secondary ranges extending east from the crest of the Washington Cascade Range (Franklin and Trappe 1963), in the Okanogan Highlands Province of northeastern Washington, and in the Blue and Wallowa Mountains of northeastern Oregon and southeastern Washington (Daubenmire 1952, Johnson 1959, Head 1959). Its lower elevational boundary is normally 1,500 meters or more in the Cascade Range and 1,300 to 1,700 meters elsewhere. The *Abies lasiocarpa* Zone is a more continental analog of the *Tsuga mertensiana* Zone with which it merges in the Cascade Range. It is the local representative of the very widespread Rocky Mountain *Abies lasiocarpa-Picea engelmannii* forests (Oosting 1956) and correlates with the Engelmann Spruce-Subalpine Fir Zone recognized in British Columbia (Krajina 1965), the *Abies lasiocarpa* Zone of the northern Rocky Mountains (Daubenmire and Daubenmire 1968), and Merriam's Hudsonian Life Zone (Barrett 1962).

The *Abies lasiocarpa* Zone may be bounded at its lower limits by forests in which *Tsuga heterophylla*, *Thuja plicata*, *Abies grandis*, or *Pseudotsuga menziesii* are climax, depending upon locale. Where mountain masses are sufficiently high, it extends upward to the subalpine-alpine ecotone and, in these cases, typically includes an area of forest-meadow parkland. As with the *Tsuga mertensiana* Zone, we will concern ourselves only with the closed forest portion of the *Abies lasiocarpa* Zone in this section.

Abies lasiocarpa-Picea engelmannii forests are often conspicuous in frost pockets and other habitats characterized by drainage and accumulation of cold air. These often occur well below the expected elevational limits for the *Abies lasiocarpa* Zone. In the Washington Cascade Range, the best developed *Abies lasiocarpa-Picea engelmannii* forests occur in glaciated valley bottoms and not on the mountain slopes.

Environmental Features

Environmental data are extremely scarce for the *Abies lasiocarpa* Zone. It is the coolest and moistest of the forested zones. Cool summers, cold winters, and development of deep winter snowpacks are more important factors than total precipitation in differentiating the *Abies lasiocarpa* from lower forested zones, however. Mean July temperatures probably range from about 13° to 16° C. (Daubenmire 1956). Climatically, the *Abies lasiocarpa* Zone is more continental than the *Tsuga mertensiana* Zone, with lower winter and higher summer temperatures and less precipitation and snowpack accumulations, on the average.

Zonal soils in the *Abies lasiocarpa* Zone are generally Cryorthods or Haplorthods (Podzols or Brown Podzolic soils) with well-developed but relatively thin mor humus layers. Regosolic and lithosolic soils are also common in some localities. Soils are more acid than in the lower forested zones—typically pH 4.5 to 5.9.

Forest Composition

Major forest tree species in the *Abies lasiocarpa* Zone are *Abies lasiocarpa*, *Picea engelmannii*, and *Pinus contorta* (fig. 151). *Pseudotsuga menziesii*, *Abies grandis*, *Larix occidentalis*, and *Pinus monticola* may be conspicuous lower in the zone; *Pinus albicaulis* and, sometimes, *Larix lyallii* are abundant higher in the zone. *Pinus ponderosa* and *Populus tremuloides* are relatively uncommon. *Abies amabilis*, *A. procera*, and *Tsuga mertensiana* may be encountered as minor stand components, although locally *Tsuga mertensiana* dominates stands disjunct from the *Tsuga mertensiana* Zone.

Daubenmire and Daubenmire (1968) recognized five *Abies lasiocarpa* associations in eastern Washington and northern Idaho: *Abies lasiocarpa/Pachistima myrsinites*, *Abies lasiocarpa/Xerophyllum tenax*, *Abies lasiocarpa/Menziesia ferruginea*, *Abies lasiocarpa/Vaccinium scoparium*, and *Pinus albicaulis-Abies lasiocarpa*. The first three associations tend to form a topographic series within the zone: the *Abies/Pachistima* occupies the lowest part of the zone; *Abies/Xerophyllum* is typical of upper south slopes and ridgetops; and *Abies/Menziesia* occupies the wettest and coolest sites—north slopes and ravines. The *Abies/Pachistima* association has representatives of the *Pachistima* union (Daubenmire 1952) as its understory; *Clintonia uniflora* and *Galium triflorum* are especially well represented. Many species occur only in this association and nowhere else in the *Abies lasiocarpa* Zone: *Acer glabrum*, *Arnica cordifolia*, *Hieracium albiflorum*, *Amelanchier alnifolia*, *Aster conspicuus*, *Mitella stauropetala*, *Actaea rubra*, *Clintonia uniflora*, *Coptis occidentalis*, *Viola glabella*, *Adenocaulon bicolor*, *Rubus parviflorus*, *Arenaria macrophylla*, *Galium triflorum*, and *Spiraea betulifolia*. McLean (1970) identifies a comparable community in southern British Columbia. In the *Abies/Xerophyllum* association, the depauperate understory is dominated by *Xerophyllum tenax* and *Vaccinium membranaceum* (see fig. 77 for an exemplary stand located outside the *Abies lasiocarpa* Zone). The *Abies/Menziesia* association has a well-developed shrub layer in which *Menziesia ferruginea* is sometimes combined with *Rhododendron albiflorum* or *Ledum glandulosum*. McLean (1970) did not find a comparable community in southern British Columbia adjacent to the Okanogan Highlands Province. In northern Idaho, Daubenmire and Daubenmire (1968) have also recognized comparable *Menziesia ferruginea* and *Xerophyllum tenax* associations in which *Tsuga mertensiana* is the characteristic tree species.

Figure 151.—Mixed stands of *Abies lasiocarpa* **and** *Picea engelmannii* **typify the subalpine forests in more continental mountain areas.**

The *Abies lasiocarpa/Vaccinium scoparium* association is widespread (Daubenmire and Daubenmire 1968, Johnson 1959) and seems particularly common in drier locales. *Vaccinium scoparium* is the major understory species; typical associates are *Juniperus communis, Vaccinium membranaceum, V. caespitosum, Carex* spp., *Hieracium albiflorum, Arnica cordifolia,* and *Aster* spp. (Illingworth and Arlidge 1960). McLean (1970) found two phases of the *Abies lasiocarpa/Vaccinium scoparium* community—the *Calamagrostis rubescens* and *Phyllodoce empetriformis* phases—to be much more common than the "typical" association as described by Daubenmire and Daubenmire (1968); including the phases, the type is more common in southern British Columbia than in eastern Washington.

The *Pinus albicaulis-Abies lasiocarpa* (Daubenmire and Daubenmire 1968) and *Abies lasiocarpa-Picea engelmannii* (Head 1959) associations described within the zone are relatively open forest types associated with the subalpine parkland.

Successional Patterns

Preforest successional developments in the *Abies lasiocarpa* Zone are poorly known. Herbaceous stages characterized by species such as *Epilobium angustifolium* may occur. Shrub communities similar to those found in lower forested zones (Daubenmire and Daubenmire 1968, Mueggler 1965) may develop on the most moderate sites; i.e., those potentially occupied by the *Abies lasiocarpa/Pachistima myrsinites* association. Elsewhere, disturbed areas are usually dominated by the same species that make up the forest understory—e.g., *Xerophyllum tenax*, *Vaccinium membranaceum*, and *Menziesia ferruginea*.

Successional relationships among tree species in the *Abies lasiocarpa* Zone are better understood. *Pinus contorta* is one of the most ubiquitous and conspicuous of the seral species. *Picea engelmannii* is also very important in many areas, and conflicting opinions are common regarding its successional status. In the Rocky Mountains, some workers have concluded *Picea engelmannii* is a climax species (Oosting and Reed 1952), others that it is subclimax (Fowells 1965). Daubenmire and Daubenmire (1968) state that *Picea engelmannii* is a major climax species only in the *Abies lasiocarpa/Vaccinium scoparium* association; it is a persistent long-lived seral species in five other associations and a minor climax component in the *Abies lasiocarpa/Pachistima myrsinites* association. McLean's (1970) work in southern British Columbia confirms Daubenmire and Daubenmire's (1968) conclusions. The subtle trend toward elimination of *Picea engelmannii* was not considered in an earlier treatment (Daubenmire 1952) in which *Picea engelmannii-Abies lasiocarpa* associations and a zone were proposed. Franklin and Mitchell (1967) concluded *Abies lasiocarpa* generally replaces *Picea engelmannii* on the eastern slopes of Washington's Cascade Range.

The relationship between *Tsuga mertensiana* and *Abies lasiocarpa* is not completely clear. Locally, *Tsuga mertensiana* is a dominant species in the northern Rocky Mountains *Abies lasiocarpa* Zone (Daubenmire and Daubenmire 1968, Habeck 1967). Sometimes it appears to be the major climax species, but in other cases reproduction of *Abies lasiocarpa* is also abundant. Factors differentiating sites with and without *Tsuga mertensiana* in this area have not been determined (Daubenmire and Daubenmire 1968). In some parts of Washington's eastern Cascade Range, and even as far south as central Oregon (Sherman 1969), disjunct populations of *Tsuga mertensiana* occur mixed with *Abies lasiocarpa* in which resolution of the successional question is not clear. Within the main *Tsuga mertensiana* Zone in the Cascade Range, *Abies lasiocarpa* is almost always seral, however (Franklin and Mitchell 1967).

Hence, *Abies lasiocarpa* is the major and often sole climax species within the closed forest portion of the *Abies lasiocarpa* Zone of eastern Washington and Oregon. Only occasionally do *Picea engelmannii* or *Tsuga mertensiana* challenge its dominance there.

Special Types

Permanent nonforest communities dominated by herbs or shrubs occur within the forest matrix. Many of these are clearly extensions of subalpine meadow types and will be discussed later. Two, more characteristic of the closed-forest portion of *Abies lasiocarpa* Zone, are grassy parks and *Artemisia tridentata* var. *vaseyana* communities.

Grassy parks or balds are frequently encountered on or near ridgetops in the *Abies lasiocarpa* Zone (Daubenmire and Daubenmire 1968, Secor 1960, Johnson 1959, Hall 1967) (fig. 152). These are typically dominated by grasses such as *Agropyron spicatum*, *Festuca idahoensis*, and *F. viridula*, although many other species may be conspicuous—e.g., *Hieracium albertinum*, *Arenaria capillaris* var. *americana*, and *Polygonum phytolaccaefolium* in northern Idaho and *Artemisia arbuscula*, *Phlox* spp., *Poa sandbergii*,

207

and *Achillea millefolium* in the Ochoco Mountains. A lower fringe of *Prunus emarginata* is sometimes present. These grasslands are associated with south-facing slopes and are more xeric than adjacent forest stands (Daubenmire 1968b). Whatever their origin, wind-transfer of moisture (snow) from these balds and soil drought seem important in maintaining them as topographic climaxes.

Artemisia tridentata var. *vaseyana,* a cold-requiring diploid ancestor of *Artemisia tridentata* var. *tridentata,* characterizes another group of nonforest communities associated with the *Abies lasiocarpa* Zone (as well as lower forest zones) (Daubenmire 1970, Johnson 1959, Hall 1967, McLean 1970). *Carex* spp., *Agropyron spicatum,* and *Festuca idahoensis* are typical associates.

Tiedemann (1972) describes an interesting mosaic of high-elevation grassland communities in parks in the *Abies lasiocarpa* Zone. The pattern is one of "biscuit-scab" microtopography and of *Stipa occidentalis* var. *minor* dominated grasslands interspersed with sparsely vegetated barrens dominated by *Madia glomerata.* Across abrupt ecotones, perennial grass and forb communities are associated with areas with a soil surface 15 to 20 centimeters lower, much bare soil, active frost heaving, and a dominance of annual forbs. Major species in the grassland are *Stipa occidentalis* var. *minor, Festuca idahoensis, Deschampsia caespitosa, Carex raynoldsii, Agoseris glauca, Senecio integerrimus, S. serra,* and *Aster foliaceus.* The annual forb communities are composed of *Madia glomerata, Polygonum sawatchense, P. lenticularis, Lithophragma bulbifera, Collinsia parviflora,* and *Cryptantha torreyana,* as well as the perennial *Sitanion hystrix, Lupinus wyethii,* and *Phacelia hastata.* Study of soil properties indicates soils under the annual community are poorer in nitrogen, sulfur, and exchangeable manganese and have higher bulk densities and lower organic matter than soils under the perennial communities.

Figure 152.—Wind transfer of snow and soil drought are important in maintaining grassy balds or parks frequently found on south slopes near ridgetops in the *Abies lasiocarpa* Zone (*photo courtesy R. Daubenmire*).

CHAPTER VIII. STEPPE AND SHRUB-STEPPE OF THE COLUMBIA BASIN PROVINCE

In the rain shadow of the Cascade Range is a large region of steppe and shrub-steppe vegetation including most of central and southeastern Washington and much of eastern Oregon. In this region of bunchgrass and sagebrush communities (Shantz and Zon 1924), typical community dominants include shrubs such as *Artemisia tridentata, Purshia tridentata, Artemisia rigida, A. arbuscula,* and *Atriplex confertifolia*; large perennial grasses such as *Agropyron spicatum, Festuca idahoensis, Elymus cinereus,* and *Stipa thurberiana*; and alien invaders such as *Bromus tectorum, Poa pratensis,* and *Elymus caput-medusae.* Forest vegetation is generally confined to mountain slopes with sufficient precipitation, either regionally (e.g., approaching the Rocky Mountains) or locally (e.g., higher elevations on interior ranges such as the Blue Mountains).

Climatically, the steppe areas can be typified as arid to semiarid with low precipitation, warm-to-hot dry summers, and relatively cold winters (table 30). Some marine influences

Table 30. — Climatic data from representative stations within steppe areas of eastern Washington and Oregon

Station	Eleva-tion	Lati-tude	Longi-tude	Temperature					Precipitation		
				Average annual	Average January	Average January minimum	Average July	Average July maximum	Average annual	June through August	Average annual snowfall
	Meters			- - - - - - - - - Degrees C. - - - - - - - - -					Millimeters		Centi-meters
Ellensburg, Wash.	526	47°02′	120°31′	8.4	−4.7	−9.3	20.6	28.9	230	33	--
Moses Lake, Wash.	368	47°07′	119°12′	9.2	−3.9	−8.3	21.3	30.7	212	35	--
Kennewick, Wash.	119	46°13′	119°08′	12.0	−.2	−4.3	24.2	33.3	190	24	34
Pullman, Wash.	776	46°46′	117°12′	8.6	−2.2	−5.3	20.1	28.1	610	75	102
The Dalles, Oreg.	31	45°36′	121°12′	11.8	.1	−2.5	22.7	30.6	389	23	--
Pendleton, Oreg.	455	45°41′	118°51′	11.2	−.7	−4.3	23.1	33.2	320	44	47
Burns, Oreg.	1,265	43°35′	119°03′	7.8	−4.1	−9.7	20.2	29.8	289	43	122
Ontario, Oreg.	654	44°03′	116°58′	10.6	−2.5	−7.3	24.5	35.1	236	24	--
Lakeview, Oreg.	1,455	42°11′	120°21′	7.8	−2.6	−7.9	19.2	29.2	364	41	136

Source: U. S. Weather Bureau (1956, 1965a,b).

are still felt (conditions are not so extreme as those in the Great Basin, for example), but continental-type climatic conditions prevail.

A great variety of soils occurs in the steppe region. The zonal sequence from the hottest, driest sites to the wettest of the steppe habitats would run through Camborthids, Haploxerolls, and Argixerolls (Sierozem, Brown, Chestnut, Chernozem, and Prairie great soil groups). From soils with light-colored, thin A horizons poor in organic matter and calcium accumulations high in the profile (Camborthids or Sierozem), the soils grade to thick, very dark-brown to black A horizons rich in organic matter in which calcium carbonate accumulations may be deep in the profile or absent (Argixerolls or Prairie). At the same time, the wide variety of intrazonal soils includes most notably those having accumulations of salts (Solonchak) and large amounts of exchangeable sodium (Natrargids or Solonetz).

Fire and grazing were apparently of limited importance in steppe vegetation before arrival of Europeans and their livestock (Galbraith and Anderson 1971, Daubenmire 1970, Heady 1968). Large herds of ungulates were never an integral part of the steppe communities in the Northwest as they were in the Great Plains. The limited grazing was confined to deer, wapiti, and antelope until arrival of the horse in the early 1700's; buffalo were never a factor in grass consumption (Galbraith and Anderson 1971). Cattle were introduced in the steppe vegetation in 1834 (Daubenmire 1970) and sheep about 1860; and the latter were generally more abundant until about 1940.[1] The periods of 1860-70 and 1892-93 were times of especially rapid expansion in cattle and sheep populations, respectively (Galbraith and Anderson 1971). Aboriginal man had little need to use fire in the steppes in contrast to forested regions or areas where he used fire as an adjunct in hunting. Daubenmire (1970) goes so far as to say that "there is no evidence that the distribution of vegetation types or species in eastern Washington is related to the past use of fire."

Man has wrought massive changes in the steppe vegetation of the Northwest by the cultivation, animals, and plants he introduced. The best lands are cultivated almost entirely for wheat, peas, and similar crops. Some of the poorer lands, cropped for a time, have since been abandoned. Additional lands lacking sufficient natural moisture have been "reclaimed" through irrigation, most notably in the Columbia Basin Province. Remaining lands have been subjected to various degrees of grazing and often overgrazed by domestic and feral livestock. Overgrazing was considered a serious problem more than 60 years ago (Griffiths 1902, 1903; Cotton 1904). Chohlis (1952) suggested that range conditions generally improved from 1900 to 1950, however, and Daubenmire (1970) was inclined to agree after examining old photographs. Man-caused range fires were common in many steppe areas and still occur occasionally.

To appreciate how grazing and wildfire can affect the natural vegetation, one should consider features of some of the major climax dominants and alien invaders (Daubenmire 1970). These features are central to any consideration of succession and successional status in the steppes.

1. Two of the major shrub species, notably *Artemisia tridentata* and *Purshia tridentata*, are fire sensitive and can be temporarily eliminated from a site by burning.[2]
2. Most of the major large perennial grasses, e.g., *Agropyron spicatum* and *Festuca idahoensis*, are not adapted to heavy grazing by ungulates. They evolved in an environment in which such animals were sparsely represented. They rarely recover to their former status after severe overgrazing but are relatively insensitive to fire.

---

[1] Gerald S. Strickler, personal communication.

[2] Steppe shrubs species are generally sprouters. *Purshia tridentata* exhibits widely varying sprouting behavior, depending on environmental conditions (Blaisdell 1953, Blaisdell and Mueggler 1956, Driscoll 1963).

3. Two alien species, *Bromus tectorum* and *Poa pratensis*, are well adapted to parts of the steppe region. They will invade or increase under heavy grazing pressure (in their respective areas) and relinquish occupied sites to native vegetation either very slowly or not at all when grazing pressures are lifted. Hence, shifts to these species are generally only reversed by human intervention. Recently, another alien species, *Elymus caput-medusae*, has entered parts of the steppe region and threatens to further alter the ecology of the region.

The vegetation of the northwestern steppes has been studied by many scientists. Early generalized accounts include those of Colville (1896), Griffiths (1902, 1903), Cotton (1904), Weaver (1917), Shantz and Zon (1924), and Aldous and Shantz (1924). In recent years, more detailed reports cover synecological aspects of steppe and shrub-steppe vegetation in eastern Washington (Daubenmire 1940, 1942, 1956, 1966, 1970, 1972; McMinn 1952; Cooke 1955; Daubenmire and Colwell 1942; Brooks 1969) and portions of eastern Oregon (Poulton 1955, Eckert 1957, McKell 1956, Tueller 1962). Anderson (1956) has divided eastern Oregon into ecological provinces; Humphrey (1945) has considered major range types in both States; and Billings (1951) constructed a general zonational approach to the Great Basin which is relevant to a part of southeastern Oregon.

From these accounts, it is possible to recognize numerous diverse community types, including many that would qualify as climatic climaxes (zonal associations) within a part of their range. In general, no single sequence of zonal belts of vegetation applies throughout the steppe region. Consequently, we focus attention in this chapter on the steppes of Washington's Columbia Basin Province, where the entire vegetational mosaic is most fully understood. Here a cross section of the communities encountered in the steppe regions of the Pacific Northwest can be illustrated and related. The other major steppe region in central and southeastern Oregon will be described in Chapter IX and similarities and differences between this region and the Columbia basin suggested.

The Columbia Basin Province of eastern Washington is a large, contiguous area of steppe and shrub-steppe vegetation—over 6,000,000 hectares (fig. 153). The vegetational mosaic has been thoroughly studied (see citations above); our discussion is abstracted almost entirely from Daubenmire's (1970) comprehensive regional account. Considered in turn are (1) zonal associations, (2) associations on specialized habitats, and (3) zootic climaxes (with comments on effects of *Artemisia* eradication).

Zonal Associations

Nine zonal associations, communities which can occur as climatic climaxes, have been recognized in the steppe region of the Columbia Basin Province (Daubenmire 1970). These are the:

Artemisia tridentata/Agropyron spicatum,
Artemisia tridentata/Festuca idahoensis,
Agropyron spicatum-Poa sandbergii,
Agropyron spicatum-Festuca idahoensis,
Festuca idahoensis/Symphoricarpos albus,
Festuca idahoensis/Rosa nutkana,
Artemisia tripartita/Festuca idahoensis,
Festuca idahoensis-Hieracium cynoglossoides, and
Purshia tridentata/Festuca idahoensis associations.

The distribution of the zones characterized by these associations is shown in fig. 153; note that this does *not* delimit the total geographic ranges of these associations. As can be seen, the last five associations are found on the periphery of the steppe region near its

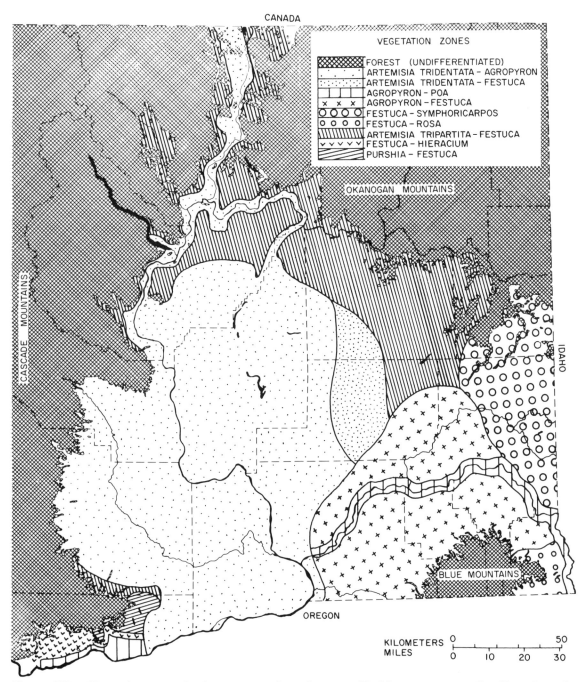

Figure 153.—Vegetation zones in the steppe region of eastern Washington; see text for discussion of the various zonal associations (*map courtesy R. Daubenmire*, **from his 1970 bulletin**).

contact with forest vegetation. These tend to be lush, meadowlike communities with conspicuous amounts of large perennial grasses and broad-leaved forbs. The term "meadow steppe" has often been applied to these types. The other four zonal associations lie in the more arid interior of the Columbia Basin Province. Vegetation is more open and forbs are less conspicuous in these communities. The contrasting physiognomy of representative steppe, shrub-steppe, and meadow-steppe communities is shown in fig. 154.

212

The nine zonal associations have differentiated in response to differences in temperature and total and seasonal distribution of precipitation. Where they occur on modal sites, as climatic climaxes, they characterize or distinguish regional units or zones of steppe vegetation. Four of these zones—the *Artemisia/Agropyron*, *Artemisia/Festuca*, *Agropyron-Festuca*, and *Festuca/Symphoricarpos* Zones—are encountered along a

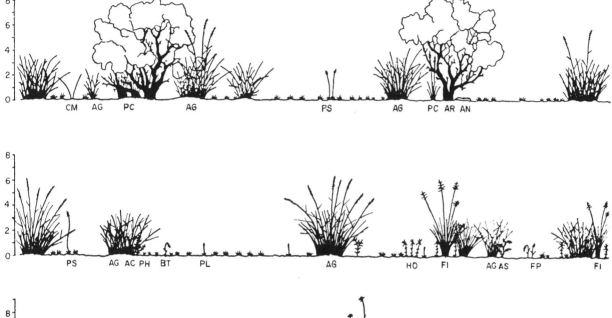

Figure 154.—**Vegetation profiles along 2- by 100-centimeter transects in a shrub-steppe** *Artemisia tridentata/Agropyron spicatum* **stand (upper), steppe** *Agropyron spicatum-Festuca idahoensis* **stand (middle), and meadow-steppe** *Festuca idahoensis/Symphoricarpos albus* **stand (lower) (from Daubenmire 1970); abbreviations are:**

Ac *Achillea millefolium*
Ag *Agropyron spicatum*
An *Antennaria dimorpha*
Ar *Artemisia tridentata*
Aa *Astragalus palousensis*
As *Astragalus spaldingii*
Ba *Balsamorrhiza sagittata*
Be *Besseya rubra*
Bj *Bromus japonicus*
Bt *Bromus tectorum*
Ce *Calochortus elegans*
Cm *Calochortus macrocarpus*
Ca *Castilleja lutescens*
Fi *Festuca idahoensis*
Fp *Festuca microstachys*
Ga *Galium boreale*

He *Helianthella uniflora douglasii*
Hi *Hieraceum albertinum*
Ho *Holosteum umbellatum*
Ir *Iris missouriensis*
Lu *Lupinus sericeus*
Mi *Microsteris gracilis*
My *Myosotis micrantha*
Ph *Phlox longifolia*
Pl *Plantago patagonica*
Pc *Poa cusickii*
Ps *Poa sandbergii*
Pt *Potentilla gracilis*
Ro *Rosa nutkana* var. *hispida* or *R. nutkana*
Se *Senecio integerrimus exaltatus*
Sy *Symphoricarpos albus*
Zy *Zygadenus venenosus gramineus*

transect beginning in the driest part of the Columbia Basin Province and extending eastward up the gentle slope of the Columbia plateau to foothills of the Rocky Mountains (fig. 153). They provide an exemplary cross section of the zonal associations. The composition of some stands found along this transect belonging to the zonal associations is shown in table 31. This tabulation includes only pristine stands found on zonal habitats (sites with gently undulating topography and deep, well-drained silt loam soils). An association table for seven of these associations (table 32) and the following descriptions amplify this tabulation and encompass all stands belonging to the association under consideration; the reader should remember that any of these zonal associations may occur as a topographic climax in one or more adjacent zones.

Table 31. — Percent canopy-coverage of species in climatic climax steppe communities along a 96-kilometer longitudinal transect extending from the center of the Columbia Basin Province (left side of table) to the Washington-Idaho border (right side of table)[1]

Species	Artemisia/Agropyron							Artemisia/Festuca			Agropyron-Festuca					Festuca/Symphoricarpos					
Stipa thurberiana	5	–	3	–	–	–	–	–	–	–	–	–	–	–	–	–	–	–	–	–	–
Poa cusickii	2	9	–	+	–	–	–	–	–	–	–	–	–	–	–	–	–	–	–	–	–
Stipa comata	2	+	13	2	9	14	–	1	–	13	6	–	–	–	–	–	–	–	–	–	–
Artemisia tridentata	18	18	9	9	16	19	11	25	13	4	–	–	–	–	–	–	–	–	–	–	–
Chrysothamnus viscidiflorus	–	–	+	–	–	2	9	+	+	8	1	–	+	–	–	–	–	–	–	–	–
Plantago patagonica	–	–	14	8	–	11	–	–	3	6	3	7	2	2	+	–	–	–	–	–	–
Phlox longifolia	+	12	–	+	7	8	12	2	+	1	–	1	3	1	5	1	–	–	–	–	–
Erigeron filifolius	–	–	–	–	3	–	5	+	+	–	–	–	–	+	–	–	–	–	–	–	–
Astragalus spaldingii	–	–	–	–	–	–	–	2	1	2	+	+	6	2	13	–	–	–	–	–	–
Poa sandbergii	40	50	29	61	55	73	36	44	38	5	19	39	16	23	45	2	+	2	+	+	+
Achillea millefolium	–	–	–	+	–	+	+	–	+	2	1	2	3	1	+	22	10	6	9	25	25
Agropyron spicatum	41	63	46	35	41	55	36	40	39	26	42	77	33	34	71	79	57	–	77	78	78
Festuca idahoensis	–	–	–	–	–	–	–	9	23	39	29	7	71	75	8	81	30	41	35	40	40
Senecio integerrimus	–	–	–	–	–	–	–	–	–	–	–	–	–	–	–	+	2	8	3	2	–
Myosotis micrantha	–	–	–	–	–	–	–	–	–	–	–	–	–	–	–	2	–	–	–	5	5
Haplopappus liatriformis	–	–	–	–	–	–	–	–	–	–	–	–	–	–	–	10	8	–	–	+	+
Koeleria cristata	–	–	–	–	–	–	–	–	–	–	–	–	–	–	–	5	3	–	4	4	4
Hieraceum albertinum	–	–	–	–	–	–	–	–	–	–	–	–	–	–	–	6	1	2	–	3	3
Lupinus sericeus	–	–	–	–	–	–	–	–	–	–	–	–	–	–	–	–	–	51	–	–	–
Festuca scabrella	–	–	–	–	–	–	–	–	–	–	–	–	–	–	–	–	8	–	–	–	–
Sidalcea oregana	–	–	–	–	–	–	–	–	–	–	–	–	–	–	–	–	+	2	9	5	5
Castilleja lutescens	–	–	–	–	–	–	–	–	–	–	–	–	–	–	–	–	5	–	–	–	+
Arnica sororia	–	–	–	–	–	–	–	–	–	–	–	–	–	–	–	29	–	–	–	–	–
Solidago missouriensis	–	–	–	–	–	–	–	–	–	–	–	–	–	–	–	–	–	41	14	51	51
Balsamorhiza sagittata	–	–	–	–	–	–	–	–	–	–	–	–	–	–	–	–	–	–	–	31	31
Helianthella uniflora	–	–	–	–	–	–	–	–	–	–	–	–	–	–	–	–	–	–	–	18	18
Astragalus palousensis	–	–	–	–	–	–	–	–	–	–	–	–	–	–	–	57	53	–	–	–	–
Poa ampla	–	–	–	–	–	–	–	–	–	–	–	–	–	–	–	1	2	–	+	1	1
Rosa nutkana + R. woodsii	–	–	–	–	–	–	–	–	–	–	–	–	–	–	–	+	29	–	2	1	1
Iris missouriensis	–	–	–	–	–	–	–	–	–	–	–	–	–	–	–	6	21	+	45	5	5
Potentilla gracilis	–	–	–	–	–	–	–	–	–	–	–	–	–	–	–	–	21	+	4	–	–
Geranium viscosissimum	–	–	–	–	–	–	–	–	–	–	–	–	–	–	–	+	11	–	–	1	1
Galium boreale	–	–	–	–	–	–	–	–	–	–	–	–	–	–	–	4	15	2	11	17	17
Symphoricarpos albus	–	–	–	–	–	–	–	–	–	–	–	–	–	–	–						

Median annual precipitation 167 mm. ←————————————————————→ 526 mm
Mean annual temperature 11.2°C. ←————————————————————→ 7.9°C.

[1] Species with maximum coverages of less than 5 percent omitted; plus sign indicates coverage of less than 1 percent (from Daubenmire 1966).

214

Table 32. — Constancy and average coverage (constancy/coverage) of selected species in seven zonal steppe associations found in the Columbia Basin Province (from Daubenmire 1970)[1]

Stratum and species	Association[2]						
	Artr/Agsp (n=15)	Artr/Feid (n=6)	Agsp/Posa (n=8)	Agsp-Feid (n=11)	Feid/Syal (n=15)	Artp/Feid (n=19)	Putr/Feid (n=10)
	Percent						
Medium shrubs:							
Artemisia tridentata tridentata	100/13	100/13	—	—	—	—	—
Chrysothamnus nauseosus albicaulis	33/tr	50/tr	88/tr	64/tr	—	26/tr	10/tr
Chrysothamnus viscidiflorus	53/tr	100/tr	25/1	18/tr	—	53/1	10/tr
Artemisia tripartita	27/1	50/1	—	—	—	100/11	20/tr
Tetradymia canescens	27/tr	83/1	—	—	—	32/2	20/tr
Purshia tridentata	7/tr	—	—	—	—	—	80/14
Low shrubs:							
Phlox longifolia	87/4	83/2	62/3	91/1	13/2	89/2	30/tr
Erigeron filifolius	53/2	67/tr	12/tr	9/tr	—	5/tr	10/tr
Eriogonum heracleoides	7/2	16/tr	—	—	—	84/2	90/10
Symphoricarpos albus	—	—	—	—	100/10	—	—
Rosa nutkana and R. woodsii	—	—	—	—	87/3	—	—
Perennial graminoids:							
Agropyron spicatum	100/49	100/40	100/68	100/59	93/62	100/38	100/66
Poa sandbergii	100/41	100/32	100/30	100/21	40/2	100/12	90/15
Stipa comata	67/4	50/8	12/3	—	—	11/13	20/2
Poa cusickii	60/5	—	—	9/tr	—	21/8	10/tr
Festuca idahoensis	—	100/34	—	100/53	100/45	100/43	100/36
Koeleria cristata	—	—	—	9/4	80/5	58/3	60/2
Poa ampla	—	—	—	—	53/21	21/1	50/3
Stipa occidentalis minor	—	—	—	—	7/4	58/4	10/tr
Carex filifolia	—	—	—	—	—	63/20	10/4
Perennial forbs:							
Achillea millefolium lanulosa	40/tr	67/2	75/2	100/2	100/8	79/1	90/2
Astragalus spaldingii	27/8	83/1	38/2	54/4	13/3	42/4	—
Balsamorhiza careyana	47/tr	50/tr	50/tr	18/tr	—	—	10/tr
Brodiaea douglasii	47/tr	16/tr	38/tr	36/1	67/2	53/1	70/1
Calachortus macrocarpus	73/tr	50/tr	38/tr	27/tr	—	79/1	30/1
Erigeron pumilus intermedius	33/tr	100/tr	62/tr	9/tr	—	42/tr	40/tr
Lithophragma bulbifera	47/2	67/tr	38/3	82/2	67/2	94/1	80/2
Lithospermum ruderale	27/tr	83/tr	12/tr	18/2	87/1	79/tr	30/tr
Lomatium macrocarpum	33/1	83/tr	75/tr	27/tr	—	—	30/1
Lomatium triternatum	40/tr	83/tr	25/tr	82/2	60/1	58/tr	30/1
Microseris troximoides	60/1	67/tr	12/tr	82/tr	—	68/1	40/tr
Senecio integerrimus exaltata	7/tr	16/tr	—	27/tr	93/2	42/1	20/2
Tragopogon dubius	—	—	62/tr	54/tr	47/tr	58/tr	10/5
Antennaria rosea	7/tr	—	—	—	—	11/1	50/6
Balsamorhiza sagittata	13/tr	—	—	—	80/25	42/5	70/16
Besseya rubra	—	—	—	9/tr	80/2	21/2	—
Castilleja lutescens	—	—	—	—	67/2	11/3	10/tr
Crepis atribarba originalis	47/tr	—	—	27/1	—	32/tr	60/1
Erigeron corymbosus	—	16/tr	—	36/3	33/2	94/3	60/2
Geranium viscosissimum	—	—	—	—	73/7	5/1	—
Geum triflorum	—	—	—	9/tr	93/13	16/5	10/tr
Haplopappus liatriformis	—	—	—	9/3	67/6	5/tr	—
Helianthella uniflora douglasii	—	—	—	—	53/13	—	—
Hieracium albertinum	—	—	—	9/tr	100/5	11/1	10/tr

Table 32. — Constancy and average coverage (constancy/coverage) of selected species in seven zonal steppe associations found in the Columbia Basin Province (from Daubenmire 1970)[1] (Continued)

Stratum and species	Association[2]						
	Artr/Agsp (n=15)	Artr/Feid (n=6)	Agsp/Posa (n=8)	Agsp-Feid (n=11)	Feid/Syal (n=15)	Artp/Feid (n=19)	Putr/Feid (n=10)
	————————————— Percent —————————————						
Perennial forbs (cont'd.):							
Iris missouriensis	—	—	—	—	67/5	—	—
Lithophragma parviflora	—	—	—	—	53/2	5/tr	20/7
Lomatium dissectum multifidum	—	—	—	—	60/5	—	—
Lupinus sericeus	—	—	—	18/tr	93/7	74/5	30/10
Potentilla gracilis	—	—	—	9/tr	100/23	11/4	—
Zigadenus venenosus gramineus	—	—	—	9/tr	73/1	63/1	30/1
Annuals:							
Bromus tectorum	80/tr	100/2	100/3	91/1	40/tr	100/2	90/1
Descurainia pinnata	60/tr	—	25/tr	9/1	—	11/tr	10/tr
Draba verna	53/tr	67/6	100/2	100/3	87/1	68/2	40/1
Epilobium paniculatum	20/tr	67/2	38/1	64/tr	93/tr	89/1	60/3
Festuca octoflora	40/1	83/tr	75/2	18/1	—	42/tr	10/1
Festuca microstachys	80/1	83/1	75/2	54/tr	33/1	63/1	30/tr
Lactuca serriola (seedlings only)	20/tr	67/tr	62/tr	82/tr	73/tr	74/tr	10/tr
Lappula redowskii	73/tr	67/tr	12/tr	27/1	—	26/1	10/tr
Microsteris gracilis	33/tr	67/tr	—	36/1	67/1	68/1	60/1
Myosurus aristatus	33/1	83/1	—	18/tr	—	42/tr	—
Plantago patagonica	20/9	83/4	100/4	64/2	7/tr	68/1	30/2
Holosteum umbellatum	—	16/1	55/4	64/1	13/tr	—	—
Bromus japonicus	—	—	12/tr	9/1	67/3	21/1	30/tr
Collinsia parviflora	20/tr	—	—	—	100/2	79/2	80/2
Collomia linearis	—	16/tr	—	18/tr	13/tr	26/1	50/1
Montia linearis	—	16/tr	—	9/2	80/1	53/1	20/1
Stellaria nitens	—	—	12/2	36/1	80/1	26/1	40/tr

[1] Average cover is based only on those plots in which a particular species occurred, not by dividing total cover by all stands included within an association; tr equals trace coverage (less than 0.5 percent).

[2] Association abbreviations are:

Artr/Agsp = *Artemisia tridentata/Agropyron spicatum*
Artr/Feid = *Artemisia tridentata/Festuca idahoensis*
Agsp/Posa = *Agropyron spicatum/Poa sandbergii*
Agsp-Feid = *Agropyron spicatum-Festuca idahoensis*
Feid/Syal = *Festuca idahoensis/Symphoricarpos albus*
Artp/Feid = *Artemisia tripartita/Festuca idahoensis*
Putr/Feid = *Purshia tridentata/Festuca idahoensis.*

Artemisia tridentata/Agropyron spicatum Association

The driest of the zones has as a climatic climax the *Artemisia/Agropyron* association (fig. 155). This zone occupies the center of the Columbia Basin Province and extends west to the foothills of the Cascade Range. Four layers are found in this association (table 32): (1) a shrub layer composed principally of *Artemisia tridentata* var. *tridentata* and 1 to 2 meters in height—very small amounts of other shrubs such as *Chrysothamnus viscidiflorus, C. nauseosus* var. *albicaulis, Artemisia tripartita* or *Grayia spinosa* may be present; (2) a layer of caespitose perennial grasses dominated by *Agropyron spicatum*—variable amounts of *Stipa comata, S. thurberiana, Poa cusickii,* or *Sitanion hystrix* may

Figure 155.—The *Artemisia tridentata/Agropyron spicatum* **association is the climatic climax in driest parts of the Columbia basin steppe region; the range pole is 4 feet tall and marked in 6-inch segments** *(photo courtesy Range Management, Oregon State University).*

be present; (3) a layer of plants within 1 decimeter of the soil surface, including species such as *Poa sandbergii, Bromus tectorum,* and *Lappula redowskii*; (4) a surface crust typically composed of crustose lichens and acrocarpous mosses (e.g., *Tortula brevipes, T. princeps,* and *Aloina rigida*). Seasonal sequences in phenology are marked with mosses, small perennials, and annuals developing earliest and larger grasses and forbs flowering in June. Shrubs remain active all summer by tapping permanent moisture supplies in the subsoil; flowering extends from late June (*Tetradymia canescens*) to October (*Artemisia tridentata*).

Some interesting spatial patterns and variations in coverage and composition are observed in *Artemisia/Agropyron* communities. *Artemisia tridentata* coverage in stands varies from 5 to 26 percent; yet this apparently is not related to past grazing. *Artemisia* coverage is not correlated with coverage of either grazing increasers or decreasers in the climax stands Daubenmire sampled. It may be that *Artemisia* coverage is reflecting subsoil variations to which the more shallow-rooted grasses are insensitive. Several species are shrub-dependent; *Poa cusickii* is found mainly under *Artemisia* or *Grayia* canopies in many stands, and *Tortula* also occurs in patches under shrub canopies. *Bromus tectorum* tends to form dense circular patches beneath shrub canopies which can be related to litter accumulation and, in some cases, mice burrowing. Composition and abundance of the annuals varies substantially, and not all species present appear every year; in two stands sampled for each of 7 years, annuals recorded were never all represented in any single year, and only three out of 10 (*Festuca microstachys, F. octoflora,* and *Bromus tectorum*) appeared every year.

The *Artemisia/Agropyron* association occurs on two major types of soils—Mollic

217

Camborthids and Calcic Haploxerolls (Sierozem, Brown, Chestnut-Brown intergrades and, in one case, Solodized Solonetz by the old soil classification). The annual soil moisture cycle has been studied for this plant community (Daubenmire 1972).

Successional changes in the *Artemisia/Agropyron* Zone are most often associated with grazing, fire, or cultivation. Grazing most seriously affects the larger perennial grasses since they are preferred and are not adapted to withstand grazing. Heavy grazing tends, therefore, to eliminate *Agropyron spicatum*, *Festuca idahoensis*, *Poa cusickii*, etc., and to increase annual grasses, particularly *Bromus tectorum*. *Poa cusickii* is preferred by horses, cattle, and sheep, with *Agropyron* as second choice. The smaller perennial *Poa sandbergii* is generally not significantly affected, particularly by sheep grazing, and *Artemisia* suffers mechanical damage only by cattle. *Bromus tectorum* will apparently relinquish ground only very slowly once grazing pressure is lifted. Fire seriously affects only one dominant, *Artemisia tridentata*. It is often completely killed by range fires, and although the remaining dominants can regenerate from subterranean organs, *Artemisia* must reoccupy the site by invasion and gradual expansion, a relatively slow process (fig. 156). A combination of both burning and overgrazing can result in development of an annual rangeland dominated by *Bromus tectorum* in which *Chrysothamnus nauseosus* may be the only significant shrub. Cropped and abandoned fields will also develop a community dominated by *Bromus tectorum*, but the tumbleweeds *Salsola kali* and *Sisymbrium altissimum* may dominate the old field for a year or two while the *Bromus* population builds up.

The *Artemisia/Agropyron* association is the most extensive element in the steppe mosaic of eastern Washington, and essentially identical communities are widely distributed elsewhere, including British Columbia (Tisdale 1947), central Oregon (Eckert 1957), southern Idaho, and Montana. Three other zonal associations (*Purshia/Festuca*, *Artemisia tripartita/Festuca*, and *Artemisia tridentata/Festuca*) can occur as topographic climaxes on moister sites within the *Artemisia/Agropyron* Zone. Conversely, the *Artemisia/Agropyron* association can occur as a topographic climax on drier sites in adjacent zones.

Figure 156.—*Artemisia tridentata* is sensitive to fire since it is a nonsprouter; the patchiness of *Artemisia* in this *Artemisia/Agropyron* community suggests the influence of past burning (*photo courtesy Range Management, Oregon State University*).

Artemisia tridentata/Festuca idahoensis Association

East along the transect, a second large perennial grass, *Festuca idahoensis*, is added to climatic climax communities, indicating presence in a second zone typified by the *Artemisia tridentata/Festuca idahoensis* association (table 31). The significance of this addition has been discussed by Daubenmire (1966); the addition indicates the moisture balance of the macroclimate in this zone is more favorable for plant growth.

The only significant floristic difference between the *Artemisia/Festuca* and *Artemisia/Agropyron* association is the addition of *Festuca idahoensis* as a major grass (tables 31 and 32). *Artemisia tridentata* remains the dominant shrub. There is, however, a general increase in the summed coverage of perennial grasses from 38 to 87 percent in the *Artemisia/Festuca* type. Soils associated with this community are mainly Calcic Haploxerolls or Lithic Mollic Camborthids with one Typic Natrixeroll (Brown and Solonetz great soil groups).

Besides its distribution as a zonal type, this community occurs as a topographic climax on north-facing slopes in the *Artemisia/Agropyron* Zone and on south-facing slopes in the adjacent cooler, moister zones. The *Artemisia/Festuca* association is also found as disjunct stands on the northeast slopes of the Blue Mountains where it is associated with Solodized Solonetz soils; perhaps soil conditions reduce the effectiveness of the much higher precipitation in this part of the association's range.

Response of the *Artemisia/Festuca* community to fire and grazing is essentially the same as that of the *Artemisia/Agropyron* association.

Agropyron spicatum-Festuca idahoensis Association

Along the transect from the central Columbia Basin Province to the east (table 31), the next zone encountered is characterized by the *Agropyron-Festuca* association (fig. 153). *Artemisia tridentata* is absent, leaving a monotonous cover of perennial bunchgrasses (fig. 157). This change again signals an improved moisture regime (Daubenmire 1966).

The *Agropyron-Festuca* association is sharply distinguished from most surrounding zonal communities—drier associations by the absence of *Artemisia tridentata*, and

Figure 157.—A monotonous, herbaceous cover dominated by perennial grasses characterizes the climatic climax of the *Agropyron spicatum-Poa sandbergii* **community** *(photo courtesy Range Management, Oregon State University).*

moister meadow steppe by the absence of *A. tripartita, Symphoricarpos albus,* and *Rosa* spp., and the scarcity of forbs. The *Agropyron-Poa* Zone, which splits the *Agropyron-Festuca* into two segments (fig. 153), has a similar zonal association except that *Festuca idahoensis* is absent.

Major dominants in the *Agropyron-Festuca* association are perennial bunchgrasses—*Agropyron spicatum, Festuca idahoensis,* and *Poa sandbergii* (table 32). There are no shrubs and few forbs of consequence. South of the Snake River, communities of this type have several distinctive features, perhaps because of the higher elevations (up to 900 meters on the Blue Mountains) and soil mosaic which incorporates abundant Solodized Solonetz. Daubenmire (1970) comments on variations in stands associated with areas where the B21 of the Solodized Solonetz lies near the soil surface. *Opuntia polyacantha* is a minor member of the climax community south of the Snake River and increases with grazing.

A wide variety of soils are associated with the *Agropyron-Festuca* association: Haploxerolls, Natrixerolls, Palexerolls, and Argixerolls (Brown, Chestnut, Chernozem, Planosol, Solodized Solonetz, and Prairie-Grumusol great soil groups).

Fire has little effect on this community since neither of the dominants is seriously affected. As with the prior associations, *Bromus tectorum* is the main increaser with grazing.

Festuca idahoensis/Symphoricarpos albus Zone

The moistest of the steppe zones are meadow steppe, and we take as our major example of this type the *Festuca/Symphoricarpos* Zone found on the eastern margin of the Columbia Basin Province (fig. 153, table 31). Here, a whole group of species are added to the zonal communities producing a luxuriant growth (fig. 154) compared with the three previously discussed associations found to the east (Daubenmire 1966).

The *Festuca idahoensis/Symphoricarpos albus* association is actually a vegetational mosaic consisting of herbaceous and shrubby components (fig. 158). The herbaceous community provides the matrix consisting of a dense herbaceous layer dominated by perennial grasses (*Festuca idahoensis, Agropyron spicatum, Koeleria cristata,* and *Poa ampla*) and a great variety of perennial forbs (fig. 159) (table 32). In this community, *Agropyron* is rhizomatous, not caespitose as is the case in drier associations. Among the more important forb associates are *Achillea millefolium, Balsamorhiza sagittata, Geum triflorum* var. *ciliatum, Hieracium albertinum, Lupinus sericeus, Potentilla gracilis, Helianthella uniflora* var. *douglasii, Iris missouriensis, Geranium viscosissimum, Astragalus palousensis,* and *Castilleja lutescens.* Dwarfed, inconspicuous sterile shrubs are scattered through the herbland including *Symphoricarpos albus* and *Rosa nutkana* or *woodsii.*

This assemblage of grasses and forbs is common to the meadow steppe which encircles the Columbia basin. It is the shrubby component which varies from one link in this associational chain to the next. In the *Festuca/Symphoricarpos* association, the shrubby islands or thickets within the herbaceous matrix are recognized as a *Symphoricarpos albus* phase. The shrub thickets are 0.5 to 3 meters in height and 4 to 25 meters in diameter. Composition may consist of *Symphoricarpos* alone, a low marginal zone of *Symphoricarpos* (0.5 meter tall) around a core of *Rosa* spp. (1 meter tall), or even a core of *Prunus virginiana* (2 to 3 meters tall) with marginal belts of *Rosa* and *Symphoricarpos.* Herbaceous species are the same as in the adjacent grassland, except highly heliophytic species may be purged; all are reduced in stature. *Agastache urticifolia* was the only native vascular plant found in thickets. The positions and sizes of the shrub thickets in the herbaceous matrix have not yet been explained. Daubenmire (1970) found no relationship to slope position or exposure or soil characteristics. There is no evidence that the thickets are spreading; grassland and shrub thickets form a stable mosaic.

Figure 158.—Complex vegetational mosaics are typical of steppe-forest ecotones, such as the *Festuca idahoensis/Symphoricarpos albus* Zone; *Physocarpus malvaceus-Symphoricarpos albus* shrub community surrounded by a *Festuca idahoensis-Agropyron spicatum* grassland with scattered *Pinus ponderosa* (Joseph Creek Canyon, Wallowa-Whitman National Forest, Oregon; photo courtesy C. Johnson).

Figure 159.—The climax grasslands of the *Festuca idahoensis/Symphoricarpos* Zone in the Columbia basin are sometimes referred to as meadow-steppe because of the dense cover of sod-forming grasses and the abundance of broad-leaved herbs (*photo courtesy R. Daubenmire*).

Soils associated with the *Festuca/Symphoricarpos* association cover a wide range: Argixerolls, Argiudolls, Palexerolls, and Haploxerolls (Chernozem, Prairie, and Planosol great soil groups). A black A1 horizon and absence or deep placement of a layer of calcium carbonate accumulation are characteristic of the soils.

Fire has little effect on the *Festuca/Symphoricarpos* community, including the balance between herbaceous and shrub phases, since all important species can regenerate from underground organs. Recovery is apparently very rapid (Daubenmire 1970, p. 24).

Heavy grazing results in major, irreversible changes, however. The perennial grasses and shrubs decline simultaneously since both groups are highly palatable. Annuals, particularly the invader *Poa pratensis*, replace the native species. *Poa pratensis* is not replaced when grazing is reduced or eliminated and will form an essentially pure sward given the opportunity by extended heavy grazing.

Artemisia tripartita/Festuca idahoensis Association

The zonal meadow-steppe association on the northern and northeastern margins of the Columbia basin is the *Artemisia tripartita/Festuca idahoensis* (fig. 153). Daubenmire (1970) points out the sharp contrasts between this association and the sometimes adjacent *Festuca/Symphoricarpos* type: *A. tripartita*, *Eriogonum heracleoides*, *Chrysothamnus*, and *Tetradymia* are absent in the *Festuca/Symphoricarpos* type, and *Symphoricarpos* and *Rosa* are essentially absent in the *A. tripartita/Festuca* association.

A dense sward of grasses and grasslike plants (*Festuca idahoensis*, *Agropyron spicatum*, *Carex filifolia*, and *Poa sandbergii*), a variety of forbs (*Lupinus sericeus*, *Erigeron corymbosus*, *Achillea millefolium*), and a discontinuous layer of *Artemisia tripartita* (average cover 11 percent) characterize this association (table 32). The *A. tripartita* is relatively short (up to 5 dm., normally 1 to 2 dm.) making it much less conspicuous than *Artemisia tridentata*. Daubenmire (1970) does mention a clinal variation in species composition along an assumed moisture gradient.

A wide variety of soils is associated with the *A. tripartita/Festuca* association. These soils fall into the Haploxeroll, Argixeroll, and Palexeroll soil great groups (Brown, Chestnut, Chernozem, Prairie, and Planosol great soil groups). Rainfall is scarcely adequate to prevent formation of a calcium carbonate layer as indicated by increasing soil pH with depth.

Fire has relatively little influence on this association since *A. tripartita* sprouts. Grazing results in reductions in the *Agropyron* and *Festuca idahoensis*, then other perennial grasses, with eventual replacement by *Bromus tectorum*, *Festuca microstachys*, and *Plantago*. *Chrysothamnus* also increases, but *A. tripartita* appears little affected by grazing.

Purshia tridentata/Festuca idahoensis Association

The *Purshia/Festuca* association is far more extensive and important than its distribution as a zonal climax would indicate (fig. 153). From about 46°30' north latitude north along the eastern slopes of the Cascade Range, it is well represented as (1) a topographic climax in the *Artemisia/Agropyron* Zone and (2) an edaphic climax alternating with *Pinus ponderosa/Purshia* communities in the Cascade foothills.

Physiognomically, the *Purshia* gives this association the appearance of a shrub-steppe (fig. 160); however, the abundance of broad-leaved forbs also makes the term meadow-steppe appropriate. *Purshia tridentata* is the only tall shrub of consequence (average 14 percent cover) with *Eriogonum heracleoides* as the major low shrub. A dense herbaceous layer includes *Agropyron spicatum*, *Festuca idahoensis*, *Poa sandbergii*, and *Balsamorhiza sagittata* as dominants (table 32).

Figure 160.—Stand representative of the *Purshia tridentata/Festuca idahoensis* **association which is abundant in the eastern foothills of the northern Washington Cascade Range; this association has physiognomic characteristics of both shrub (***Purshia***) and meadow-steppe (abundance of broad-leaved forbs) (near Ellensburg, Washington).**

Soils associated with the *Purshia/Festuca* association are Typic Haploxerolls (Chestnuts) in the area where it is the zonal climax and Vitrandepts (Chernozem-Regosol intergrades and Regosols).

Purshia tridentata in this association is deciduous and nonsprouting. Consequently, fire produces a major physiognomic change by eliminating the *Purshia* which is killed outright. This produces a stand resembling an *Eriogonum/Festuca* or *Artemisia tripartita/ Festuca* type. Mice play an important role in reinvasion of *Purshia* into burned areas. *Purshia* is an important source of winter browse for deer and elk. Grazing of this association by domestic livestock reduces the *Agropyron* and *Festuca*, increases *Stipa occidentalis* var. *minor* and *Poa sandbergii* during early stages in the retrogression, and ultimately leads to dominance of *Bromus tectorum*.

Daubenmire (1970) recognizes a closely related *Purshia/Agropyron spicatum* community from which *Festuca idahoensis* is absent.

Other Zonal Associations

There remain three minor zonal associations of which one is a pure steppe type (*Agropyron spicatum/Poa sandbergii* association) and two are peripheral meadow-steppe types (fig. 153).

The *Agropyron/Poa* association is basically like the *Artemisia/Agropyron* except for the absence of *Artemisia tridentata* (table 32). *Agropyron spicatum*, *Poa sandbergii*, and scattered individuals of *Chrysothamnus nauseosus* constitute most of the community. Soils are Camborthids or Haploxerolls (Sierozem, Brown, and Chestnut). This association occurs in two locations: along the Snake River (fig. 153) and along the Columbia River east of the Columbia Gorge. As with related associations, *Agropyron* decreases with grazing and *Bromus tectorum* increases; sheep grazing may favor *Poa sandbergii* over the *Bromus*, however.

The *Festuca idahoensis/Rosa nutkana* association is very similar to the *Festuca/Symphoricarpos* type (table 32). *Symphoricarpos albus* is absent from the *Festuca/Rosa* type which is confined to the rain shadow of the Blue Mountains (fig. 153). The shrubby phase of this association is related to swales in the topography.

The *Festuca idahoensis-Hieracium cynoglossoides* association is poorly known. A meadow-steppe type, it lacks any distinguishing shrub. *Festuca idahoensis* and *Agropyron spicatum* dominate in combination with an abundance of perennial herbs. This association is found in the vicinity of Goldendale, Washington, and abuts *Quercus garryana* woodland in the foothills of the Cascade Range.

Associations on Specialized Habitats

Within the steppe region are a wide variety of habitats which have soils sufficiently unusual in physical or chemical properties to develop climax communities not assignable to any of the zonal associations. Some of these are associated with particular zones as defined by the nine zonal associations; many are not. Consequently, some of the associations typical of such sites are handled separately in this section.

Stipa comata Associations of Deep Soils Dominated by Gravel or Sand

A series of communities, generally dominated by *Stipa comata* and a shrub associate, are found on regosolic soils of the type being considered here. The communities are:

> *Artemisia tridentata/Stipa comata,*
> *Purshia tridentata/Stipa comata,*
> *Stipa comata-Poa sandbergii,* and
> *Artemisia tripartita/Stipa comata.*

These deep soils, dominated by gravels or sands, have low moisture-holding capacity; surprisingly, soils of strongly weathered volcanic ash appear to be ecologically equivalent despite their higher moisture-holding capacity as indicated by comparable communities. Because of this, Daubenmire (1970) suggests low fertility rather than moisture-holding capacity may be the key environmental factor differentiating the *Stipa* communities from the zonal associations. A comparison of soil properties under *Stipa* and *Agropyron* communities shows that moisture equivalent, cation exchange capacity, and exchangeable magnesium differ significantly between the two groups. It is notable that the shrub associates seem insensitive to environmental differences reflected so strongly in the herbaceous layer (i.e., between the *Agropyron* and *Stipa* groups).

The *Artemisia tridentata/Stipa comata* association is very similar to the *Artemisia tridentata/Agropyron spicatum* physiognomically and floristically except for the substitution of *Stipa* for *Agropyron*. It is found throughout the *A. tridentata/Agropyron* and *A. tridentata/Festuca* Zones. Soils include Haploxerolls, Xeropsamments, Xerorthents, Camborthids, and Vitrandepts (Sierozem, Brown, and Chestnut-Regosol and Brown-Regosol intergrades).

The *Purshia tridenta/Stipa comata* association intergrades extensively with the *Artemisia tridentata/Stipa comata* association differing only in the substitution of *Purshia* for *Artemisia*. In fact, the *Purshia/Stipa* can be seral to the *Artemisia/Stipa* association. It is found in the *Artemisia tridentata/Agropyron* Zone on Camborthids, Xeropsamments, and Vitrandepts (Sierozem, Regosol, and Brown-Regosol intergrades).

224

The *Stipa comata-Poa sandbergii* association is dominated by these two herbs with widely scattered *Chrysothamnus nauseosus* as the only shrub. It is found within the *Agropyron-Poa* and *Agropyron-Festuca* Zones on sandy soils (Camborthids, Haploxerolls, and Torriorthents or Sierozems and Brown soils). An *Eriogonum niveum* phase of this association (*Eriogonum* with 6- to 10-percent cover) is recognized which occurs in the *Artemisia tridentata/Agropyron* and *Artemisia tripartita/Festuca* Zones.

Associations on Shallow Soils

Lithosolic sites are those where soils are stony and extremely shallow to bedrock.[3] These sites provide an extremely rigorous plant environment, with heat and drought in the summer and intense frost action (the result of excess water) during the winter. The lithosolic series of associations have in common a carpet of *Poa sandbergii*, a crust of mosses and lichens, and a lithosolic substrate (fig. 161). Nearly all associations have a taller layer of shrubs, but the species vary—*Eriogonum niveum, E. sphaerocephalum, E. douglasii, E. compositum, E. thymoides, E. microthecum,* or *Artemisia rigida.* Many stands are outstandingly dominated by a single shrub species, but others have dominance divided among several (table 33). Several plant groups—*Allium, Eriogonum, Lomatium,* Cruciferae—have their best steppe representation in the lithosolic associations.

Daubenmire (1970) recognizes a whole series of associations based on shifts in the dominance of the shrubs (table 33) but cannot satisfactorily relate their occurrence to soil factors or macroclimate; often stands of several associations are in close proximity. He suggests that variation in the fracture system of the underlying basalt may hold the key to explaining variation in shrub dominants.

Figure 161.—Lithosolic sites in the Columbia basin steppe region are occupied by a series of associations of which *Eriogonum niveum/Poa sandbergii* is one (*photo courtesy Range Management, Oregon State University*).

[3] Daubenmire's (1970) investigations suggest that soil depth is not critical (i.e., soils are not "shallow") until the mean depth of soil declines to 10 to 30 cm.

Table 33. — Canopy cover of principal low shrubs, *Poa sandbergii* and *Agropyron spicatum*, in stands belonging to various lithosolic associations in the Columbia basin; stands interpreted as intergrades are not named (from Daubenmire 1970)

Association	Species[1]									
	Posa	Arri	Erth	Hast	Phho	Ersp	Erdo	Erni	Ermi	Agsp
	----------------------Percent----------------									
Artemisia rigida/Poa sandbergii	33	29	—	—	—	—	—	—	—	—
	13	27	—	—	—	—	—	—	—	—
	32	25	—	—	—	—	—	—	—	3
	31	14	—	tr	—	—	—	—	—	—
	31	10	—	—	—	—	—	—	—	—
	2	34	—	—	—	—	—	tr	—	—
	34	10	—	—	tr	—	tr	—	—	—
Intergrades	20	12	5	—	—	—	—	—	—	—
	36	19	4	—	—	tr	—	tr	—	—
	26	13	tr[2]	1	2	—	—	—	—	—
	2	6	2	tr	1	—	tr	—	—	3
	26	17	1	6	tr	3	—	—	—	—
	14	9	tr	9	2	tr	—	—	—	—
Eriogonum thymoides/Poa sandbergii	9	—	4	4	tr	—	—	—	—	—
	36	—	3	3	tr	3	—	—	—	—
	29	—	14	4	—	—	6	1	—	—
	38	—	1	1	—	—	—	—	—	—
	21	—	3	8	—	—	—	—	—	—
	9	—	—	tr	6	—	—	—	—	—
Eriogonum sphaerocephalum/Poa sandbergii	28	—	—	6	2	8	—	—	—	—
	22	—	—	2	—	4	—	—	—	11
	58	—	tr	—	—	13	—	—	—	—
Eriogonum douglasii/Poa sandbergii	25	—	—	—	—	—	6	—	—	—
	23	—	—	—	10	—	23	—	—	—
Eriogonum niveum/Poa sandbergii	37	—	—	—	—	—	—	tr	—	—
	24	—	—	—	—	—	—	2	—	—
	26	—	—	—	—	—	—	7	—	tr
	34	—	—	—	—	—	—	5	—	tr
	26	—	—	—	—	—	—	1	—	tr
	36	—	—	—	—	—	—	14	—	—
Eriogonum microthecum-Physaria oregana	—	—	—	—	—	—	—	tr	17	2
	2	—	—	—	—	—	—	tr	14	tr
	—	—	—	—	—	—	—	tr	15	—
	—	—	—	—	—	—	—	12	12	—
Agropyron spicatum-Poa sandbergii (lithosolic phase)	18	—	—	—	—	—	—	tr	—	70
	5	—	—	—	—	—	—	—	—	47
	29	—	—	—	—	—	—	—	—	69
	29	—	—	—	—	—	—	—	—	39

[1] Species abbreviations: Posa = *Poa sandbergii*; Arri = *Artemisia rigida*; Erth = *Eriogonum thymoides*; Hast = *Haplopappus stenophyllus*; Phho = *Phlox hoodii*; Ersp = *Eriogonum sphaerocephalum*; Erdo = *Eriogonum douglasii*; Erni = *Eriogonum niveum*; Ermi = *Eriogonum microthecum*; Agsp = *Agropyron spicatum*.

[2] tr indicates trace amounts (less than 0.5 percent cover).

These lithosolic associations are less subject to modification by grazing than any other of the steppe associations because of the low stature of the vegetation and stony substrate.

The *Artemisia rigida/Poa sandbergii* association is the most widespread member of this series. It occurs on basalt throughout the Columbia basin steppe and into the forests of the Cascade Range and Blue Mountains as far as the *Pseudotsuga menziesii* Zone. Sharp ecotones are encountered between this association and the *Eriogonum niveum/Poa sandbergii* type.

Noteworthy features of the remaining lithosolic associations are: (1) the tendency of *Eriogonum niveum* to invade recently bared gravelly soils and to occur mainly in pure stands rather than in mixture with other low shrubs; (2) the confinement of the *Eriogonum douglasii/Poa* association to the eastern flank of the Cascade Range; (3) the showy nature of the *Eriogonum thymoides/Poa sandbergii* association in the spring due to a large number of plants with conspicuous flowers; and (4) the distinctive substrate (coarse, unstable material from hydrothermally altered basalt) and group of character species (*Eriogonum microthecum laxiflorum*, *Physaria oregana*, and *Mentzelia laevicaulis parviflora*) of the *Eriogonum microthecum/Physaria oregana* association as well as the absence of *Poa sandbergii* and a moss-lichen crust.

Distichlis stricta Associations on Saline-Alkali Soils

A carpet of *Distichlis stricta* links the associations found on saline-alkali soils which are high in salts, sodium ion, and pH:

Distichlis stricta,

Elymus cinereus-Distichlis stricta, and *Sarcobatus vermiculatus/Distichlis stricta* associations.

Distichlis is a low (herbage less than 20 cm. high), perennial, strongly rhizomatous grass. Superimposed on this sward may be an open stand of either *Sarcobatus* (a deciduous, succulent-leaved shrub) or *Elymus cinereus* (a large, coarse bunchgrass). A notable common feature of these associations is the bareness of the fine-textured soils between the vascular plants, i.e., the absence of a surface crust of mosses and lichens.

Vegetation has almost always been disturbed by grazing because green foliage is available all through the summer. Heavy grazing leads to increases in annuals (*Lepidium perfoliatum*, *Bromus tectorum*, and *Bassia hyssopifolia*), although *Distichlis* is highly tolerant of grazing.

A generalized and zonational pattern around a playa would show *Distichlis* occupying the lowest lying central areas and either the *Elymus/Distichlis* or *Sarcobatus/Distichlis* types forming an outer ring. *Distichlis* and *Sarcobatus* appear equally tolerant of high salinity and pH, both being superior to *Elymus* in this regard. The pure *Distichlis* sward in playa centers may relate to a greater tolerance of poor soil aeration than either of its associates.

Crataegus douglasii Associations and Related Riparian Types

On moister sites within the *Festuca/Symphoricarpos* and *Festuca/Rosa* Zones, the *Crataegus douglasii-Symphoricarpos albus* and *Crataegus douglasii/Heracleum lanatum* associations may occur. Both have a nearly complete woody plant cover 5 to 7 meters tall which is dominated by *Crataegus douglasii*. The *Crataegus-Symphoricarpos* association is typically found in moist draws and includes *Symphoricarpos albus* and many herbs common to the *Festuca/Symphoricarpos* association. Some examples are *Galium boreale*, *Geranium viscosissimum*, *Iris missouriensis*, and *Potentilla gracilis*. *Spiraea betulifolia*, *Crataegus columbiana*, *Prunus virginiana*, and *Amelanchier alnifolia* are often present.

The *Crataegus/Heracleum* association is found on aggraded valley floors. Understory dominants are *Heracleum lanatum*, *Hydrophyllum fendleri* var. *albifrons*, and *Urtica dioica*, singly or collectively. Only *Lomatium dissectum* is shared with the *Festuca/Symphoricarpos* or *Festuca/Rosa* associations. However, many elements more common in forested zones may be present, such as *Circaea alpina*, *Elymus glaucus*, *Geum macrophyllum*, and *Pteridium aquilinum*.

Both *Crataegus* associations have been profoundly affected by human activities. Cattle heavily graze the major shrubs and herbs, including any *Crataegus* foliage within reach. The native understory is replaced by a *Poa pratensis-Poa compressa* sward which incorporates exotic forbs such as *Cirsium vulgare*, *Dipsacus sylvestris*, and *Taraxacum officinale*. The *Crataegus* itself is killed by herbicides. As a consequence, little remains of these types.

Populus tremuloides phases of these associations, in which *Populus* and *Crataegus* alternate their dominance, are recognized. *Populus* grows through the *Crataegus* canopy and overtops it, resulting in reduced vigor of *Crataegus*. The short-lived *Populus* eventually dies back releasing the *Crataegus*, and new *Populus* sprouts start the cycle over again.

Two related riparian types briefly described by Daubenmire (1970) are: the *Populus trichocarpa/Cicuta douglasii* association which replaces the *Crataegus/Heracleum* type in drier portions of the steppe; and the *Alnus rhombifolia* forests which occur in some riparian habitats, sometimes in association with *Populus trichocarpa*.

Shrub/*Poa sandbergii* Associations

Within the lowest, driest part of the *Artemisia/Agropyron* Zone is a series of three associations found on reasonably deep, loamy soils which are drier than those on associated zonal habitats. These are the *Artemisia tridentata/Poa sandbergii*, *Grayia spinosa/Poa sandbergii*, and *Eurotia lanata/Poa sandbergii* associations. These associations lack a large perennial grass; and, in fact, the species used in the names are often nearly the only vascular plants present. These communities are associated with highly calcareous regosolic soils.

Artemisia tridentata has a higher density (average cover, 24 percent; range, 8 to 35 percent) in the *Artemisia/Poa* association than in any other type of undisturbed vegetation in the Columbia basin (fig. 162). *Poa sandbergii* is the only significant herb (average cover, 47 percent; range, 41 to 51 percent), and only *Bromus tectorum*, *Descurainia pinnata*, and *Microsteris gracilis* have high constancy. Daubenmire (1970) has substantial evidence that this community is not a consequence of grazing (i.e., elimination of larger grasses and forbs from *Artemisia/Agropyron* communities). Soils are Mollic Camborthids (Sierozems).

The *Grayia spinosa/Poa sandbergii* association is very similar except that the *Artemisia* is replaced by *Grayia*, a highly palatable shrub.

The *Eurotia lanata/Poa sandbergii* association is dominated by *Eurotia*, a white-woolly shrub about 3 decimeters tall (fig. 163). *Poa sandbergii* and *Bromus tectorum* are the only significant associates. The *Bromus* reflects the heavy grazing which has occurred in this community because of year-round availability of highly palatable forage (the *Eurotia* leaves and twigs). Ecotones with adjacent communities are usually sharp (fig. 163), but the reasons are not clear.

Sporobolus cryptandrus and *Aristida longiseta* Associations

Sporobolus cryptandrus and *Aristida longiseta* each dominate a grassland community with *Poa sandbergii* as the major associate. Daubenmire (1970) notes that both of these

Figure 162.—Stand of *Artemisia tridentata/Poa sandbergii* **association;** *Artemisia* **coverage is high, and larger perennial grasses and forbs are essentially absent in this association which is found only in the warmest, driest parts of the Columbia basin (Rattlesnake Hills Research Natural Area, Washington).**

Figure 163.—Mosaic of *Eurotia lanata/Poa sandbergii* **and** *Artemisia tridentata/Poa sandbergii* **stands; in this area, the** *Eurotia* **occupies convex surfaces (shoulders of small ridges) with** *Artemisia* **occupying intervening areas (Rattlesnake Hills Research Natural Area, Washington).**

grasses may be seral to *Agropyron spicatum* on some habitats but concludes that because of undetermined soil abnormalities, each is also dominant in an edaphic climax community.

The *Sporobolus/Poa* association is generally found along the Snake River. These two species form nearly all the perennial grasses, and widely scattered *Chrysothamnus nauseosus*, the only shrub. *Sporobolus* develops later in the spring than the other grasses and, for that reason, is heavily grazed by cattle and horses.

The *Aristida/Poa* association also is found mainly along the Snake River and its tributaries. Again *Aristida* and *Poa* are the major perennial grasses, and widely scattered *Chrysothamnus* is the only shrub.

Plant Associations on Colluvium, Alluvium, and Talus

Three associations typified by *Rhus glabra* are found on colluvial or alluvial soils in canyons in the *Agropyron-Poa* and *Artemisia/Agropyron* Zones. Most *Rhus glabra* stands are heavily grazed, and *Bromus tectorum* and annuals such as *Erodium cicutarium* usually are understory dominants. When a dominant perennial grass is present, it can be *Agropyron spicatum*, *Sporobolus cryptandrus*, or *Aristida longiseta*. Daubenmire (1970) therefore recognizes a series of three hypothetical climaxes based on the *Rhus* and each grass but concludes that grazing has effectively reduced them to a *Rhus glabra/Bromus tectorum* zootic climax. He also concludes that *Rhus* is neither an increaser (or invader) or decreaser with grazing.

A *Celtis douglasii/Bromus tectorum* zootic climax typifies colluvial cones and aprons in the *Agropyron-Poa* Zone (along the Snake River). A *Grayia spinosa/Bromus tectorum* community can be found on talus in some areas. Shrub "garlands" of *Philadelphus lewisii*, *Prunus virginiana* var. *melanocarpa*, *Amelanchier alnifolia*, and *Rosa* spp. are typical of talus margins in the interior steppe zones (fig. 164).

Figure 164.—Shrub garlands of *Philadelphus lewisii, Prunus virginiana,* **and** *Amelanchier alnifolia* **are typically found around talus in the interior of the Columbia basin** (*photo courtesy R. Daubenmire*).

Vegetation on Sand Dunes

Sand dunes are a common phenomenon near the Columbia River in the lowest part of the Columbia Basin Province. *Psoralea lanceolata* is found on active windward slopes of some of these dunes. *Elymus flavescens* can be found on high dune summits, and *Rumex venosus* or *Agropyron dasystachyum* on slip faces of active dunes. Stabilized dune surfaces can have communities of *Chrysothamnus nauseosus* and *C. viscidiflorus, Agropyron dasystachyum, Oryzopsis hymenoides, Koeleria cristata, Poa sandbergii, Achillea millefolium* var. *lanulosa, Microsteris gracilis, Descurainia pinnata,* and *Holosteum umbellatum* (fig. 165). The *Artemisia tridentata/Stipa comata* association has been noted on very old dune surfaces. *Juniperus scopulorum* savanna is found on some of the sand dunes.

Figure 165.—Stabilized dune community dominated by *Oryzopsis hymenoides* **and** *Hymenopappus filifolius* **with a variety of associates (near Vernita Bridge, Washington).**

Pond Vegetation

Numerous ponds are encountered in the Columbia basin, particularly in the Channeled Scablands. Harris (1954) described vegetational sequences in some small intradune ponds:

(1) *Spartina gracilis*, the highest zone of hydrophytic vegetation,

(2) *Distichlis stricta-Scirpus nevadensis*, barely above the water surface,

(3) *Juncus balticus-Carex douglasii-Scirpus americanus-Eleocharis palustris*, in water 0 to 3 decimeters deep,

(4) *Scirpus acutus-Typha latifolia*, in water 3 to 9 decimeters deep, and

(5) *Potamogeton pectinatus-Ceratophyllum demersum-Myriophyllum spicatum*, in deep water.

Sequences 2 and 3 formed extensive marshes in some areas.

Zootic Climaxes and *Artemisia* Eradication

At the beginning of this chapter, we mentioned the peculiar roles of *Bromus tectorum* and *Poa pratensis* in the steppe vegetation of the Pacific Northwest. We would like to review this further.

The normal situation after disturbances is the successional return to a given climax vegetation when disturbance (e.g., grazing) is eliminated. In the steppes of the Northwest, however, the introduction of two aggressive widely adapted aliens have altered the natural successional patterns.

Bromus tectorum is the most important invading species in the drier zones. This winter annual, which is usually 2 to 3 decimeters tall (maximum 1 meter), was introduced in Washington about 1890. Since that time, it has become ubiquitous and is the most common invader of excessively grazed or abandoned farmland except in the *Festuca/Symphoricarpos* and *Festuca/Rosa* Zones. On abandoned fields, a brief 1- to 2-year stage of dominance

231

by *Salsola kali* or *Sisymbrium altissimum* is followed by permanent *Bromus tectorum* dominance. Other annuals and *Chrysothamnus nauseosus* may be associated with the *Bromus*; there are suggestions that *Elymus caput-medusae* may prove a superior competitor on at least some sites. And several other alien *Bromus* spp. may compete well with *Bromus tectorum* in moister portions of the steppe. Daubenmire (1970) also points out that high coverage of *Chrysothamnus* may represent a further stage in site degradation, since animals will graze *Bromus*, especially in the spring, but avoid *Chrysothamnus*.

In most cases, however, the perennial grasses and forbs are increasingly replaced by *Bromus tectorum*, and there is little evidence that it will relinquish a site once occupied. Daubenmire (1970) notes a *Bromus* stand surrounded by *Agropyron-Poa* which was created by cultivation 50 years ago, only lightly grazed since, and is still covered by *Bromus*. This competitive ability seems to be related to the early spring development of *Bromus* which takes full advantage of the warm spring and early summer when soils are still moist; in laboratory tests, Harris and Wilson (1970) found *Bromus* in flower while *Agropyron* plants were still vegetative. The long root system of *Bromus*, developed through the winter months, is also an element in competition; Harris and Wilson (1970) report maximum root depth penetrations of 97 centimeters for *Bromus*, 74 centimeters for *Elymus caput-medusae*, and 56 centimeters for *Agropyron*.

The consequence of all this is that communities dominated by *Bromus tectorum* are a permanent and widespread feature of the landscape. Where a major shrub dominant has not been eliminated by cultivation, fire, or grazing, a shrub/*Bromus* community may result (fig. 166); otherwise, essentially pure *Bromus tectorum* stands may be encountered (note the exception for *Chrysothamnus* mentioned above), regardless of wide variations in environmental conditions so clearly indicated by the original climax communities.

Daubenmire (1970) points out that *Bromus*, in small quantities, must also be considered an element of the climax vegetation on undisturbed sites. Population densities are low, and plants are dwarfed, but they are present even in the most inaccessible tracts of steppe.

Poa pratensis plays a similar role in the moistest steppe zones, the *Festuca/Symphoricarpos* and *Festuca/Rosa*. Although *Poa* is grazed, it is very tolerant of grazing; and as the native perennial grasses are gradually reduced, *Poa* takes possession of the space and

Figure 166.—Heavy grazing in drier portions of the Columbia basin has produced herbaceous layers dominated by *Bromus tectorum* over extensive steppe areas; fire in the *Artemisia tridentata/Bromus* community would eliminate the *Artemisia*, leaving the *Bromus* as the sole dominant (near Vernita Bridge, Washington).

retains it. A nearly pure sward of *Poa* can result, and even alien *Bromus* spp. (*B. brizaeformis*, *B. japonicus*, *B. mollis*, and *B. tectorum*) which colonize bare soil ultimately give way to the *Poa*.

One other major human disturbance should be considered—the elimination of *Artemisia tridentata* by herbicide spraying. Such activities are frequently proposed by ranchers as a means to increase grass production since *Artemisia* is not valuable to domestic livestock. Daubenmire (1970) has pointed out several key points which raise doubts as to the desirability or efficacy of this practice:

1. There is no evidence that increases in grass production will be sustained—effects of green manure or fire or both may be responsible for much of the short-term gain which has been measured.

2. *Artemisia* protects the perennial grasses and may be the key reason why as much grass remains today as does occur on depleted rangelands.

3. In the summer, *Artemisia* uses only water which has percolated through the upper soil profile, out of the reach of grass roots. Removing the shrub would suspend activity for several months when herbs have aestivated, reducing the total productivity, leaving a moisture resource unused and reducing the extent of the soil profile involved in nutrient cycling if not the amount of cycling itself.

4. *Artemisia* provides valuable habitat for birds.

5. Herbicidal spraying may eliminate or damage the perennial broad-leaved forbs, some of which are valuable forage.

6. *Artemisia* often promotes uniform accumulation of snow and delays its melting.

It appears that elimination of *Artemisia* is focused only on the production of perennial grass primarily valuable for one type of animal (cattle) and has many undesirable side effects for the system as well as reducing future management options.

It seems clear to us that the introduction of domestic livestock, cultivation, wildfire, and herbicides has and will continue to change the diverse steppe landscape to a much simpler array of communities. These communities thoroughly obscure much of the environmental diversity apparent in the original climax vegetation.

CHAPTER IX. STEPPE AND SHRUB-STEPPE OF CENTRAL AND SOUTHEASTERN OREGON

Steppe and shrub-steppe communities are widespread in Oregon, dominating the entire southeastern quarter of the State and a strip around the northwestern and western margins of the Blue Mountains. The communities found in the latter area, near the Columbia River, are very similar to those described in Chapter VIII on the Columbia Basin Province (Poulton 1955, Anderson 1956). Therefore, we will focus this chapter on the shrub-steppe of southeastern Oregon; i.e., the High Lava Plains, Basin and Range, and Owyhee Upland Provinces (fig. 2).

In some respects, the steppes of southeastern Oregon are similar to those of the Columbia Basin Province: (1) both areas have hot, dry summers and cold winters (table 30); and (2) there are similar communities, e.g., the *Artemisia tridentata/Agropyron spicatum* and *Artemisia tridentata/Festuca idahoensis* associations. Contrasts between the two areas are notable, however: (1) southeastern Oregon shrub-steppes average much higher in elevation; (2) deep, loamy soils are not common in southeastern Oregon; (3) desert or salt desert shrub communities are common enough to appear on regional vegetation maps; (4) *Juniperus occidentalis* and *Cercocarpus ledifolius* occur in association with shrub-steppe; and (5) meadow-steppes of sod-forming grasses and dicotyledonous herbs, which ring much of the Columbia Basin Province, are nearly absent. Hence, although the physiognomy and dominants of many communities are often similar, the environmental and community mosaics are different.

The steppe and shrub-steppe communities of southeastern Oregon have not been comprehensively treated. Nevertheless, from the generalized accounts of Poulton (1962), Anderson (1956), and Shantz and Zon (1924) and detailed but localized work by Eckert (1957), Dean (1960), Tueller (1962), Culver (1964), Dealy (1971), and McKell (1956), many of the important community types in the vegetational mosaic emerge.

Artemisia Communities

Artemisia communities dominate nearly every vegetational mosaic in southeastern Oregon's shrub-steppe (fig. 167). There are four major *Artemisia* species, each of which characterizes particular habitats:

Artemisia tridentata on the deeper soils,
Artemisia arbuscula on shallow, stony soils,
Artemisia rigida on very shallow soils (Lithosols), and
Artemisia cana on moister habitats.

Associations of *Artemisia tridentata, A. arbuscula,* and *A. rigida* have been described in detail (Eckert 1957, Culver 1964, Hall 1967, Dealy 1971); lists of characteristic species for six of these associations are provided in table 34.

Table 34. — Characteristic species for *Artemisia* associations found in the High Lava Plains (Eckert 1957), Owyhee Upland (Culver 1964), and southern Blue Mountains (Hall 1967) Provinces

Association	Species group	High Lava Plains	Owyhee Upland	Southern Blue Mountains
Artemisia tridentata/ Agropyron spicatum	Shrubs Grasses Herbs	*Artemisia tridentata* *Agropyron spicatum* *Poa sandbergii* *Phlox diffusa* *Aster scopulorum* *A. canescens* *Chaenactis douglasii* *Collinsia parviflora* *Phlox gracilis* *Lappula redowskii* *Gayophytum ramosissimum*	*Artemisia tridentata* *Agropyron spicatum* *Poa sandbergii* *Lupinus sericeus* *Lomatium triternatum* *L. macrocarpum* *Zigadenus paniculatus* *Microseris troximoides* *Astragalus filipes* *A. lentiginosus*	
Artemisia tridentata/ Festuca idahoensis	Shrubs Grasses Herbs	*Artemisia tridentata* *Chrysothamnus viscidiflorus* *Symphoricarpos rotundifolius* *Ribes cereum* *Juniperus occidentalis* *Festuca idahoensis* *Agropyron spicatum* *Poa sandbergii* *Koeleria cristata* *Phlox diffusa* *Antennaria corymbosa* *Calochortus nitidus*	*Artemisia tridentata* *Chrysothamnus viscidiflorus* *Festuca idahoensis* *Agropyron spicatum* *Poa sandbergii* *Sitanion hystrix* *Bromus tectorum* *Elymus cinereus* *Balsamorhiza sagittata*	
Artemisia tridentata/ Elymus cinereus	Shrubs Grasses Herbs	(mentioned but not described)	*Artemisia tridentata* *Elymus cinereus* *Poa sandbergii* *Agropyron spicatum* *Bromus tectorum* *Penstemon speciosus* *P. cusickii* *Thlaspi arvense* *Eriogonum umbellatum*	
Artemisia arbuscula/ Agropyron spicatum	Shrubs Grasses Herbs	*Artemisia arbuscula* *Eriogonum sphaerocephalum* *Juniperus occidentalis* *Agropyron spicatum* *Poa sandbergii* *Festuca idahoensis* *Phlox diffusa* *Erigeron linearis* *Collinsia parviflora*	*Artemisia arbuscula* *Agropyron spicatum* *Poa sandbergii* *Sitanion hystrix* *Bromus tectorum* *Penstemon aridus* *Lagophylla ramosissima*	*Artemisia arbuscula* *Purshia tridentata* *Agropyron spicatum* *Poa sandbergii* *Sitanion hystrix* *Trifolium macrocephalum*
Artemisia arbuscula/ Festuca idahoensis	Shrubs Grasses Herbs	*Artemisia arbuscula* *Juniperus occidentalis* *Festuca idahoensis* *Agropyron spicatum* *Poa sandbergii* *Phlox diffusa* *P. hoodii* *P. longifolia* *Microseris troximoides* *Antennaria dimorpha* *Astragalus stenophyllus* *Lupinus saxosus* *Trifolium gymnocarpon* *T. macrocephalum*	*Artemisia arbuscula* *Festuca idahoensis* *Agropyron spicatum* *Poa sandbergii* *Arabis holboellii* *Phlox diffusa* *Erigeron linearis* *Astragalus miser* *Balsamorhiza hookeri* *Agoseris heterophylla* *Achillea millefolium* *Haplopappus stenophyllus*	*Artemisia arbuscula* *Festuca idahoensis* *Agropyron spicatum* *Poa sandbergii* *Phlox douglasii* *Balsamorhiza serrata*
Artemisia rigida/ Poa sandbergii	Shrubs Grasses Herbs		*Artemisia rigida* *Poa sandbergii* *Bromus tectorum* *Festuca microstachys* *Agropyron spicatum* *Sitanion hystrix* *Mimulus nanus* *Zigadenus paniculatus*	*Artemisia rigida* *Poa sandbergii* *Sitanion hystrix* *Phlox douglasii*

235

Figure 167. — *Artemisia* communities dominate the shrub-steppe of southeastern Oregon; in this High Lava Plains Province landscape, *Artemisia tridentata* communities occur on the hills, *Artemisia arbuscula* communities on the shallow soils of basalt flows, and *Artemisia cana* communities in seasonally ponded valley bottoms *(photo courtesy Range Management, Oregon State University).*

Artemisia tridentata Associations

Artemisia tridentata/Agropyron spicatum Association

The *Artemisia tridentata/Agropyron spicatum* association is very similar to the Columbia Basin Province association of the same name (fig. 168); it is the most widespread association in southeastern Oregon. Eckert (1957) and Culver (1964) recognized the *Artemisia tridentata/Agropyron* association as the climatic climax in their areas (High Lava Plains and Owyhee Upland Provinces, respectively). Eckert (1957) found it a highly variable community in which *Artemisia tridentata* was the only important shrub (average coverage 10 percent). *Agropyron spicatum* was the dominant grass and *Poa sandbergii* the typical associate in both localities. The component of perennial and annual herbs differed considerably between the High Lava Plains and Owyhee Upland Provinces, however (table 34). Eckert (1957) noted an abundance of annual herbs. *Tortula ruralis* was a conspicuous mass in the leaf-fall area of the shrubs. *Festuca idahoensis* and *Stipa thurberiana* are typically minor elements, but *Festuca* and *Stipa* phases (as well as *Purshia tridentata* phases) of the association have been described or listed (Culver 1964, Eckert 1957). The *Festuca* phase occupies more mesic sites and the *Stipa* phase more xeric sites than the typical association.

Tueller (1962) compared grazed and ungrazed *Artemisia tridentata/Agropyron spicatum* stands along fence lines and in enclosures. The important perennial grasses—*Agropyron spicatum*, *Festuca idahoensis*, and *Stipa thurberiana*—decreased with grazing (fig. 169). *Poa sandbergii* and many of the herbs increased with grazing. *Artemisia tridentata* was indeterminant; significant changes in its status with grazing were not found. *Chrysothamnus* spp. varied in response, increasing in some cases and appearing indeterminant in others. *Bromus tectorum* or *Chrysothamnus* spp., or both, dominate many severely overgrazed areas as well as abandoned farmland (fig. 170).

Soils associated with the *Artemisia/Agropyron* association were Haploxerolls (Brown great soil group), and were usually moderately deep (60 to 80 cm. to cemented layers or bedrock), and usually had well-developed, clay, B horizons.

Figure 168.—The *Artemisia tridentata/Agropyron spicatum* community is probably the most widespread single type in southeastern Oregon and contains the same dominants as in the Columbia basin; the range pole is 2½ feet tall and marked in 6-inch segments *(photo courtesy Range Management, Oregon State University).*

Figure 169.—Larger perennial grasses have been found to decrease under grazing pressure, although *Poa sandbergii* may increase, and *Artemisia tridentata* appears little affected; left, the area to the left of the fence (lacking *Agropyron spicatum*) has been continuously and heavily grazed, and the area to the right has been protected for 8 years; right, an *Artemisia tridentata/Poa sandbergii* community which has developed under heavy cattle grazing *(photos courtesy Range Management, Oregon State University).*

237

Artemisia tridentata/Festuca idahoensis Association

The *Artemisia tridentata/Festuca idahoensis* association is a topographic or topoedaphic climax on sites more mesic than those occupied by the *Artemisia tridentata/Agropyron spicatum* association (Culver 1964, Eckert 1957). *Artemisia tridentata* is the dominant shrub, although *Chrysothamnus viscidiflorus* is an important minor component. *Festuca idahoensis*, *Poa sandbergii*, and *Agropyron spicatum* are dominant grasses. The grass and perennial herb associates differ in the two study areas (table 34). The more mesic nature of Eckert's stands is indicated by occurrence of rhizomatous *Agropyron spicatum*, *Symphoricarpos rotundifolius*, *Ribes cereum*, and *Juniperus occidentalis* as typical constituents.

In the Owyhee Upland Province, Culver (1964) found the *Artemisia tridentata/Festuca idahoensis* association on Brown soils. In the High Lava Plains Province (Eckert 1957), it was the only association found on Chestnut as well as Brown soils. Both these old great soil groups are, of course, within the Haploxeroll soil great group within the new classification. In any case, these soils had higher moisture storage capacities than those supporting adjacent *Artemisia tridentata/Agropyron spicatum* communities.

Artemisia tridentata-Purshia tridentata/Festuca idahoensis Association

An *Artemisia tridentata-Purshia tridentata/Festuca idahoensis* association is a dominant shrub-steppe community in the only portion of the Basin and Range Province which has been studied (Dealy 1971) rather than the *Artemisia/Agropyron* and *Artemisia/Festuca* types already mentioned. Although *Artemisia-Purshia* communities have been identified in the High Lava Plains and Blue Mountain Provinces (Eckert 1957, Tueller 1962), they apparently do not achieve the prominence found in the Silver Lake area.

The *Artemisia-Purshia/Festuca* association is located just below the edge of the *Pinus ponderosa* and *Cercocarpus ledifolius* woodlands (Dealy 1971). *Artemisia tridentata* and *Festuca idahoensis* dominate the shrub layer with a crown cover ratio of 2 to 1 (12 and 7 percent, respectively). *Chrysothamnus viscidiflorus* is consistently present in small amounts. Important grasses are *Festuca idahoensis, Sitanion hystrix,* and *Bromus tectorum,* all with approximately equal coverage. Additional consistent grass and grasslike plants are *Carex rossii, Stipa occidentalis, Agropyron spicatum, Poa sandbergii,* and *Koeleria cristata.* Forbs are of relatively minor importance, composing less than 1 percent of the community in the aggregate; *Lupinus* and *Astragalus* spp. are the most consistent perennials, and *Collinsia parviflora, Cryptantha ambigua,* and *Epilobium minutum* are the consistent annuals.

Interestingly, the soil found under this association is a well-drained sandy loam derived primarily from pumice (Dealy 1971) rather than the finer textured soils typically found under many *Artemisia tridentata* communities.

Artemisia tridentata/Elymus cinereus Association

Culver (1964) describes and Eckert (1957) mentions an *Artemisia tridentata/Elymus cinereus* association which occurred primarily on moist alluvial bottom lands. Culver (1964) and Anderson (1956) mention occurrence of such a type on uplands as well. *Artemisia tridentata* and *Chrysothamnus viscidiflorus* are the characteristic shrubs. *Elymus cinereus* is always conspicuous and sometimes dominates the ground layer. Culver (1964) mentions several herbs which were restricted to this association in the Owyhee Upland Province (table 34).

Artemisia arbuscula Associations

Artemisia arbuscula has a much lower growth form (1 to 4 dm.) than *A. tridentata* and also dominates large portions of the eastern Oregon shrub-steppe. Two widespread *Artemisia arbuscula* associations have been recognized: *Artemisia arbuscula/Agropyron spicatum* and *A. arbuscula/Festuca idahoensis* (fig. 171) (Eckert 1957, Culver 1964, Hall 1967, Dealy 1971). *Artemisia arbuscula, Agropyron spicatum,* and *Poa sandbergii* are the major species in the *Artemisia arbuscula/Agropyron spicatum* association. A variety of associates is present depending upon locale (table 34). *Artemisia arbuscula, Festuca idahoensis, Poa sandbergii, Agropyron spicatum,* and a variety of perennial herbs distinguish the *Artemisia arbuscula/Festuca idahoensis* association (table 34).

These associations typically occur with *Artemisia tridentata* communities as edaphic climaxes on shallow, stony phases of the zonal Haploxerolls (Brown soils). The *Agropyron* type occupies more xeric sites than the *Festuca* type in the steppe area (Culver 1964, Eckert 1957). Where they occur within a forest mosaic (Ochoco Mountains), the *Agropyron* type is on about 35 centimeters of clayey soil derived from basic igneous rock and the *Festuca* type is on comparable soils from acid igneous rock (Hall 1967). The *Artemisia arbuscula* communities are common as openings or parks within the lower forest zones throughout much of the central and southeastern Oregon area (fig. 172).

Dealy (1971) describes an interesting mosaic of communities in the rocky "scab flats" of the Silver Lake area. Three *Artemisia arbuscula* communities and a *Cercocarpus* woodland are involved: *A. arbuscula/Festuca idahoensis, A. arbuscula/Koeleria cristata, A. arbuscula/Danthonia unispicata,* and *Cercocarpus ledifolius/Festuca idahoensis-Agropyron spicatum.* The *Cercocarpus* woodland occurs around the edges of the scab flats between the *A. arbuscula* communities and the forest. *A. arbuscula* is the only shrub species in the three sage communities, forming a canopy of about 20 centimeters above a mixture of herbaceous species. Compositional data for each *A. arbuscula* community are in table 35; canopy cover of the *Artemisia* is 14, 8, and 26 percent in the *Festuca, Danthonia,* and *Koeleria* communities, respectively.

Figure 171.—*Artemisia arbuscula* **typically dominates communities on shallower, stonier soils than those found under** *Artemisia tridentata*; **this is an** *Artemisia arbuscula/ Festuca idahoensis* **community in the High Lava Plains Province of central Oregon** (*photo courtesy Range Management, Oregon State University*).

Figure 172.—Communities of *Artemisia arbuscula* **perennial grasses (***Poa sandbergii* **and** *Festuca idahoe* **in this case) are common on shallow soil opening "parks" within the lower forested zones in the Mountains and Basin and Range Provinces (Good Mountain Research Natural Area, Fremont Nati Forest, Oregon).**

Each of these communities appears to be an edaphic climax occupying a distinctive habitat (Dealy 1971). Soil-site factors typical of each are as follows:

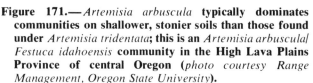

Community	A-horizon depth	Depth to restrictive layer	B-horizon texture	Stoniness in soil	Drainage
	(Centimeters)			(Percent)	
Artemisia arbuscula/ Festuca idahoensis	15	46	Clay loam	50	Moderately well drained
A. arbuscula/ Danthonia unispicata	18	46	Clay loam	60	Imperfectly or somewhat poorly drained
A. arbuscula/ Koeleria cristata	10	38	Clay	5	Imperfectly or somewhat poorly drained

The *A. arbuscula/Festuca* community occupies the better drained soils; associated soils also have a thicker A horizon and less restrictive B horizon than those found under *A. arbuscula/Koeleria* communities. The *A. arbuscula/Danthonia* community occupies lower parts of the landscape where surface runoff is channeled or collects to form temporary

240

Table 35. – Community composition in three *Artemisia arbuscula* communities which form a mosaic in the "scabland" areas of the Basin and Range Province (after Dealy 1971)

| Group and species | Community | | |
	A. arbuscula/ Festuca	A. arbuscula/ Danthonia	A. arbuscula/ Koeleria
	------------ Percent ------------		
Shrubs:			
Artemisia arbuscula	26	13	59
Grasses:			
Festuca idahoensis	32	–	–
Danthonia unispicata	16	28	–
Poa sandbergii	13	23	13
Sitanion hystrix	4	4	5
Koeleria cristata	2	2	7
Stipa thurberiana	1	–	–
Perennial forbs:			
Eriophyllum lanatum	3	–	–
Arenaria congesta	.9	–	–
Lomatium triternatum	.5	1	2
Erigeron bloomeri	.4	1	1
Agoseris sp.	.2	1	.2
Microsteris gracilis	.2	3	–
Lomatium nudicaule	.1	1	.2
Antennaria dimorpha	–	17	2
Annual forbs:			
Polygonum douglasii	.4	.4	4
Navarretia tagetina	.2	.1	2
Polemonium micranthum	–	2	–
Collomia tenella	–	–	3

pools. Both water and root penetration are very restricted in soils associated with *A. arbuscula/Koeleria* communities.

Tueller (1962) studied effects of grazing in habitats characterized by the *Artemisia arbuscula/Festuca idahoensis* association. He found the larger perennial grasses—*Festuca idahoensis, Sitanion hystrix, Agropyron spicatum,* and *Koeleria cristata*—decreased with grazing. *Chrysothamnus viscidiflorus, Poa sandbergii,* and a variety of herbs increased, whereas amounts of *Artemisia arbuscula* remained constant under grazing.

Artemisia rigida/Poa sandbergii Association

Artemisia rigida is the dominant shrub of the *Artemisia rigida/Poa sandbergii* association (fig. 173). This association is widespread on lithosolic sites (Culver 1964, Hall 1967, Daubenmire 1970). *Artemisia rigida* is normally the only shrub present; average coverage was about 20 percent in the Owyhee area (Culver 1964). *Poa sandbergii* is the dominant grass (cover 40 percent). Other typical grasses and herbs are listed in table 34.

This association is always found on very shallow, stony soils. Hall (1967) reported 15 centimeters of soil as typical of *Artemisia rigida* habitats compared with 35 centimeters on *Artemisia arbuscula* habitats.

Figure 173.—*Artemisia rigida* **dominates the** *Artemisia rigida/Poa sandbergii* **association widespread on lithosolic sites throughout the steppes and drier forests of eastern Oregon and Washington** *(photo courtesy Range Management, Oregon State University).*

Other *Artemisia* Associations

Several other *Artemisia*-dominated communities have been mentioned by various authors. These include:

Community	Location	Source
Artemisia tridentata-Chrysothamnus spp.	High Lava Plains Province	Eckert (1957)
Artemisia tridentata-Chrysothamnus nauseosus/Stipa thurberiana	Sand hills; Owyhee Upland Province	Culver (1964)
Artemisia tridentata-Grayia spinosa	Steep, south-exposed talus; Owyhee Upland Province	Culver (1964) Dean (1960)
Artemisia tridentata-Sarcobatus vermiculatus/Stipa thurberiana	Slopes; High Lava Plains and Owyhee Upland Provinces	Eckert (1957) Dean (1960)
Artemisia tridentata/Stipa comata-Carex spp.	Sandy soils; High Lava Plains Province	Eckert (1957)
Artemisia tridentata/Stipa occidentalis-Lathyrus bijugatus	Basin and Range Province	Dealy (1971)
Artemisia cana/Muhlenbergia richardsonis	Basin and Range Province	Dealy (1971)

Other Oregon Steppe Communities

Cercocarpus ledifolius Communities

Communities dominated by *Cercocarpus ledifolius* are most often found in the ecotone between lower edge of the *Pinus ponderosa* forest and upper edge of the southeastern and central Oregon *Artemisia* shrub-steppe (fig. 174). *Cercocarpus* also forms pure stands at high elevations in some of the mountain ranges of extreme southeastern Oregon, particularly in the Mahogany Mountains and may even occur, in small amounts, as inclusions within the *Pinus ponderosa* forest (Dealy 1971).

Dealy (1971) recognizes two *Cercocarpus* communities in the Silver Lake area: *C. ledifolius/Festuca idahoensis* and *C. ledifolius/Festuca-Agropyron spicatum*. The former differs from the latter in its much denser stand of *C. ledifolius* (66-percent vs. 36-percent crown cover), fewer shrubs, and the dominance of *Festuca idahoensis* in the understory. Both communities typically have scattered individuals of *Pinus ponderosa* and *Juniperus occidentalis*. Shrubs scattered through the *Cercocarpus* openings are mainly *Artemisia tridentata* and *Chrysothamnus viscidiflorus* in the *Cercocarpus/Festuca* community. Small amounts of *Sitanion hystrix*, *Agropyron spicatum*, *Koeleria cristata*, and *Poa*

Figure 174.—*Cercocarpus ledifolius* communities are common in the forest-steppe ecotone in central Oregon; *Chrysothamnus nauseosus* is the major associate in this stand (Deschutes National Forest, Oregon).

sandbergii are associated with the *Festuca*. Forbs are not important, but the most common are *Hieracium cynoglossoides* and *Microsteris gracilis*. In the *Cercocarpus/Festuca-Agropyron* community, *Artemisia tridentata* and *Chrysothamnus viscidiflorus* are again the most common shrubs. *Festuca idahoensis* and *Agropyron spicatum* share understory dominance, with *Bromus tectorum*, *Sitanion hystrix*, *Poa sandbergii*, *Koeleria cristata*, *Balsamorhiza sagittata*, and *Achillea millefolium* as additional frequent associates.

These communities are important browse types for deer which often leave browse lines or "highline" the mature *Cercocarpus* and severely hedge young plants (Dealy 1971). This apparently has little effect on maintenance or survival of the species, as seedlings escape detection in the tall bunchgrass and snow until they are well established.

Purshia tridentata Communities

Purshia tridentata is a dominant in little known but widespread communities, particularly near the foothills of the Cascade Range. Dealy (1971) has described two of these: *Purshia tridentata-Artemisia arbuscula/Stipa thurberiana* and *Purshia tridentata/Festuca idahoensis* communities which are found near the forest-steppe ecotone in central Oregon. Similar communities are also found in the southern Blue Mountains (F. C. Hall, personal communication) and southeastern Wallowa Mountains (Johnson 1959). As mentioned earlier, *Purshia* is codominant with *Artemisia tridentata* on some sites within the steppe proper.

Dealy (1971) characterizes his *Purshia/Festuca* community by a dominance of *Purshia tridentata* with *Artemisia tridentata* and *A. arbuscula* commonly occurring throughout the stand. Occasionally, *Juniperus occidentalis* and *Cercocarpus ledifolius* are minor components. *Festuca idahoensis* dominates the herbaceous layer but *Poa sandbergii*, *Stipa thurberiana*, *Sitanion hystrix*, and *Agropyron spicatum* occur consistently and prominently and *Antennaria rosea*, *Erigeron* sp., *Astragalus purshii*, and *Arabis* sp. are consistent forbs. Soils are well drained, moderately deep loams.

The *Purshia-Artemisia arbuscula/Stipa* community is actually a vegetational mosaic in which *Purshia tridentata* appears scattered through an *Artemisia arbuscula* stand (Dealy 1971). In fact, the two species are growing on different microsites, with the *A. arbuscula* found on soils having a root-restricting layer at 40 to 50 centimeters and *Purshia* on soils lacking such a layer.

Grassland Communities

Steppe communities lacking a major shrub constituent, such as the grassy types so abundant in the *Agropyron-Festuca* and *Festuca/Symphoricarpos* Zones of the Columbia Basin Province (Daubenmire 1969), are not common in southeastern Oregon (fig. 175). Johnson (1959) described an *Agropyron spicatum-Poa sandbergii* Zone on the southeastern slopes of the Wallowa Mountains in which two steppe communities were recognized:

> *Agropyron spicatum-Poa sandbergii*, the most extensive community with *Balsamorhiza sagittata* and *Eriogonum heracleoides* as associates, and
>
> *Festuca idahoensis-Agropyron spicatum*, an uncommon community with *Balsamorhiza* and *Bromus brizaeformis* as typical constituents.

The *Agropyron-Poa* community also occurred on lithosolic soils in the adjacent *Pseudotsuga menziesii* Zone (Johnson 1959). A similar community was noted by Dean (1960) in a part of the Owyhee River canyon and by F. C. Hall (personal communication) in the Blue Mountains.

Figure 175.—Steppe communities lacking a major shrub dominant are not common in southeastern Oregon; where they do occur, burning may have been responsible, as in this *Agropyron spicatum-Lygodesmia spinosa* **community located near Fort Rock, Oregon** *(photo courtesy Range Management, Oregon State University).*

Hall also described a *Poa sandbergii-Danthonia unispicata* association which occurs on lithosolic soils throughout much of eastern Oregon. This scabland community generally replaces the *Artemisia rigida-Poa sandbergii* type on less permeable bedrock.

Desert or Salt Desert Shrub Communities

Communities variously designated as desert shrub, salt desert shrub, shadscale (*Atriplex confertifolia*), salt sage (*Atriplex nuttallii*), or saltbush-greasewood (*Atriplex-Sarcobatus*) have been mapped but described in only general terms (Shantz and Zon 1924, Poulton 1962, Küchler 1964, Hansen 1956). These communities are on saline soils and often intermingled with upland communities dominated by *Artemisia tridentata*. Salt desert shrub communities are most common in the Basin and Range Province, where interior drainage and old lakebeds are typical.

Important shrubs in these communities can include *Grayia spinosa*, *Atriplex confertifolia*, *A. nuttallii*, *Eurotia lanata*, *Artemisia spinescens*, and *Sarcobatus vermiculatus*. Grasses sometimes associated with these shrubs include *Elymus cinereus*, *E. triticoides* (which may dominate on ancient lakebeds), and *Distichlis stricta*. An *Atriplex confertifolia/ Sitanion hystrix* community with some *Artemisia spinescens*, *Eurotia lanata*, *Poa sandbergii*, and *Oryzopsis hymenoides* is one of the most common communities. *Sarcobatus vermiculatus/Distichlis stricta* communities are also typical of some of the moister saline habitats.

The desert shrub communities are much better developed to the south and east where they dominate extensive areas (Shantz and Zon 1924; Billings 1949, 1951).

Riparian and *Populus tremuloides* Communities

A variety of little known community types is found on riparian or other moist sites within the southeastern Oregon shrub-steppe. These include *Salix-Crataegus*, *Salix-Prunus*,

245

and *Elymus cinereus* communities (Dean 1960, Hansen 1946) and wet *Carex* meadows (Hansen 1956). There are extensive tule marshes in the Klamath Lake area in which *Scirpus validus* is an important dominant (Shantz and Zon 1924) but, despite their importance to migratory waterfowl, they are still virtually undescribed.

Populus tremuloides is a wide-ranging species and not particularly characteristic of the shrub-steppe. Nevertheless, in some locales *Populus tremuloides* communities or colonies are common on moist sites near the forest-steppe ecotone. Dealy (1971) describes colonies of this type which have a heavy cover of grasses and forbs; livestock and deer make heavy use of such areas.

Populus tremuloides is particularly conspicuous between 1,950- and 2,400-meter elevations on Steens Mountain where it occurs just above a belt of *Juniperus occidentalis/ Artemisia* vegetation (fig. 176) (Hansen 1956, Faegri 1966). The understory beneath *P. tremuloides* consists largely of *Artemisia arbuscula* in openings and of *Symphoricarpos rotundifolius* and a rich variety of large herbs elsewhere (Faegri 1966). Faegri's description is reminiscent of the *Populus tremuloides/Artemisia tridentata/Bromus carinatus* community found in the lower forests near Silver Lake (Dealy 1971).

Figure 176.—*Populus tremuloides* **is conspicuous between elevations of 1,950 and 2,400 meters on Steens Mountain in southeastern Oregon;** *Populus* **groves are typically associated with wet meadows in this landscape mosaic.**

246

Status of *Juniperus occidentalis*

Although it occurs well into forested regions (e.g., see Canyon Creek and Ochoco Divide Research Natural Areas in Franklin et al. 1972) and characterizes a zone in its own right, *Juniperus occidentalis* is sometimes associated with *Artemisia tridentata* and *A. arbuscula* communities throughout much of central and southeastern Oregon. In the High Lava Plains Province, its occurrence is related to more mesic microhabitats (Eckert 1957); *Juniperus* is typical of escarpments and rock outcrops, mesic northerly slopes (with *Artemisia arbuscula/Festuca idahoensis* communities), and intermittent drainageways in this area. Soil depths are commonly greater under trees than under adjacent *Artemisia*. In some locales of the High Lava Plains Province, *Juniperus occidentalis* is sufficiently common that *Juniperus* associations or phases of shrub-steppe associations are recognized; unique microcommunities and soil properties are associated with the trees (Eckert 1957). A *Juniperus occidentalis* belt is recognized at 1,750- to 1,950-meter elevations on Steens Mountain (Hansen 1956, Faegri 1966); *Artemisia arbuscula* is its most typical associate there. *Juniperus* also occurs along the Owyhee River canyon (Head 1959).

CHAPTER X. TIMBERLINE AND ALPINE VEGETATION

On the highest mountain ranges of Oregon and Washington are subalpine parklands and alpine meadows. The parklands constitute an ecotone in which tree dominance is gradually giving way under the increasingly harsh alpine environment. Typically, the area between forest line[1] and scrub line is a mosaic of tree patches and meadow communities (fig. 177), the former gradually being reduced in area and in stature as elevations increase. This belt was referred to as the Hudsonian Zone by Merriam (Bailey 1936). In western Washington and British Columbia, it has been split into two units by Krajina (1965) and Franklin and Bishop (1969) depending on occurrence of trees as a climatic or topographic climax. In this fashion, a part is considered the upper segment of the *Tsuga mertensiana* Zone (Parkland Subzone), and the other constitutes the lower part of the Alpine Zone.

Some scientists have chosen to confine the term "subalpine" to this region of meadow-forest mosaic (e.g., Douglas 1970 and 1972). We feel this ignores the closed forest sub-zones usually found below the parklands which are clearly subalpine in character.

The subalpine meadow-forest mosaic or parkland is extensively developed in the mountains of the Pacific Northwest, perhaps to a greater extent than anywhere else in the world. Whereas in many mountain ranges the forest-tundra ecotone is reasonably

Figure 177.—Timberline regions are typically mosaics of tree groups, meadows, snow patches, and rock outcrops (Jefferson Park, Willamette National Forest, Oregon).

---

[1] We recognize three types of timberline: *forest line*, the general upper limit of contiguous closed forest; *tree line*, the upper limit of erect arborescent growth; and *scrub line*, the general upper limit of krummholz (= elfinwood or wind-timber line) (Arno 1966, Habeck and Hartley 1968). Our timberline region in this discussion covers the entire area from forest line to scrub line.

sharp and occupies a relatively narrow elevational band, parklands extending over an elevational span of 300 to 400 meters or more are not unusual in the Cascade Range. We suspect that deep, late-lying snowpacks are a major reason for this broad, extended forest-meadow ecotone. In any case, a great diversity of subalpine parkland communities as well as large expanses of an exceptionally attractive landscape are present.

The entire forest-meadow mosaic of the region between forest and scrub line is the major topic of this chapter. We will consider first the meadow communities which are the dominant feature, then the trees, tree groups, and dynamic relations between trees and meadows, and conclude with a section on alpine communities. Published information on this region above forest line has expanded tremendously during the last 5 years, allowing a much more thorough treatment than was possible in our earlier work (Franklin and Dyrness 1969). Among the recent valuable contributions are those of Douglas in the northern Cascades (Douglas 1970, 1971, 1972; Douglas and Ballard 1971); Henderson (1973) and Hamann (1972) at Mount Rainier; Kuramoto and Bliss in the Olympic Mountains (Kuramoto 1968, Kuramoto and Bliss 1970, Bliss 1969); Campbell (1973) in the Oregon Cascades; Lowery (1972) on development of subalpine tree groups; Van Ryswyk (1969) and Bockheim (1972) on subalpine soils in relation to vegetation; Arno (1970) and Arno and Habeck (1972) on the ecology of *Larix lyallii*; and Brooke et al. (1970) on the ecology of the entire mosaic on adjacent coastal British Columbia.

Occurrence of Timberline Regions

Elevations sufficient to develop true timberline conditions are encountered generally in the northern Cascade Range and Olympic Mountains in Washington and in the Wallowa Mountains of Oregon. Further south in the Cascade Range, major peaks such as Mount Rainier, Mount Hood, and Three Sisters have sufficient elevation to develop a climatic timberline and alpine regions. Timberline environments are occasionally encountered in the Blue, Steens, and Warner Mountains of Oregon and the Okanogan Highlands and Rocky Mountain outliers of eastern Washington.

Average elevation of forest and scrub lines at representative Oregon and Washington locations is listed in table 36. It can be seen that the timberline mosaics generally involve a 300- to 500-meter elevational band. Elevations of forest and scrub lines vary with exposure on an order of ± 150 meters, dropping on cool, northerly exposures and rising on warmer, southerly exposures (Bailey 1936, Arno 1966). Timberline elevations decrease notably with increasing latitude; Daubenmire (1954) indicates the general tendency is for timberline to drop about 110 meters per degree of increase in latitude under a given climatic regime. The forest and scrub lines are markedly lower in coastal mountain regions (dominated by a maritime climate) than they are further inland. For example, forest line is about 500 meters higher at the eastern edge of the Cascade Range than it is at the same latitude on the western edge; and, of course, it is lower (as much as 2,000 meters) in the Cascade Range than in the main body of the Rocky Mountains at the same latitude.

Subalpine Meadow Communities

The variety and richness of the meadow flora and communities make the subalpine parkland attractive to scientists and laymen alike. Many of the species (and communities in the broad sense) are circumpolar. The mosaic of meadow communities is an intricate

Table 36. — Average elevation of forest and scrub lines at selected locations in Oregon and Washington

Area	Latitude	Longitude	Forest line	Scrub line
			- - - - - - - Meters - - - - - - -	
Mount Baker	48°45′	121°50′	1,400	1,750
Wenatchee Mountains	47°30′	120°45′	2,000	2,440
Mount Rainier (except northeast)	47°10′	121°40′	1,580	2,100
Mount St. Helens	46°15′	122°10′	1,340	—
Mount Hood	45°20′	121°45′	1,680	1,980
Three Sisters	44°10′	121°50′	1,980	2,290
Mount McLoughlin	42°50′	122°20′	2,130	2,440
Olympic Mountains (central)	47°45′	123°30′	1,460	1,890
Olympic Mountains (northeast)	47°50′	123°20′	1,680	1,980
Wallowa Mountains	45°10′	117°20′	—	2,700

Source: Partially from Arno (1966), Bailey (1936), and Brockman (1949).

and often sharp response to local variations in substrate, moisture conditions, and duration of winter snowpack (fig. 178).

Even within the Pacific Northwest, the dominant meadow communities in the parkland mosaic and lower alpine vary from area to area just as the tree species do. Many communities occur throughout and retain their same basic character over a wide geographic range, but their importance in the mosaic changes. We will consider first the meadow communities found in a cooler and moister maritime region, the western slopes of Washington, and then outline some different meadow types characteristic of southern Oregon and of the interior mountain ranges.

Western Washington

The meadow communities found in the subalpine parklands of western Washington and adjacent British Columbia are the best known in the Pacific Northwest. Among the relevant studies are those of Douglas (1970, 1971, 1972), Douglas and Ballard (1971), and Bockheim (1972) in the northern Cascade Range; Henderson (1973) and Hamann (1972) at Mount Rainier; Kuramoto (1968) and Kuramoto and Bliss (1970) in the Olympic Mountains; and Krajina (1965), Brooke (1965), Brooke et al. (1970), Peterson (1965), and Archer (1963) in coastal British Columbia.

Figure 178.—Several varied communities are frequently in juxtaposition in subalpine landscapes, reflecting a sharp response to differences in environmental conditions, such as depth and duration of snowpack; in this landscape, *Phyllodoce empetriformis-Vaccinium deliciosum* (heather-huckleberry in foreground), *Carex nigricans* (dwarf sedge type around small pond), and *Valeriana sitchensis* (lush herbaceous type on slopes of peak) communities are visible along with patches of forest and tree groups (Green Mountain, Mount Baker National Forest, Washington).

A basic spectrum or pattern of meadow communities is repeated in each of the study areas (table 37). These communities can be arranged in five major type groups (adapted from Henderson 1973): (1) *Phyllodoce-Cassiope-Vaccinium* (heath shrub or heather-huckleberry) group, (2) *Valeriana sitchensis-Carex spectabilis* (lush herbaceous) group, (3) *Carex nigricans* (dwarf sedge) group, (4) rawmark and low herbaceous group, and (5) *Festuca viridula* (grass or dry grass) group which is conspicuously absent in the more northerly study areas (table 37). We will key our discussion to the communities at Mount Rainier where the entire spectrum of communities is richly represented.

Phyllodoce-Cassiope-Vaccinium (Heath Shrub) Communities

The heather communities of *Phyllodoce empetriformis*, *Cassiope mertensiana*, and *Vaccinium deliciosum* are a distinctive and conspicuous type group in the coastal mountains of the Pacific Northwest and are often used to characterize the parkland subzone. The group is well represented in each of the study areas (table 37). Ericaceous shrubs dominate the relatively dense, closed vegetative canopies of the communities in this group.

Phyllodoce empetriformis-Vaccinium deliciosum Community

The major community in this type group is the *Phyllodoce empetriformis-Vaccinium*

Table 37. — Major subalpine meadow communities recognized in four different localities in the Pacific Northwest arranged to show analogies

Community type group	Locale and author			
	Mount Rainier (Henderson 1973)	Northeastern Olympic Mountains (Kuramoto and Bliss 1970)	Western north Cascade Range, Washington (Douglas 1972)	Coastal British Columbia (Brooke et al. 1970)
Phyllodoce-Cassiope-Vaccinium (heath shrub or heather-huckleberry)	*Phyllodoce empetriformis-Vaccinium deliciosum* *Vaccinium deliciosum* *Phyllodoce empetriformis/ Lupinus latifolius*	Heath shrub type	*Cassiope mertensiana-Phyllodoce empetriformis* *Vaccinium deliciosum*	*Phyllodoce empetriformis Cassiope mertensiana* *Vaccinium deliciosum*
Valeriana sitchensis-Carex spectabilis (lush herbaceous)	*Valeriana sitchensis-Veratrum viride* *Valeriana sitchensis-Lupinus latifolius* *Lupinus latifolius-Polygonum bistortoides* *Lupinus latifolius-Carex spectabilis* *Polygonum bistortoides-Carex spectabilis* *Mimulus lewisii*	Moist *Valeriana* forb type Tall sedge type *(Carex spectabilis)* Moist *Saussaurea* forb type	*Valeriana sitchensis-Veratrum viride* *Carex spectabilis* *Rubus parviflorus/ Epilobium angustifolium*	*Leptarrhena pyrolifolia*
Carex nigricans (dwarf sedge)	*Carex nigricans* *Carex nigricans-Pedicularis groenlandica* *Carex nigricans-Caltha biflora*	Dwarf sedge type	*Carex nigricans*	*Carex nigricans*
Rawmark and low herbaceous	*Saxifraga tolmiei* *Luetkea pectinata* *Eriogonum pyrolaefolium-Spraguea umbellata* *Antennaria lanata* *Aster alpigenus-Antennaria lanata*	Cushion plant type	*Saxifraga tolmiei* *Luetkea pectinata*	*Saxifraga tolmiei*
Festuca viridula (grass or dry grass)	*Festuca viridula-Lupinus latifolius* *Festuca viridula-Aster ledophyllus*	Mesic grass type Dry grass-forb type		

deliciosum (fig. 179) (Henderson 1973) (*Phyllodoce-Cassiope mertensiana* of some authors). *Phyllodoce empetriformis*, *Cassiope mertensiana*, and *Vaccinium deliciosum* are the dominant species (tables 38 and 39), but there are a number of other relatively constant associates such as *Luetkea pectinata*, *Antennaria lanata*, *Lycopodium sitchense*, and *Deschampsia atropurpurea*. There are clearly shifts in the proportions of *Phyllodoce* and *Cassiope* from stand to stand; Douglas (1972) shows an average preponderance of *Cassiope*, and Kuramoto and Bliss (1970) and Henderson (1973) show slightly higher average values for *Phyllodoce*. Henderson (1973) indicates an elevational shift with *Phyllodoce* generally dominant in stands from 1,650 to 1,830 meters and *Cassiope* above that elevation; at about 2,075-meter elevation, the communities shift gradually into a more alpine heather type dominated by *Phyllodoce glanduliflora*.

The *Phyllodoce-Vaccinium* community occupies a variety of sites. Henderson (1973) found it most common on gentle slopes and moist, moderately well-drained soils; Douglas (1972) indicates moist slopes and ridges with poor to well-drained soils. According to Brooke et al. (1970) a snow-free period of 3 to 4 months is typical, and the topography occupied shifts from concave areas at lowest elevations to convex surfaces near its upper limits. The diverse soils apparently associated with *Phyllodoce-Vaccinium* communities include podzolic types (Alpine Podzols or Cryohumods), with 0-A2-B2-C horizon sequences and rankers with O-A-C horizon sequences. Bockheim (1972) found Humic or Humic Lithic Cryorthods (Alpine Turf) as the soil types present under heather-huckleberry communities in the Mount Baker area; these had 8- to 10-centimeter A1 horizons underlain by 8- to 12-centimeter B2h horizons. He describes podzolic soil types (Cryohumods) only from under tree clumps.

The *Phyllodoce empetriformis/Lupinus latifolius* community is a variant of the heather-huckleberry group which has been described from Mount Rainier (Henderson 1973). It differs mainly in the abundance of *Lupinus* (table 38).

Figure 179.—A *Phyllodoce empetriformis-Vaccinium deliciosum* **(or** *Phyllodoce-Cassiope***) community typifies the heather-huckleberry or heath shrub type group in the subalpine parklands of the Cascade Range and Olympic Mountains (western slopes of Mount Baker, Washington).**

Table 38. – Mean dominance ratings for the major species in some subalpine meadow communities on Mount Rainier (from Henderson 1973)[1]

Species	Phem-Vade	Vade	Phem/Lula	Vasi-Vevi	Vasi-Lula	Lula-Pobi	Cani	Anla	Lupe	Sato	Fevi-Asle	Fevi-Lula	Fevi-Pofl
						Communities[2]							
Phyllodoce empetriformis	4.0	–	3.7	–	–	–	+	+	–	+	–	–	–
Cassiope mertensiana	3.9	–	2.8	–	–	–	+	1.2	–	+	–	–	–
Vaccinium deliciosum	3.7	3.0	+	–	2.4	+	1.0	+	–	–	–	–	–
Lupinus latifolius	+	3.5	4.0	1.0	4.2	4.5	–	–	+	+	2.1	4.0	1.6
Gentiana calycosa	+	–	+	–	–	–	–	–	–	–	–	–	–
Pedicularis ornithorhyncha	1.7	+	2.4	–	–	–	–	–	–	–	–	–	–
Tauschia stricklandii	+	–	–	–	–	–	+	–	–	–	–	–	–
Valeriana sitchensis	–	1.8	–	4.8	4.1	1.1	–	–	–	–	+	–	1.5
Veratrum viride	–	–	–	3.5	1.0	1.1	–	–	–	–	–	–	–
Ligusticum grayi	–	2.0	–	+	2.4	2.0	–	–	–	–	2.3	2.4	1.6
Carex spectabilis	+	1.2	1.5	1.2	1.2	3.3	1.5	+	1.2	1.2	1.7	1.7	1.6
Potentilla flabellifolia	+	1.8	+	+	2.0	2.4	1.5	1.5	–	–	+	1.6	3.5
Polygonum bistortoides	–	2.8	1.6	+	2.9	3.8	+	1.2	–	–	2.0	2.3	2.5
Castilleja parviflora	1.2	2.2	1.5	1.0	3.0	2.3	+	–	–	+	+	1.0	1.8
Erythronium spp.	1.3	–	+	1.5	1.6	+	–	–	–	–	–	–	–
Carex nigricans	–	+	1.3	–	+	+	5.0	3.3	2.8	+	–	–	–
Aster alpigenus	+	1.8	–	–	1.4	+	1.2	2.0	–	–	–	–	–
Luetkea pectinata	1.6	1.2	1.7	–	–	–	+	1.3	3.6	+	–	–	–
Antennaria lanata	1.4	2.3	1.2	–	+	1.3	+	4.5	1.2	+	+	+	1.4
Saxifraga tolmiei	–	–	–	–	–	–	–	–	+	2.8	–	–	–
Festuca viridula	–	2.0	–	+	1.2	+	–	–	–	–	4.2	4.4	4.2
Anemone occidentalis	–	1.8	+	1.2	+	2.3	+	–	–	+	1.7	1.4	2.4
Erigeron peregrinus	–	–	1.0	–	2.4	+	+	–	–	–	1.5	+	+
Vernoica cusickii	+	1.5	1.1	+	1.5	1.2	–	–	+	–	1.7	2.4	2.2
Aster ledophyllus	–	+	–	–	+	+	–	–	–	–	4.2	1.4	+
Phlox diffusa	–	–	–	–	–	–	–	–	–	–	1.5	1.2	+
Castilleja miniata	–	–	–	–	–	–	–	–	–	–	1.6	+	–
Microseris alpestris	–	–	–	–	–	–	–	–	–	–	+	1.1	+

[1] The dominance rating scheme is on a subjective 5-point scale ranging from 1 (species which can be seen only by searching) to 5 (species which dominates the layer); if several species are codominant, they are assigned a 4 rating. A plus (+) indicates an average rating of less than 1.0. Italicized numbers are the dominants.

[2] Abbreviations for communities are:
Phem-Vade = *Phyllodoce empetriformis-Vaccinium deliciosum*
Vade = *Vaccinium deliciosum*
Phem/Lula = *Phyllodoce empetriformis/Lupinus latifolius*
Vasi-Vevi = *Valeriana sitchensis-Veratrum viride*
Vasi-Lula = *Valeriana sitchensis-Lupinus latifolius*
Lula-Pobi = *Lupinus latifolius-Polygonum bistortoides*
Cani = *Carex nigricans*
Anla = *Antennaria lanata*
Lupe = *Luetkea pectinata*
Sato = *Saxifraga tolmiei*
Fevi-Asle = *Festuca viridula-Aster ledophyllus*
Fevi-Lula = *Festuca viridula-Lupinus latifolius*
Fevi-Pofl = *Festuca viridula-Potentilla flabellifolia.*

Table 39. — Mean prominence values for selected shrubs and herbs[1] in nine subalpine meadow communities in the northern Cascade Range of Washington (from Douglas 1972)[2]

Species	Communities[3]								
	Came-Phem	Vade	Vasi-Vevi	Casp	Rupa-Epan[4]	Cani	Lupe	Lupe (raw)	Sato
Cassiope mertensiana	441	8	—	—	—	11	11	1	tr
Phyllodoce empetriformis	386	78	—	—	—	2	15	1	1
Vaccinium deliciosum	92	689	—	tr	—	tr	21	tr	1
Luetkea pectinata	73	71	—	2	—	21	502	462	—
Lycopodium sitchense	16	2	—	—	—	tr	16	—	—
Deschampsia atropurpurea	9	5	tr	tr	—	15	44	6	1
Polygonum bistortoides	1	10	8	30	27	tr	9	—	—
Valeriana sitchensis	5	14	305	15	127	—	47	2	tr
Carex spectabilis	1	4	52	782	5	6	41	80	2
Mitella breweri	—	1	32	tr	4	—	1	—	—
Veratrum viride	—	tr	290	2	58	—	tr	—	—
Lupinus latifolius subalpinus	4	—	59	42	25	—	—	—	—
Rubus parviflorus	—	—	1	—	256	—	—	—	—
Pteridium aquilinum pubescens	—	—	—	—	154	—	—	—	—
Epilobium angustifolium	—	tr	12	8	161	—	—	—	—
Viola glabella	—	—	5	7	105	—	—	—	—
Heracleum lanatum	—	—	6	tr	80	—	—	—	—
Thalictrum occidentale	—	—	—	—	78	—	—	—	—
Hydrophyllum fendleri	—	—	—	—	71	—	—	—	—
Carex nigricans	2	4	—	16	—	803	13	6	1
Epilobium alpinum	—	—	tr	tr	—	32	10	6	3
Hieracium gracile	3	8	—	1	—	3	36	16	4
Luzula wahlenbergii	—	—	—	—	—	2	10	29	47
Potentilla flabellifolia	—	1	4	7	—	1	13	tr	—
Castilleja parviflora albida	tr	2	—	—	—	tr	16	5	4
Anemone occidentalis	tr	—	—	2	—	—	11	—	tr
Juncus drummondii subtriflorus	—	—	—	—	—	5	6	11	1
Juncus mertensianus	—	—	—	—	—	—	—	10	—
Saxifraga tolmiei	—	—	—	—	—	tr	—	15	78

[1] All species with prominence of at least 10 in at least one community.

[2] Douglas calculated prominence values by multiplying the average percent of cover by the square root of the species frequency; tr = trace values.

[3] Abbreviations for communities are:
Came-Phem = *Cassiope mertensiana-Phyllodoce empetriformis*
Vade = *Vaccinium deliciosum*
Vasi-Vevi = *Valeriana sitchensis-Veratrum viride*
Casp = *Carex spectabilis*
Rupa-Epan = *Rubus parviflorus-Epilobium angustifolium*
Cani = *Carex nigricans*
Lupe = *Luetkea pectinata* (residual or regosolic phase)
Lupe (raw) = *Luetkea pectinata* (rawmark phase)
Sato = *Saxifraga tolmiei*.

[4] Additional species in the *Rubus parviflorus-Epilobium paniculatum* community, which are unique to that community and have prominence values of over 10, are: *Lathyrus nevadensis* (27), *Elymus hirsutus* (24), *Aster engelmannii* (23), *Disporum smithii* (21), *Angelica arguta* (15), *Saussurea americana* (15), *Bromus carinatus* (13), *Galium triflorum* (11), and *Arenaria macrophylla* (10).

Vaccinium deliciosum Community

A *Vaccinium deliciosum*-dominated community has been recognized in most subalpine areas (fig. 180). Associated species differ somewhat depending upon the investigator's definition of this community. Henderson (1973) explicitly excludes stands with significant amounts of heather and identifies a variety of herbs as the major associates (table 38). *Phyllodoce empetriformis* and *Luetkea pectinata* are major associates in Douglas' (1972) community (table 39).

There appears to be considerable disagreement on the successional status of this community, i.e., its relationship to the *Phyllodoce-Vaccinium* or *Phyllodoce-Cassiope* types. Brooke et al. (1970) feel that the *Vaccinium deliciosum* community replaces their *Phyllodoce-Cassiope* type successionally; this is based partially on its frequent position as a ring around tree groups above (inside) the heather communities. Douglas (1970) indicates it can either succeed or precede a heather community. Henderson (1973) finds the *Vaccinium deliciosum* community seral to the *Phyllodoce-Vaccinium* type and concludes it is mainly a fire-derived type. He cites Douglas and Ballard's (1971) study which shows that fires in heather-huckleberry communities have little effect on the *Vaccinium* but greatly reduce or eliminate the heathers. We suspect that the *Vaccinium deliciosum* communities play several successional roles, i.e., as a pioneer community on burned habitats where it may be succeeded either directly by forest or by *Phyllodoce-Vaccinium* communities and as a climax or near-climax type around tree groups.

Figure 180.—A *Vaccinium deliciosum*-dominated community typical of those found lower in the subalpine parklands of the Cascade Range.

Valeriana sitchensis-Carex spectabilis (Lush Herbaceous) Communities

This group of subalpine meadow communities is the showiest, the richest floristically, and among the most diverse (fig. 181). This group does not appear to be as common in subalpine parklands outside western Washington and seems to have an affinity with the

256

tall herb lands or mountain meadows such as are found within a forest matrix in the Rocky Mountains. Several distinctive subgroups are obvious: the *Valeriana sitchensis-Veratrum viride* community, communities with *Carex spectabilis* or *C. albonigra* as major components, riparian types (e.g., *Mimulus lewisii*), the *Rubus parviflorus-Epilobium angustifolium* community, and a *Saussurea americana* community. As will be seen, most of these are closely interrelated, and communities often grade from one type into another as well as with the *Festuca viridula* or grass type group.

Figure 181.—Lush herbaceous communities, typified here by a *Valeriana sitchensis-Veratrum viride* **community, are among the showiest and most floristically rich of the subalpine types (Green Mountain, Mount Baker National Forest, Washington).**

Valeriana sitchensis-Veratrum viride Community

This lush herbaceous community is dominated by *Valeriana*, *Veratrum*, and a rich assemblage of associated herbs which form a stand 1 meter or more in height (fig. 182). Among the associates are *Lupinus latifolius*, *Carex spectabilis*, *Castilleja parviflora*, *Erythronium* spp., *Anemone occidentalis*, *Polygonum bistortoides*, *Mitella breweri*, *Epilobium angustifolium*, and *Heracleum lanatum* (tables 38 and 39).

This community is most commonly found on steep, well-watered ("fresh") slopes of varying exposure. Such habitats are frequently subject to recurring avalanches. Thornburgh

257

Figure 182.—*Veratrum viride*, **one of the major dominants which grows 1 meter or more in height in the lush** *Valeriana sitchensis* **communities found on well-watered parkland slopes in the Pacific Northwest.**

(personal communication) suggests that this community is favored over a *Phyllodoce-Vaccinium* type as a consequence, since all important constituent species survive the winter in underground parts protected from snow creep and avalanching. According to Douglas (1972), weakly developed A-B-C soil horizon sequences are typical.

Henderson (1973) recognizes a closely related community type (*Valeriana sitchensis-Lupinus latifolius*) which differs in increased abundance of *Lupinus*, *Ligusticum grayi*, *Potentilla flabellifolia*, the addition of several new associates (especially *Vaccinium deliciosum* and *Erigeron peregrinus*), and a substantial reduction in *Veratrum viride* (table 38). This is a conspicuous community on the southern and western slopes of Mount Rainier, and it occupies gentler topography than the *Valeriana-Veratrum* type. Intergrades between the *Valeriana-Lupinus* and *Festuca-Lupinus* communities are common.

Carex spectabilis Communities

The *Valeriana-Veratrum* communities appear to grade into another series of somewhat more attenuated herbaceous types which have *Carex spectabilis* as one of the dominants; the *Carex albonigra* reported earlier by Kuramoto and Bliss (1970) has since been re-evaluated as *Carex spectabilis*. These communities are typically not as tall or productive as the *Valeriana-Veratrum* type but can still be considered "lush herbaceous" communities.

Douglas (1970, 1972) and Kuramoto and Bliss (1970) each recognize only a single community in this category (the "Tall Sedge" of the Olympic Mountains and *Carex spectabilis* of the northern Cascade Range). Henderson recognizes three related communities (table 37) based upon shifts in dominance. In all cases, however, *Carex spectabilis*, *Polygonum bistortoides*, and *Lupinus latifolius* are present as the three dominants.[2] Associates vary including *Potentilla flabellifolia* (all areas), *Castilleja parviflora* (Mount Rainier), *Ligusticum grayi* (Mount Rainier), *Carex nigricans* (all areas), *Anemone occidentalis* (Cascade Range), *Valeriana sitchensis* (Cascade Range), and *Erigeron peregrinus* (Mount Rainier and Olympic Mountains).

Rubus parviflorus-Epilobium angustifolium Community

This community has been described as a subalpine meadow type only by Douglas (1970, 1972). A tall, dense, rich community (fig. 183), it occupies steep slopes and may cover an altitudinal range of 600 meters. The snow-free season is reportedly long (commencing in April or May). Douglas (1972) indicates a total species tally of 70 for this community type with an average of 32 per stand. Nineteen major species (prominence values greater than 10) are listed in table 39. Aspection is conspicuous in this community beginning with *Claytonia lanceolata* and *Erythronium grandiflorum*, dominance by *Pteridium aquilinum* and *Epilobium angustifolium* in midsummer, and late summer development of *Solidago canadensis*, *Artemisia ludoviciana*, and *A. norvegica*.

The *Rubus parviflorus-Epilobium angustifolium* community is really not a part of the subalpine parkland mosaic; its center of distribution is well below the normal elevational line of closed forest. Furthermore, it appears to have close affinities with mountain meadow communities which occupy high ridges in otherwise forested locations (e.g., the *Abies amabilis* and *Abies concolor* Zones) as far south as 43° north latitude in the Cascade Range. Perhaps it can be considered a subalpine variant of a much broader *Pteridium aquilinum-Rubus parviflorus* synecological unit. The *Rubus-Epilobium* type may intergrade with *Valeriana-Veratrum* communities at its upper elevational limits, however.

Figure 183.—The *Rubus parviflorus-Epilobium angustifolium* is a rich, dense herbaceous community found in the lower subalpine zone (Green Mountain, Mount Baker National Forest, Washington).

---

[2] Except *Lupinus* is inconspicuous or absent in Henderson's (1973) *Polygonum-Carex spectabilis* community type.

Saussurea americana Community

Another rich herbaceous community is the "moist *Saussurea* forb type" recognized by Kuramoto (1968) and Kuramoto and Bliss (1970) in the Olympic Mountains; Henderson (personal communication) noted a similar type at Mount Rainier. As in the case of the *Rubus parviflorus-Epilobium angustifolium* community, this appears to be a lower elevation (under 1,500 meters) mountain meadow type rather than a true constituent of the subalpine parkland subzone. Again, it is closely related to the *Valeriana-Veratrum* community, however.

The *Saussurea americana* community is described as floristically rich (average 21 species per stand) with three layers (tall herbs, understory herbs, and creepers). Major species are *Saussurea, Heracleum lanatum, Hydrophyllum occidentale, Senecio integerrimus, Thalictrum occidentale,* and *Delphinium glaucum.* Other associates include *Pedicularis bracteosa, Lupinus latifolius, Artemisia douglasiana, Elymus glaucus, Bromus sitchensis* var. *aleutensis, Melica subulata, Viola glabella, Aster foliaceus, Epilobium alpinum, Mitella breweri,* and *Vicia americana.* This community is found primarily in shallow gullies and has young, poorly developed soils with A1-B1-B2 horizon sequences.

Mimulus lewisii and Other Riparian Communities

A final group of lush herbaceous communities are those found along streams (fig. 184). Only one such community has been described from the Washington Cascade Range, the *Mimulus lewisii* type (Henderson 1973). *Mimulus lewisii, M. tilingii, Epilobium alpinum, Petasites frigidus,* and *Carex nigricans* are listed as dominants. Our observations indicate that *Leptarrhena pyrolifolia, Parnassia fimbriata, Caltha biflora, C. leptosepala, Juncus drummondii, Philonotis fontana, Lupinus latifolius, Epilobium latifolium, Drepanocladus aduncus,* and *Erigeron peregrinus* may also be characteristic species on such habitats. Brooke et al. (1970) also recognize a lush hydric community which they call the *Leptarrhena pyrolifolia-Caltha leptosepala* community. *Erigeron peregrinus* and *Parnassia fimbriata* are additional constant dominants. *Equisetum palustre* is another dominant in some stands. *Petasites frigidus* and *Drepanocladus exannulatus* occur closest to the stream.

Figure 184.—*Mimulus lewisii-Epilobium latifolium* **community typical of the showy and varied meadow types found on riparian habitats in the subalpine parklands.**

Habenaria dilatata, Mimulus lewisii, Mitella pentandra, and *Tofieldia glutinosa* occupy slightly raised habitats.

Henderson (1973) recognizes several other riparian or hydric communities which are not showy herbaceous types: the *Carex nigricans-Caltha biflora, Carex nigricans-Kalmia polifolia*, and *Eriophorum polystachion/Sphagnum* spp. types.

Carex nigricans Communities

Carex nigricans communities typify sites with a short growing season due to late-persisting snowbanks and cold, wet soil (fig. 185). They occur in every area studied (table 37) and exhibit little variation from locale to locale (tables 38 and 39). The communities are dominated by the short *Carex nigricans* which forms a dense mat. Small amounts of *Phyllodoce empetriformis, Cassiope mertensiana, Vaccinium deliciosum, Carex spectabilis, Luetkea pectinata, Aster alpigenus, Deschampsia atropurpurea, Hieracium gracile, Epilobium alpinum*, and *Antennaria lanata* may be present. Kuramoto and Bliss (1970) mention that *Erythronium montanum* attains best development in *Carex nigricans* communities in the Olympic Mountains and is important there. *Polytrichadelphus lyallii, Polytrichum norvegicum*, and *Pogonatum alpinum* are typical bryophytes.

As mentioned, *Carex nigricans* communities are associated with snow bed habitats. Brooke et al. (1970) indicate the snow-free period is less than 3 months. Most authors

Figure 185.—*Carex nigricans* communities typify sites with a short growing season due to late-persisting snowbanks and cold wet soil (southeastern slopes of Mount Baker, Washington).

indicate A-C soil horizon sequences are typical; depositional strata (of organic matter or of different inorganic materials such as pumice) are often conspicuous in soil profiles.

Some *Carex nigricans* communities occur along streams and have long growing seasons as well as typically hygric constituents such as *Caltha* spp.

Rawmark and Low Herbaceous Communities

Grouped here are a series of poorly developed, pioneer communities associated with a variety of habitats from pumice slopes to snow beds.

The *Saxifraga tolmiei* community is consistently recognized (table 37) as a pioneer type on recently exposed sandy and gravelly substrates receiving abundant seepage water. Snow-free seasons are short, and substrate is unstable. Total plant cover is low; Douglas (1972) and Brooke et al. (1970) report about 20 percent. *Saxifraga tolmiei* and *Luzula wahlenbergii* are the most important vascular plants. Douglas (1972) mentions *Polytrichadelphus lyallii* as the only important cryptogam, but Brooke et al. (1970) emphasize *Marsupella brevissima* (= *Gymnomitrium varians*) and *Oligotrichum hercynicum* and list several other important bryophytes. The *Marsupella* is apparently an early colonizer of harsh habitats in its own right as well as a component of the *Saxifraga tolmiei* community.

Luetkea pectinata communities are recognized by both Henderson (1973) and Douglas (1970, 1972) (tables 38 and 39). Generally it is considered to be a pioneer type, but Douglas recognizes two phases depending on the habitat: "rawmark" and "residual or regosolic" depending on substrate. These differ primarily in the occurrence of elements of the *Saxifraga tolmiei* community (table 39) and of bare soil and rock in the rawmark phase. Major associates of the dominant *Luetkea pectinata* appear to be *Carex nigricans, C. spectabilis, Luzula wahlenbergii, Deschampsia atropurpurea, Castilleja parviflora, Hieracium gracile, Epilobium alpinum, Valeriana sitchensis,* and small amounts of *Phyllodoce empetriformis, Cassiope mertensiana,* and *Vaccinium deliciosum*. The prominence values of associates in the regosolic phase of the *Luetkea* community suggest successional relationships with several other types, particularly the *Carex nigricans* and *Phyllodoce-Vaccinium*.

The *Antennaria lanata* community type, as described by Henderson (1973), occupies gentle or flat topography similar to the *Carex nigricans* type but is drier and has a longer snow-free season. *Antennaria lanata* and *Carex nigricans* dominate (table 38). We mention this community not because of its extent in western Washington but because of its resemblance to some of the communities reported in the Okanogan Highlands Province (Van Ryswyk 1969). Related communities include Henderson's (1973) *Aster alpigenus-Antennaria lanata* and *Trisetum spicatum* communities.

Raw, gentle to moderately steep, pumice slopes are typically colonized by low, shrubby Polygonaceae. This will be discussed later in this chapter. Only one such community—*Eriogonum pyrolaefolium-Spraguea umbellata*—has been recognized in western Washington (Henderson 1973).

A cushion plant community has been noted in the Olympic Mountains on steep, southerly exposed slopes which are free of snow after late April or early May (Kuramoto and Bliss 1970). *Phlox diffusa*, a mat-former, is the most important species with an average cover of 6 percent (43 percent of the total plant cover). Species unique to this habitat are *Arabis cobrensis, Collinsia parviflora,* and *Douglasia laevigata*. Other important components are *Allium crenulatum, Lomatium martindalei, Geum triflorum, Achillea millefolium, Festuca viridula, Arenaria capillaris, Campanula rotundifolia,* and *Polygonum bistortoides*.

There are numerous other pioneer communities which occur on particular habitats. One of the more common types encountered in subalpine scree and talus is a nearly pure community of *Athyrium distentifolium* var. *americanum.* Douglas (1970) mentions several cryptogams which are pioneers on rock outcrops. *Juniperus communis, Arctostaphylos uva-ursi, Arenaria capillaris, Pachystima myrsinites, Phlox diffusa* var. *longistylis,* and *Sibbaldia procumbens* are pioneering vascular plants on dry rock outcrops. Dwarfed *Tsuga mertensiana* may follow since growing seasons are long.

Festuca viridula Communities

Grassy meadows dominated by *Festuca viridula* are conspicuous elements of the parkland mosaic at Mount Rainier (Henderson 1973) and in the Olympic Mountains (Kuramoto 1968, Kuramoto and Bliss 1970).[3] These generally occupy warmer, drier habitats and are best developed in western Washington subalpine parklands which are in a rain-shadow area (fig. 186). These are clearly related to the *Festuca* communities so common on some interior mountains which will be discussed later in this chapter. The *Festuca viridula* community complex has been studied most extensively at Mount Rainier where three types are recognized. Two of these intergrade and represent different segments of a moisture gradient: the *Festuca viridula-Lupinus latifolius* (moist) and *Festuca viridula-Aster ledophyllus* (dry) communities (Henderson 1973). The third community is the *Festuca viridula-Potentilla flabellifolia.*

The *Festuca-Potentilla* community is found in the lower portion of the subalpine zone, and stands are often surrounded by trees. A total ground cover of 85 percent is dominated by *Festuca* and *Potentilla* (table 38). Associates include *Polygonum bistortoides, Anemone occidentalis, Veronica cusickii, Lupinus latifolius, Ligusticum grayi,* and *Carex spectabilis.* Henderson (1973) suggests that it may be seral to forest vegetation and is closely related to an *Anemone occidentalis-Castilleja parviflora* community which differs only in the absence of *Festuca.*

The *Festuca-Lupinus* community occurs throughout Mount Rainier National Park (see Butter Creek Research Natural Area in Franklin et al. 1972) but is most common northeast of the mountain (fig. 186). *Festuca viridula* and *Lupinus latifolius* dominate, but substantial amounts of *Polygonum bistortoides, Ligusticum grayi, Veronica cusickii, Carex spectabilis,* and *Potentilla flabellifolia* are also present (table 38). Early in the season

Figure 186.—*Festuca viridula* **meadows are found in the Olympic Mountains and Cascade Range, especially in the rain shadows on the eastern slopes of the major volcanoes in the latter;** *Festuca viridula-Lupinus latifolius* **community on the northeastern slopes of Mount Rainier (Sunrise Ridge).**

---

[3] We assume the *Festuca* in meadows in the Olympic Mountains is actually *Festuca viridula* and not *Festuca idahoensis* as reported by Kuramoto (1968) and Kuramoto and Bliss (1970). In any case, it plays an identical role with *Festuca viridula* in the Cascade Range.

(shortly after snowmelt), the *Potentilla*, *Anemone occidentalis*, *Claytonia lanceolata*, and *Ranunculus eschscholtzii* are prominent. Pocket gophers are a significant influence in the loose, friable soil. The "Mesic Grass Type" of Kuramoto and Bliss (1970) appears to be most closely related to Henderson's *Festuca-Lupinus* type. One major difference would appear to be a greater importance of *Phlox diffusa* in the Olympic Mountain community.

The *Festuca-Aster* community is the driest of the subalpine meadow types Henderson (1973) recognizes at Mount Rainier. It is generally found on steep, south-facing slopes with coarse-textured soils. *Festuca viridula* and *Aster ledophyllus* dominate with *Ligusticum grayi* and *Polygonum bistortoides* as major associates. Kuramoto's (1968) and Kuramoto and Bliss' (1970) "Dry Grass-Forb Type" appears closely related, based on habitat and *Festuca* dominance. *Aster* is absent, however; *Delphinium glareosum*, *Eriophyllum lanatum*, *Lathyrus nevadensis*, and *Lomatium martindalei* have taken its place as the *Festuca* associate.

Henderson (1973), Kuramoto (1968), and Kuramoto and Bliss (1970) feel that fire has played a role in the creation of the *Festuca*-dominated meadows.

Environmental and Successional Relationships of the Meadow Communities

From the analyses of Henderson (1973), Douglas (1970, 1972), Kuramoto and Bliss (1970), and Brooke et al. (1970), it is possible to see some relationships between the communities in terms of floristics, environment, and succession. We have already alluded to many of these relationships.

Two dimensional ordinations suggest some of the vegetative relationships between the various community types (figs. 187 and 188A and B). They also indicate general environmental relationships. Henderson's (1973) ordination uses cold wet and warm dry community types as the end points on the X-axis and mesic, deep-soiled, short growing season and mesic shallow-soiled, long growing season communities as end points on the Y-axis (fig. 187). Kuramoto and Bliss (1970) indicate growing season temperatures and soil moisture on their ordination axes (fig. 188A). Douglas (1972) indicates relationships with type of snowmelt (fig. 188B). Brooke et al. (1970) discuss environmental controls at length and emphasize time of snowmelt (length of growing season) as one of the key controllers of community patterns; some of these relationships are suggested in their excellent diagrams showing topographic sequences, one of which is reproduced as figure 188C.

Successional relationships among the meadow communities, and between meadow and forest, have been a subject of conjecture for many subalpine ecologists (Douglas 1970, Henderson 1973, Hamann 1972, and Brooke et al. 1970). The reader should bear in mind that the subalpine parkland is considered by most authors to be potentially forested habitat, i.e., the climatic climax vegetation is forest. Consequently, most hypothesized successional sequences lead toward forest and, in the Clementsian sense, all meadows are seral to forest. We will have more to say about forest-meadow relationships later in this chapter.

For the moment, we will consider only relationships between the meadow communities themselves. Henderson's (1973) diagram (fig. 189A) shows the greatest diversity in pioneer communities and successional "routes," probably reflecting (in part) the greater diversity of meadow habitats in the Mount Rainier area. He also feels fire has had a role in creating and maintaining subalpine communities and offers a second successional diagram which incorporates some of its effects (fig. 189B). Douglas (1970) has constructed a successional diagram for subalpine meadow habitats in the northern Cascade Range (fig. 190).

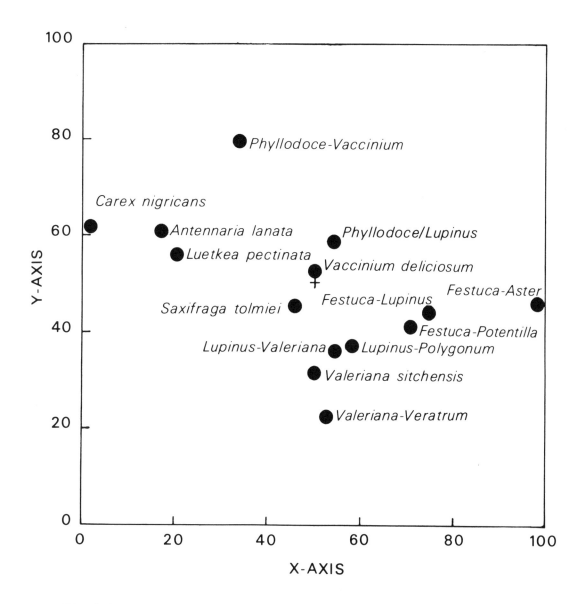

Figure 187.—Ordination of major community types at Mount Rainier in which *Carex nigricans* **(Cani; cold wet habitat),** *Festuca viridula-Aster ledophyllus* **(Fevi-Asle; warm dry habitat),** *Phyllodoce empetriformis-Vaccinium deliciosum* **(Phem-Vade; mesic shallow-soiled, short growing season habitat), and** *Valeriana sitchensis-Veratrum viride* **(Vasi-Vevi; mesic, deep-soiled, long growing season habitat) communities are used as the four end points; ordination procedures follow those of Dick-Peddie and Moir (1970) (from Henderson 1973).**

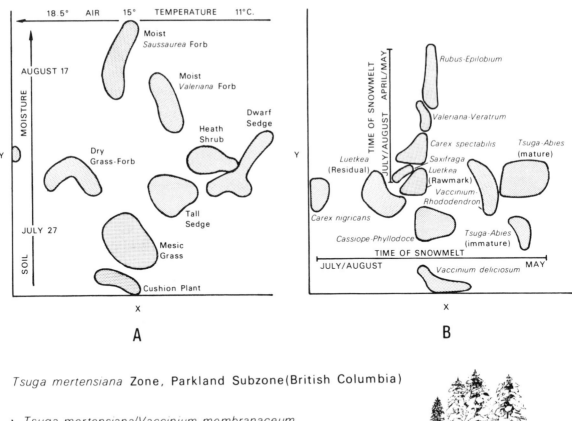

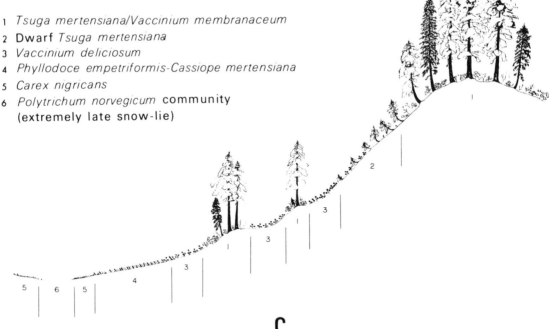

Figure 188.—Ordinations of the meadow communities in *A,* **the Olympic Mountains (X axis is mean daily air temperature at 15 centimeters above ground during July and August 1967; Y axis is date at which permanent wilting percentage was reached at a soil depth of 20 centimeters) (from Kuramoto and Bliss 1970); and** *B,* **northern Cascade Range, showing floristic and environmental relationships (from Douglas 1972).** *C* **shows topographic relationships between several communities in the parkland subzone of the British Columbia** *Tsuga mertensiana* **Zone (from Brooke et al. 1970).**

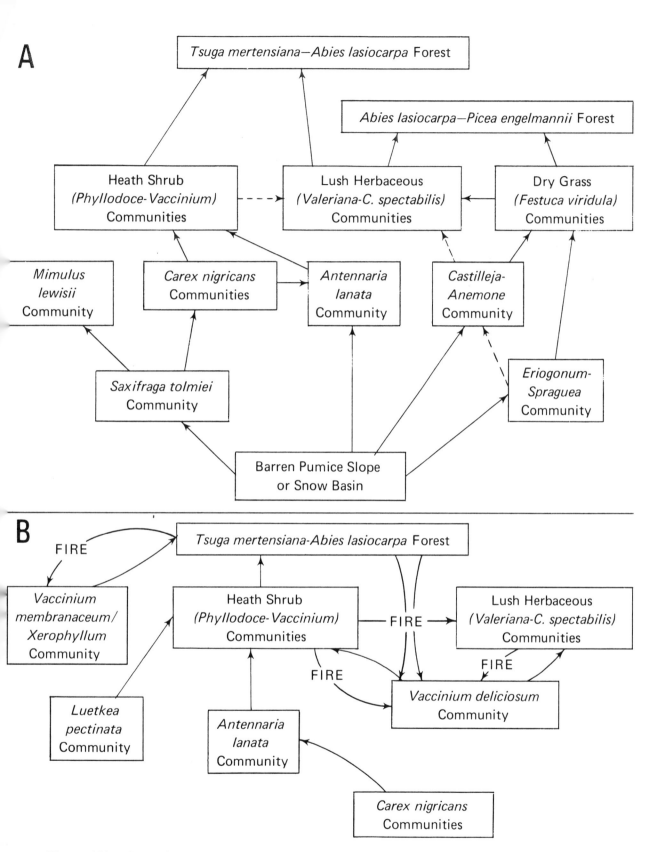

Figure 189.—Successional patterns as hypothesized for *A*, high elevation pumice slopes or snow basins, and *B*, for areas influenced by fire, in the parkland subzone of Mount Rainier National Park (from Henderson 1973).

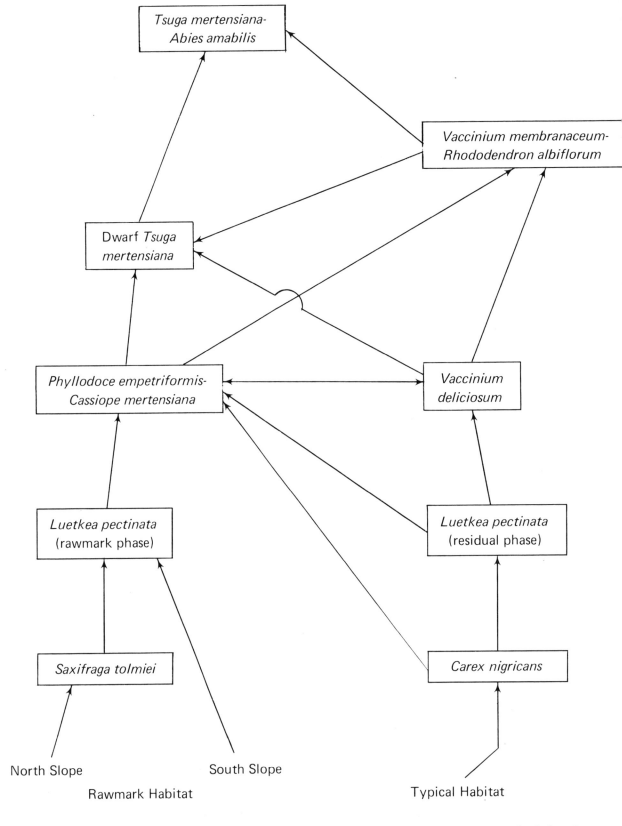

Figure 190.—General successional trends for subalpine meadow habitats as hypothesized for the northern Cascade Range (from Douglas 1970).

High Cascades of Oregon

Subalpine parklands in the central and southern Oregon Cascade Range appear considerably sparser than the densely vegetated regions just discussed. The High Cascades are the consequence of Pleistocene vulcanism, and the resulting substrates combined with a somewhat warmer and drier climate are important factors in the sparser vegetation. Pumice and cinder fields, outwash flats, and lava fields are common in timberline areas and provide extremely rigorous environments for plant growth.

Subalpine herbaceous communities have been studied on a broad scale on Three Sisters (Van Vechten 1960) and near Mount Jefferson (Swedberg 1961). Lower in the zone, *Lupinus latifolius* meadows appear on moister sites and include *Castilleja parviflora*, *Potentilla flabellifolia*, *Epilobium alpinum*, *Aster ledophyllus*, *Senecio triangularis* and *Ligusticum grayi* as constituents. On somewhat drier sites, *Trisetum canescens*, *Carex nigricans*, *Juncus drummondii*, and *Aster alpigenus* appear as dominants. Dry, south slopes are dominated by *Lupinus latifolius* or *Festuca viridula* or both. *Phyllodoce empetriformis*, *Cassiope mertensiana*, and *Vaccinium scoparium* are characteristic species on rockier sites and on cool north slopes, where they are joined by *Cardamine bellidifolia*, *Saxifraga tolmiei*, *Luetkea pectinata*, and *Castilleja* spp. This grouping and the *Carex nigricans-C. spectabilis* communities, associated with some wet meadows, are similar to subalpine communities previously discussed. *Phyllodoce* and *Cassiope* are also characteristic of ridges and slopes higher in the timberline region; VanVechten (1960) mentions *Luetkea pectinata*, *Castilleja parviflora*, *Hieracium gracile*, *Carex* spp., *Vaccinium scoparium*, and *Anemone occidentalis* among the associates.

Campbell's (1973) work is the only fine-scale study of subalpine meadow vegetation in the Oregon Cascades. She recognizes two broad classes of meadow communities in a small valley near Mount Jefferson—"typical" and "atypical" or hydric. The typical community types are:

Phyllodoce empetriformis-Cassiope mertensiana,
Vaccinium deliciosum,
Potentilla flabellifolia-Carex nigricans,
Carex nigricans-Polytrichum norvegicum,
Carex nigricans-Aster alpigenus,
Senecio triangularis, and
Alpine bryophyte.

Four of the five major groups of communities in our earlier discussion of western Washington subalpine vegetation can be recognized in this list. The *Phyllodoce-Cassiope* and *Vaccinium* communities are so similar to those described earlier no additional description is required. Campbell (1973) does mention the occurrence of *Vaccinium deliciosum* communities as a band along the margin of mature *Tsuga mertensiana* forest and common invasion by *Tsuga* and *Abies lasiocarpa* seedlings, mostly since 1934. The *Potentilla-Carex* community is found on well-drained sites between the *Phyllodoce-Cassiope* and *Carex nigricans* types. Major species, in order of percent cover, are *Potentilla flabellifolia*, *Carex nigricans*, *Ligusticum grayi*, *Carex spectabilis*, *Hieracium gracile*, *Castilleja parviflora* var. *oreopola*, and *Polytrichum juniperinum*. Both *Carex nigricans* communities occupy poorly drained bottoms, but the *Carex-Polytrichum* has a significantly shorter snow-free season than the *Carex-Aster* type; it is also a much more depauperate community (three vs. eight vascular plants, and one vs. six bryophytes). The *Senecio triangularis* community is a lush, tall-herb type. Thirty-two vascular plants are constituents of which *Senecio*, *Aster occidentalis*, *Potentilla flabellifolia*, *Viola glabella*, *Carex spectabilis*, *Lupinus latifolius*, and *Dodecatheon jeffreyi* have over 1-percent coverage. The alpine bryophyte community is a pioneer type found on areas with extremely persistent snow-banks. Six bryophytes comprise the community list: *Bryum alpinum*, *Pohlia ludwigii*,

Barbilophozia lycopodioides, *Polytrichum norvegicum*, *Moerckia blyttii*, and *Dicranella heteromolla*.

The hydric communities indentified by Campbell (1973) are mainly new ones:
Carex rostrata-Sphagnum squarrosum,
Eleocharis pauciflora-Aulacomnium palustre,
Carex scopulorum, and
Carex sitchensis.

The *Carex-Sphagnum* community is a bog type occurring below main seepage areas; moving water is found at the level of the moss layer all year. *Carex rostrata* is the major dominant averaging 5 decimeters in height, but *Eleocharis pauciflora* is also important and *Salix commutata* is an invader. Other species with over 1-percent cover are *Polygonum bistortoides*, *Epilobium alpinum*, *Dodecatheon jeffreyi*, *Saxifraga oregana*, *Pedicularis groenlandica*, *Carex illota*, *Sphagnum squarrosum*, and *Drepanocladus exannulatus*. The *Carex scopulorum* community is the most widespread hydric type. The sites occupied are rarely flooded, but soils are saturated throughout the year. *Carex scopulorum* is the major dominant growing up to 60 centimeters in height. On seepage areas, *Caltha biflora* and *Carex luzulina* assume the importance of codominants. Other important constituent species are *Polygonum bistortoides*, *Carex illota*, *Carex interrupta*, *Potentilla flabellifolia*, *Calamagrostis canadensis*, *Ligusticum grayi*, and *Philonotis americana*. The *Eleocharis-Aulacomnium* community dominates small areas of stagnant shallow water, usually as patches within a *Carex scopulorum* meadow community. *Carex rostrata* and *C. scopulorum* are important associates of the *Eleocharis*. *Aulacomnium palustre* and *Philonotis americana* dominate the moss layer, with *Sphagnum squarrosum* a minor component. The *Carex sitchensis* community is subject to inundation for up to 3 weeks during the spring snowmelt period. *Polygonum bistortoides* and *Calamagrostis canadensis* are consistent associates of the 6- to 8-decimeter-tall *Carex sitchensis*. Other significant species are *Caltha biflora*, *Dodecatheon jeffreyi*, *Carex illota*, *Salix commutata*, and *Boykinia major*.

Campbell (1973) reports that the relationships between time of snowmelt and community types in her study area are similar to those reported for the Washington Cascade Range. She also shows a relationship between floristics and time of snowmelt within the *Carex nigricans* communities.

Large, nearly barren pumice flats are conspicuous features of the subalpine parkland from Mount Jefferson south in the Cascade Range (Van Vechten 1960, Swedberg 1961, Horn 1968). Typical colonizers are low compact perennials with large taproots. Characteristic species in the north are *Eriogonum marifolium*, *E. pyrolaefolium*, *Lupinus lepidus*, *Penstemon procerus*, *Raillardella argentea*, *Spraguea umbellata*, *Polygonum newberryi*, *Juncus drummondii*, *Aster alpigenus*, *Carex breweri*, *Castilleja arachnoidea*, and *Lomatium angustatum*. VanVechten (1960) studied one area (the Cinder Desert) on which *Eriogonum pyrolaefolium* was the only species of significance; occasional plants of *Spraguea umbellata*, *Draba aureola*, *Smelowskia calycina*, *Hulsea nana*, and *Carex breweri* were also present. Outwash flats apparently have similar poorly developed communities. VanVechten (1960) mentions *Eriogonum pyrolaefolium*, *E. marifolium*, *Raillardella argentea*, *Senecio fremontii*, *Spraguea umbellata*, and *Aster alpigenus* as characteristic species.

Horn's (1968) study of the Pumice Desert at Crater Lake National Park illustrates the sparsity of vegetation on these pumice and cinder flats. Coverage of all vascular plants totaled only 4.5 percent. *Eriogonum marifolium*, *Carex breweri*, *Stipa occidentalis* var. *californica*, *Arenaria pumicola*, *Spraguea umbellata*, and *Polygonum newberryi* were the most important of 14 vascular species present on the desert. Horn (1968) conducted some environmental studies and concluded that a severe climatic regime (wide diurnal temperature fluctuations) and low soil fertility were responsible for the sparsity of the vegetation. Soil moisture was apparently available throughout the short, intense growing season.

270

Interior Mountains

Grasslands are the dominant meadow types in many timberline areas of eastern Oregon and Washington (fig. 191). Included here are subalpine areas on the east side of the Washington Cascade Range and in the Wallowa and Blue Mountains in Oregon.

A bunchgrass, *Festuca viridula*, characterizes pristine communities of this type

Figure 191.—Grasslands dominate the subalpine parklands in drier mountain ranges of eastern Oregon and Washington; top, climax *Festuca viridula* grassland in virgin condition, and bottom, overgrazed *Festuca* grassland which has deteriorated into a community of weeds (e.g., *Eriogonum* spp.) and *Stipa* spp. (Wallowa Mountains, northeastern Oregon).

(Pickford and Reid 1942) except in the Blue Mountains. Since we are considering a group of communities covering a wide geographic area, there are many associates. Some of these are:

Grasses and grasslike: *Stipa lettermanii, S. occidentalis* var. *minor, Agrostis rossiae* (Cascade Range), *Carex geyeri, C. hoodii, Sitanion hystrix, Phleum alpinum, Agropyron* spp., *Bromus carinatus, Poa* spp., and *Trisetum spicatum.*

Forbs: *Eriogonum heracleoides, E. piperi, Gilia nuttallii, Lupinus leucophyllus, L. latifolius, Polemonium pulcherrimum, Penstemon rydbergii, Erigeron speciosus, E. peregrinus, Arenaria formosa, Hieracium gracile, Potentilla glandulosa, Phlox diffusa, Polygonum phytolaccaefolium, P. newberryi,* and *P. bistortoides.*

Shrubs: *Artemisia tridentata* var. *vaseyana, Potentilla fruticosa, Ribes* spp., *Phyllodoce empetriformis,* and *Vaccinium scoparium* (all minor).

Most of the *Festuca viridula* grasslands have been grazed, many of them to excess, and are in some deteriorated grass-forb stage. In 1938, Pickford and Reid (1942) started a study of changes in composition, productivity, and soil erosion associated with grazing of these types in the Wallowa Mountains. Recently, Strickler (1961) continued this study and made interesting photographic and analytic comparisons of conditions in 1938 and 1956-57. *Festuca viridula, Agropyron caninum, Melica bulbosa, Stipa lettermanii, S. occidentalis* var. *minor,* and *Lupinus leucophyllus* composed the virgin communities; total plant coverage was high. "Mixed grass-and-weed" and "weed-needlegrass" (*Stipa* spp.) communities with a high proportion of bare soil characterized the overgrazed sites. Amounts of *Eriogonum* spp., *Gilia nuttallii, Penstemon rydbergii, Arenaria formosa, Artemisia tridentata,* and *Stipa* spp. were generally higher in these communities. Eventual return of these communities to their climax state (dominance of *Festuca viridula*) appears questionable.

Tree Species, Tree Groups, and Forest-Meadow Dynamics

The other component of the parkland subzone is made up of individual trees, patches of forest, and tree groups or clumps. We will consider the major timberline tree species found in Oregon and Washington, the communities associated with forest or tree clumps in the parkland subzone, and, finally, some successional or dynamic aspects of the parkland forest-meadow ecotones.

Timberline Tree Species

A great many tree species occur in timberline areas, some of which are listed in table 40. *Abies lasiocarpa, Tsuga mertensiana, Pinus albicaulis,* and *Larix lyallii* are characteristic, however. All these species except *L. lyallii* are in one of Clausen's (1963) "most tolerant timberline tree species complexes," and it was apparently omitted by accident or oversight. *Abies lasiocarpa* is the most widespread, occurring in all timberline areas except in parts of southern Oregon. Near forest line, it is usually abundant as an erect tree dominating the islands of forest (fig. 192). It is reduced to a shrubby krummholz form at higher elevations where it often forms dense mats by layering.

Tsuga mertensiana is widely distributed at timberline throughout all but the most xeric portions of the Cascade Range and Olympic Mountains. In the most maritime portions of these ranges, it is usually more important than *Abies lasiocarpa;* timberline on Mount Baker is completely dominated by *Tsuga mertensiana,* and both *Abies lasiocarpa*

Table 40. — Tree species typically found between forest line and timberline in selected parts of Washington and Oregon

Species	Cascade Range			Olympic Mountains, Washington	Wallowa Mountains, Oregon
	Northwestern Washington	Northeastern Washington	Central Oregon		
Abies amabilis	m	—	m	m	—
Abies lasiocarpa	M	M	M	M	M
Abies magnifica var. *shastensis*	—	—	m	—	—
Chamaecyparis nootkatensis	m	—	—	m	—
Larix lyallii	—	M	—	—	—
Picea engelmannii	—	m	—	—	m
Pinus albicaulis	m	M	M	m	M
Pinus contorta	—	m	m	m	m
Tsuga mertensiana	M	—	M	M	—

Note: M = major species; m = minor species.

and *Pinus albicaulis* are very rare. Its timberline growth behavior is like that of *Abies lasiocarpa* (fig. 193). Clausen (1965) believes that both *T. mertensiana* and *Pinus albicaulis* have evolved distinct "subalpine erect" and "alpine horizontal" races, but this has not been confirmed experimentally.

Pinus albicaulis is present in most timberline areas. However, it is unquestionably most important in more xeric regions—eastern and southern parts of the Cascade Range and Okanogan and Wallowa Mountains, or locally, on the eastern rain-shadow slopes of the major volcanoes such as Mount Rainier and Mount Hood. *Pinus albicaulis* is able to grow erect to higher elevations than either *Abies lasiocarpa* or *Tsuga mertensiana*, although it will form krummholz at its upper limits. It sometimes functions as a pioneer tree species in invasion of meadow areas (Franklin and Mitchell 1967). The wingless seeds of *Pinus albicaulis* are distributed primarily by the Clark's nutcracker (*Nucifraga columbiana*), a large jay, which consumes a portion of the seed crop and hoards the rest. Reproduction develops from the forgotten hoards (fig. 194). Their relationship is the same as that described for *Pinus cembra-Nucifraga caryocatactes* in the European Alps and for *Pinus pumila-Nucifraga caryocatactes* var. *japonicus* in Japan.

Figure 192.—*Abies lasiocarpa* is the most widespread of the timberline tree species; pictured are groups of this species near timberline in the eastern Cascade Range (Wenatchee National Forest, Washington).

Figure 193.—*Tsuga merten-siana* is an important timberline tree species throughout most of the Cascade Range and Olympic Mountains.

Figure 194.—*Pinus albicaulis* is a major timberline tree species in drier mountain areas; reproduction of the species is largely dependent upon the hoarding habits of a large jay, *Nucifraga columbiana,* which give rise to groups of seedlings and saplings (Sunrise Ridge, Mount Rainier National Park, Washington).

Larix lyallii is, with rare exception, limited to the eastern half of the northern Cascade Range and Okanogan Mountains in Washington. It occurs only at or near timberline in these areas and appears exceptionally well adapted to the environment (fig. 195). *Larix lyallii* typically grows to higher elevations than any of its associates and maintains an erect habit when other species are unable to grow or occur only as krummholz beneath the *Larix* (Arno 1970, Arno and Habeck 1972). As a result, *Larix lyallii* stands often form a distinctive forest belt between forest and alpine; this belt is particularly conspicuous during the fall as a swath of brilliant orange yellow.

Many other species do occur in localized timberline areas. *Chamaecyparis nootkatensis* is a major krummholz species in parts of the Washington Cascade Range and northeastern Olympic Mountains. *Pinus contorta* is typical in more xeric timberline tracts in association with *Pinus albicaulis. Picea engelmannii* is occasional at timberline, generally to the east of the Cascade crest (fig. 196). *Abies amabilis* is common below the tree line in the most maritime portions of the Cascade Range and Olympic Mountains. *Abies magnifica* var. *shastensis* and *Pinus monticola* occupy similar sites in the southern Oregon Cascade Range (e.g., at Crater Lake).

Figure 196.—*Picea engelmannii* is one of several species which form true krummholz in upper subalpine and lower alpine regions in more continental parts of the Pacific Northwest; this particular specimen is on the northeast slopes (rain-shadow side) of Mount Rainier, however (Mount Rainier National Park, Washington).

Figure 195.—Although limited in distribution, *Larix lyallii* appears exceptionally well adapted to the timberline environment, often growing to higher elevation than any arborescent associates (Wenatchee National Forest, Washington).

Forest and Tree Groups

Tree groups or clumps are one of the most distinctive and picturesque features of subalpine parkland (fig. 197). The characteristics of these tree groups, specifically their composition and development, have been the subject of considerable study in wetter portions of the Cascade Range (Franklin and Mitchell 1967; Douglas 1970, 1972; Lowery 1972; Brooke et al. 1970). In addition, there is a substantial body of information about *Larix lyallii* communities (Arno 1970, Arno and Habeck 1972).

276

Figure 197.—Tree groups are one of the more characteristic and picturesque aspects of subalpine parklands; these groups are predominantly *Abies lasiocarpa* and *Pinus albicaulis* (Sunrise Ridge, Mount Rainier National Park, Washington).

Tree Clumps or Groups

Most of the study of subalpine tree groups has been conducted in the parkland subzone of the *Tsuga mertensiana* Zone in wet coastal British Columbia and western Northern Cascades Province (Lowery 1972; Douglas 1970, 1972; Brooke et al. 1970).

The tree groups may range in size from a few trees to 0.1 hectare or more in area. They typically (but not exclusively) occur on slight hummocks, ridges, or other convex topography indicating a relationship with areas in the landscape with longer snow-free seasons. Four tree species are constituents—*Chamaecyparis nootkatensis*, *Tsuga mertensiana*, *Abies lasiocarpa*, and *Abies amabilis*. These species can all occur in a single tree group, but there are some indications of different amplitudes. *Abies lasiocarpa* is more characteristic of warmer, drier slopes which probably reflects an ability of its seedlings to become established on droughty sites superior to those of *Tsuga* or *Abies amabilis* (Lowery 1972). *Abies lasiocarpa* also tends to be more common in the alpine krummholz formations. *Abies amabilis* is more common in mesic and lower elevation tree groups; as will be mentioned, it also is often a later arrival in group development. Lowery (1972) found *Chamaecyparis nootkatensis* a rather infrequent constituent of subalpine tree groups in his study area (mainly on moist, low-elevation habitats); but in some other areas, we have observed it to be rather common even extending as krummholz mats into the Alpine Zone.

Distinctive floras (composed largely of species from the closed forest subzone) and vegetational belts are frequently associated with subalpine tree groups (Douglas 1970, 1972; Brooke et al. 1970). Brooke et al. (1970) recognized two mature tree communities in coastal British Columbia: *Tsuga mertensiana/Vaccinium membranaceum* (mesic habitats) and *Tsuga mertensiana/Cladothamnus pyrolaeflorus* (moderately dry habitats). *Vaccinium membranaceum* and *Rhododendron albiflorum* are the most characteristic tall shrubs in the *Tsuga/Vaccinium* community and occur closely clumped around the bases of isolated trees or clumps of trees in exposed habitats (i.e., forming marginal "rings" of tall shrubs around the groups). Most of the lesser associated species are typical of adjacent meadows, e.g., *Phyllodoce empetriformis, Cassiope mertensiana,* and *Vaccinium deliciosum.* The *Tsuga/Cladothamnus* community has an even ranker growth of tall shrubs (average cover, 65 percent) with *Cladothamnus, Menziesia ferruginea, Vaccinium alaskaense,* and *V. membranaceum* as the dominants. Shrublike *Chamaecyparis nootkatensis* and sapling *Tsuga* and *Abies amabilis* are associated with these tall shrubs. Again, meadow species characterize the lesser vegetation.

Douglas (1970, 1972) recognizes only a single tree group community, the "Mature *Tsuga mertensiana-Abies amabilis*"; however, he does recognize both an "open" and "closed" phase of this community. The major shrubs are *Vaccinium membranaceum* and (in the open phase) *Menziesia ferruginea.* Low shrubs and herbs in the closed phase are dominantly *Luzula wahlenbergii* and *Rubus lasiococcus.* In the open phase, *Luetkea pectinata, Phyllodoce empetriformis,* and *Rubus pedatus* are conspicuous. Douglas (1970, 1972) also recognizes a shrub community which occurs on the margin of tree groups— *Vaccinium membranaceum-Rhododendron albiflorum.* The close relationship of this community with that of Brooke et al. (1970) is obvious. Douglas (1970, 1972) maintains that this community develops after the forest is established (rather than preceding it) as a consequence of the lengthened growing season wrought by the trees. Major species, in order of importance, are *Rhododendron albiflorum, Vaccinium membranaceum, Sorbus sitchensis grayi, Menziesia ferruginea, Phyllodoce empetriformis, Vaccinium deliciosum, Rubus pedatus,* and *Vaccinium ovalifolium.* Douglas (1970, 1972) and Brooke et al. (1970) recognize a similar dwarfed *Tsuga mertensiana* community which may or may not be associated with a mature tree group. Douglas (1970, 1972) interprets this as an early stage in forest or tree group development and calls it an "immature" *Tsuga-Abies* type.

Apparently, podzolic soil types are characteristic under most subalpine forest or tree groups. The best study conducted to date is that of Bockheim (1972) in the Mount Baker area. Two subalpine tree groups, as well as an alpine krummholz stand, had well-developed Podzols (Lithic Cryohumod, Humic Cryorthod, and Humic Lithic Cryorthod) with 3- to 9-centimeter A2 horizons; 5- to 17-centimeter B21hir horizons; and 9- to 25-centimeter B22hir or B22h horizons.

The very appearance of subalpine tree groups suggests their establishment as one or a few trees at a central point followed by gradual expansion (fig. 198); the tallest and largest

Figure 198.—Subalpine tree groups in early stages of development (Sunrise Ridge, Mount Rainier National Park, Washington). Left: Group initiated by *Pinus albicaulis* with expansion taking place primarily by layering of *Abies lasiocarpa*; some *Tsuga mertensiana* seedlings have become established within the protection of the larger trees. Right: Similar group initiated by *Abies lasiocarpa*; note the confinement of *Pinus* saplings to the margin of the group.

diameter trees are generally in the central area with increasingly smaller trees toward the peripheries. This general hypothesis has been examined by several scientists but most carefully by Lowery (1972). His dissection of two clumps at Mount Baker shows clearly that they have expanded radially around the first trees established at the site (fig. 199). Furthermore, the height and diameter growth rates are slower on the peripheries of the clumps than near the centers. *Abies lasiocarpa* was the major pioneer in both clumps.

Similar tree groups or clumps have been recognized in subalpine parklands throughout Oregon and Washington and, indeed, the entire mountain west. Franklin and Mitchell (1967) studied subalpine tree groups around Mount Rainier and hypothesized a developmental sequence (fig. 200) which has been partially confirmed (Franklin et al. 1971). Douglas (1969) has described other types of subalpine tree groups.

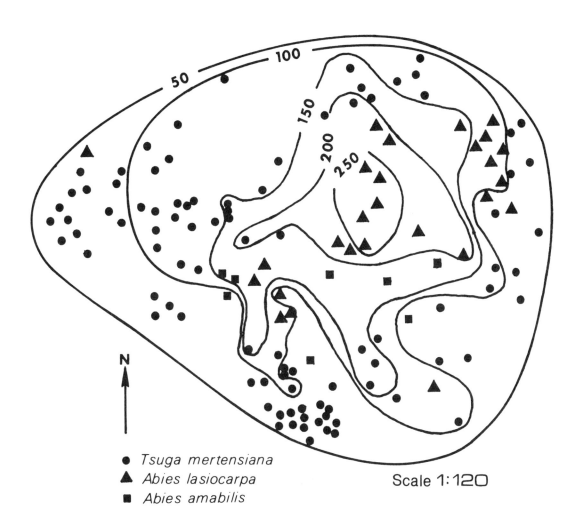

Figure 199.—Isolines of total tree age for tree clump in subalpine parkland near Mount Baker; all trees within an isoline are that age or older except for two *Abies amabilis* **within the 150-year-age isoline which were 35 years old and two very large dead** *Abies lasiocarpa* **in the center of the clump (from Lowery 1972).**

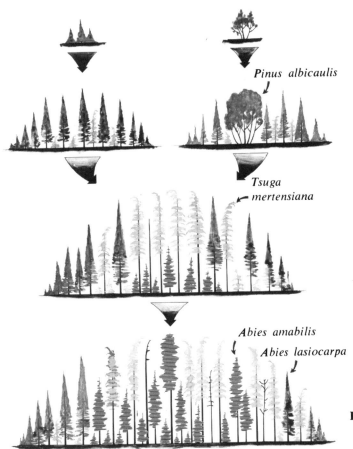

Pinus albicaulis

Tsuga mertensiana

Abies amabilis

Abies lasiocarpa

Figure 200.—Schematic diagram showing one possible sequence in development of subalpine tree groups in the Cascade Range.

To summarize, a general successional sequence in tree clump development would appear to be as follows. The group is initiated by a hardier subalpine tree species—often *Abies lasiocarpa* or *Pinus albicaulis* in drier habitats or more continental regions, *Tsuga mertensiana* in moister subalpine habitats. As the invaders grow larger, they begin to exert an influence on their microenvironment, i.e., lengthening the growing season because of their black body effect which results in earlier snowmelt. Consequently, additional trees of the same or other species become established on the peripheries by layering of branches and seed germination. As the group enlarges, trees in the central area die, either leaving an unfilled hole (Lowery 1972) and an "atoll" form of tree group or being replaced by seedlings of the same or a more tolerant species (Franklin and Mitchell 1967). The pioneer species generally continues to be represented at the peripheries (Franklin and Mitchell 1967); it is notable, however, that at Mount Baker *Tsuga mertensiana* was best represented in the peripheries of groups pioneered by *Abies lasiocarpa* (Lowery 1972). Nonetheless, given sufficient time for tree clump development and a constant climate, the successional sequence in expanding tree groups should follow the same general pattern as in the adjacent closed forests.

The reader should note that successional status is not often considered among timberline tree species since all are climax in a sense. None are in danger of elimination from the parkland since they can migrate to other open areas if more tolerant species become too competitive. The same is, of course, true of meadows being replaced by forests. Nonetheless, gradual changes do take place, and timberline tree species (and meadows) are displaced from sites they colonize.

A succession of tree species can apparently occur even in alpine krummholz stands. Archer (1963), for example, suggests a sequence of *Chamaecyparis nootkatensis*, *Abies lasiocarpa*, and finally, *Tsuga mertensiana* may occur in krummholz stands at Garibaldi Park in British Columbia.

Larix lyallii and Other Interior Timberline Forests

As mentioned, *Larix lyallii* is one of the most interesting of the subalpine trees, forming upright stands at higher elevations than any other associated tree species. It is a continental species found only on the eastern slopes of the northern Cascade Range in Washington in our region where it forms a distinctive zone.

The efforts of Arno (1970) and Arno and Habeck (1972) provide descriptions of the numerous *Larix lyallii* communities found in this area. *Larix lyallii* occurs with a wide variety of tree associates, of which *Picea engelmannii*, *Abies lasiocarpa*, and *Pinus albicaulis* are most important in more continental areas; maritime species such as *Tsuga mertensiana* and *Abies amabilis* become more important on approaching the crest of the Cascade Range. These associates are confined to a krummholz form in many of the stands which are at 1,800 to 2,400 meters. The ameliorating influence of the *Larix* appears to allow typical lower elevation subalpine understory species and trees to ascend to higher elevations than would otherwise be possible. Major understory associates in the Cascade Range are *Vaccinium scoparium*, *V. deliciosum*, *Rhododendron albiflorum*, *Luetkea pectinata*, *Luzula glabrata*, *Phyllodoce empetriformis*, *P. glanduliflora*, *Cassiope mertensiana*, *Juniperus communis*, *Carex rossii*, and *Carex nigricans*. Each of these dominates the understory in at least one of the Cascade Range stands that was sampled. Arno (1970) and Arno and Habeck (1972) do not attempt a classification of *Larix lyallii* communities, however, and suggest several reasons why community types or patterns are difficult to define.

Daubenmire and Daubenmire (1968) mention another interior timberline community, the *Pinus albicaulis-Abies lasiocarpa* type. Dwarfed and wind-deformed trees of these two species are scattered singly or in small groves and share climax status. Ground cover commonly consists of *Vaccinium scoparium*, *Xerophyllum tenax*, *Carex* spp., *Luzula* sp., *Erigeron peregrinus*, and *Polygonum bistortoides*.

Dynamics at Forest-Meadow Ecotones and Timberline

Timberline areas are tension zones—dynamic ecotones between tree and treeless regions. As at lower elevations, directional changes are constantly taking place in response to allogenic (e.g., long- and short-term climatic changes) and autogenic (changes in environment brought about by organisms) factors. Successional considerations are particularly complex in subalpine parkland since they involve relationships between tree species, various meadow communities, and forest and meadow communities.

Some of the successional relationships, mostly of the gradual developmental type, have been discussed in earlier sections: succession between meadow community types and development of subalpine tree groups and their gradual encroachment on meadows. We would like to consider three more dramatic phenomena in a little more detail here: (1) extensive invasion (replacement) of meadows by trees, (2) the role of fire, and (3) upward movement of timberline.

Massive Invasions of Subalpine Meadows

Gradual changes in the forest-meadow ecotone are to be expected, with a net loss of meadow areas during times of more favorable climate or for a substantial period after termination of a major glacial epoch under a stabilized climate. Most authors consider forest the climatic climax in at least the lower portions of the parkland subzone of the Pacific Northwest. Expansion of tree clumps would fall into this "gradual" category.

During the last half century, however, massive invasion of meadows has taken place at many locations in the Pacific Northwest (fig. 201) (Brink 1959; VanVechten 1960; Fonda 1967; Kuramoto and Bliss 1970; Brooke et al. 1970; Douglas 1970, 1972; Henderson 1973; Franklin et al. 1971; Lowery 1972; and others). This in no sense resembles a gradual change; it is rapid and recent. Several authors have examined age class distributions of the seedlings invading meadows (e.g., Lowery 1972, Henderson 1973, Franklin et al. 1971) and found similar patterns with most trees having become established 20 to 50 years ago (fig. 202). Exact patterns of species and ages vary from area to area; but, in general, the period of invasion is centered on the 1930's, and both young seedlings and trees older than 40 to 50 years are uncommon. The simultaneous, apparently temporary conditions, favorable for conifer invasion over such a wide geographic area, suggest a climatic flux, i.e., lessened duration of snow cover or lengthened growing season. Franklin et al. (1971) suggest that the warmer, drier climate experienced between the late 1800's and mid-1940's is responsible. Henderson (1973) suggests that the explanation is more complex, actually involving good seed years followed by summers of above normal precipitation and temperature; he also indicates that significant meadow invasion has continued to the present, a phenomenon not confirmed by other authors.

Intensity of tree invasion and growth rate and form of trees after establishment vary from one meadow community to another (Henderson 1973, Franklin et al. 1971). Henderson (1973) indicates some of these relationships:

Figure 201.—Changes in the subalpine forest-meadow ecotones have typically been gradual; however, massive invasions of meadow areas by tree species have taken place in the last 50 years all over the Cascade and Olympic Mountains (Paradise Valley, Mount Rainier National Park).

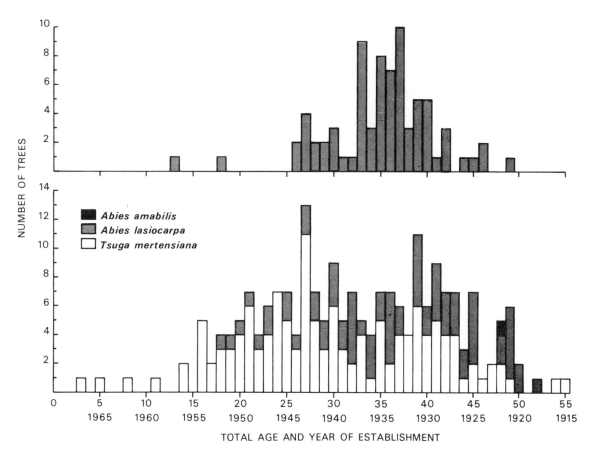

Figure 202.—Age distributions of invading seedlings at Paradise Valley on Mount Rainier (upper) and in the Mount Baker area (lower, from Lowery 1972).

Meadow community type	Intensity of tree invasion	Spatial pattern	Growth rate	Growth form
Carex nigricans	Very low	Singly	Very low	Distorted
Phyllodoce-Vaccinium	High	Widespread	Low	Distorted
Valeriana-Veratrum (Lush herbaceous group)	Moderate	Singly or small groups	Moderate	Straight
Festuca viridula group	Low	Singly or copses	Moderate	Straight

It seems clear that, in the *Festuca viridula* communities, drought and possibly heat retard forest establishment; trees grow well once established, however. The wet, cool environment of the *Phyllodoce-Vaccinium* group places different constraints on tree seedling establishment, and it is here that recent meadow invasion has been most intense; short growing seasons and heavy snow loads severely retard growth, however.

The climate has entered a cooler, moister period since the mid-1940's; and, as mentioned, most investigators note a drastic decline in tree establishment since that time. It will be interesting to observe whether the deep snowpacks and short growing seasons experienced during recent years (especially 1971 and 1972) result in mortality of established trees, i.e., reclamation of meadows.

Role of Fire

Fire is considered an important factor in creation and maintenance of subalpine meadows by several scientists (e.g., Kuramoto and Bliss 1970, Henderson 1973, and Bockheim 1972). Henderson (1973) incorporates fire effects into one of his successional diagrams (fig. 189B). Bockheim (1972) describes charcoal from all of his subalpine soils, even under meadows, and indicates it would probably be necessary to have forest vegetation before a significant amount of charcoal could be produced. A fire-created meadow might be maintained by other factors, such as herbaceous competition, for many centuries without fire recurrence.

Unquestionably, fire has played an important role, particularly in the lower part of the subalpine parkland where forest and tree groups are more abundant. Recent wildfires, experienced by the authors, show clearly that fire will carry from tree clump to tree clump often skipping over intervening meadows. Douglas and Ballard (1971) show how fire in krummholz stands eliminates the trees and moves succession back to a previous meadow stage (*Phyllodoce-Vaccinium* in their case). Henderson (1973) recognizes several meadow communities which he feels are seral to forest and may be products of fire. His *Vaccinium membranaceum/Xerophyllum tenax* community is certainly a product of fire in the subalpine forest. In many of these seral meadow communities, however, a fire origin is obvious from rotten wood, old logs and snags, presence of several "forest-site" herbs and shrubs, and an abundance of charcoal.

We would suggest caution in ascribing the majority of subalpine meadows to past fires, however. The evidence suggests other historic and environmental factors are dominant in determining the present vegetational mosaic in wetter mountain areas. In drier regions, particularly the interior mountains, fire has probably had more of a key role.

Changes in Timberline

Changes in the overall elevation of forest line or scrub line have not been reported in this region with one exception—on Mount St. Helens, a relatively low (2,948 m.) peak in southwestern Washington. The existing cone (fig. 203) is of very recent origin (less than 2,000 years ago) and its slopes are mantled with coarse pumice. Timberline is very low compared with other peaks in the vicinity (about 1,340 m.) and composed of species not normally found at timberline—*Pinus contorta* (dominant), *P. monticola, Pseudotsuga menziesii, Populus trichocarpa, Abies procera,* and *Tsuga heterophylla,* for example. Lawrence (1938) concluded that timberline is advancing at a discernible rate on Mount St. Helens after studying some photographs taken in 1897 and suggests present timberline is a consequence of edaphic conditions.

Alpine Communities

The Alpine Zone is not well developed in the mountains of Oregon and Washington. Habitats capable of supporting alpine plant communities are not extensive in contrast to the large alpine acreages in the Rocky Mountains. Most of the Alpine Zone is steep and rugged, occupied by glaciers, snowfields, bare rock, and rubble (fig. 204). Numerous observers have commented upon the narrow elevational belt between the timberline (upper limit of upright tree growth) and line of permanent snow and ice (see Troll 1955 and Knapp 1965, for example). Furthermore, much of the Alpine Zone occurs on only recently dormant Pleistocene volcanoes of the Cascade Range where substrates are youthful and steep slopes help perpetuate them in that state.

Figure 203.—Timberline is low (1,340 m.) but advancing on Mount St. Helens, a Cascade Range volcano of very recent origin (less than 2,000 years old). This photograph was taken prior to the eruption of Mount St. Helens in May 1980.

Figure 204.—The Alpine Zone is not well developed in Oregon and Washington; areas above timberline are occupied primarily by glaciers, snowfields, bare rock, and rubble (Mount Rainier from Sunrise Ridge).

Since the broad, snow-free alpine tracts typical of the Rocky Mountains are absent, it is not surprising that alpine vegetation in the Pacific Northwest is neither well known nor extensively developed. Krajina (1965) feels that the *Phyllodoce-Cassiope* association is the climatic climax in the Alpine Zone of adjacent British Columbia. Scott (1962) lists a Saxifrage-Heather association as climax in western Washington's "Arctic-Alpine Zone"; this he considers to lie between 2,300 and 2,600 meters with essentially no vegetation above that level. We suspect that some type of herbaceous turf community may ultimately be identified as the zonal climax in the upper part of the Alpine Zone with an ericaceous shrub type possibly characterizing the lower part of the zone.

Despite limited extent, the basic features of alpine regions throughout most of the temperate zone are apparently present; wind and snow depth and duration combine with edaphic features to determine the structure and composition of the communities. Bliss (1969) concludes that the primary factors controlling community patterns in the Olympic Mountains are time of snowmelt and accumulation of snow in relation to slope aspect. Patterned ground is consistently reported from the alpine areas including frost boils, stone nets, and stone stripes, as well as solifluction terraces (fig. 205) (Bliss 1969, Hamann, 1972, Van Ryswyk 1969).

Figure 205.—Alpine landscape mosaic; note the stone stripes on the background slope (Mount Fremont, Mount Rainier National Park, Washington).

Community descriptions are available from three alpine regions: the Olympic Mountains (Bliss 1969), Mount Rainier (Hamann 1972), and the Okanogan Highlands (Van Ryswyk 1969, McLean 1970). It seems notable that all three study areas are relatively dry, more continental tracts than is typical of most of the coastal mountains; the focal points of the studies at Mount Rainier and in the Olympic Mountains are on the northeastern slopes in the lee of the mountain massifs. We have observed extensive development of alpine vegetation only in such areas; the heavy snows and ice prevent development of substantial tracts of alpine vegetation in wetter areas. The reader should also bear in mind that many of the meadow communities characteristic of the subalpine parkland extend into at least the lower reaches of the Alpine Zone, although they may be somewhat modified in form and composition (see Archer 1963, for example). *Phyllodoce empetriformis-Cassiope mertensiana*, *Valeriana sitchensis*, *Carex nigricans*, and *Luetkea pectinata* communities are examples.

Hamann (1972) recognizes seven alpine plant associations on the northern slopes of Mount Rainier:

Association	Remarks
Empetrum nigrum/Lupinus lepidus	On north-facing fellfields
Arctostaphylos uva-ursi/Solidago spathulata	On south-facing fellfields
Arenaria obtusiloba-Lupinus lepidus	Pioneer cushion plant community on gentle slopes
Phyllodoce glanduliflora/Aster alpigenus	Moistest alpine environments
Pedicularis contorta-Carex spectabilis	Forms well-developed turf
Erigeron aureus-Lupinus lepidus	
Phlox diffusa/Arenaria capillaris	Seral to *Pedicularis-Carex* type

The *Empetrum/Lupinus* association is a low stature community which is relatively rich in species (table 41); however, only the *Empetrum*, *Lupinus*, *Erigeron aureus*, and *Pedicularis contorta* have high constancies. Vegetative coverage is generally 40 to 75 percent in the block-fields or fellfields it occupies. Snow patches persist until July or August, and the deeper winter snowpack provides a more protected winter environment for this north-slope community. The *Arctostaphylos/Solidago* association is the south-slope fellfield analog. It is also rich in species with *Juniperus communis* and *Potentilla fruticosa* joining *Arctostaphylos* as shrubby components. Numerous herbaceous species have high constancies (table 41). Patterned ground is frequently encountered in this association which occurs on a relatively dry and exposed alpine habitat. Cushions of *Arenaria obtusiloba* and a high percentage of bare ground are characteristic of the *Arenaria-Lupinus* association (fig. 206). *Arenaria obtusiloba* and *Lupinus lepidus* provide over half the cover with the remaining associates (table 41) typically present in only low coverages; it is the richest of the alpine associations. *Phlox diffusa* and *Silene acaulis* help, along with the *Arenaria*, to produce a physiognomy of mounds or cushions. Frost action is characteristic, producing a variety of phenomena. The *Erigeron-Lupinus* is a closely related association occurring between active nonsorted stone stripes; average cover is only 9 percent. It is relatively poor floristically, although *Erigeron aureus* is best developed here. The *Phyllodoce/Aster* association has a high plant coverage, although it consists primarily of three species—*Phyllodoce glanduliflora*, *Aster alpigenus*, and *Lupinus lepidus*. On the moist habitats occupied by this association, vegetation has stabilized the soil. The *Pedicularis-Carex* association forms a well-developed turf. High constancy of several moisture-requiring species suggests a moist habitat, at least early in the season. *Carex* spp. make up 25 percent or more of the cover. Some of the best developed

Table 41. – Constancy class (1 = low to 5 = high) for major species in six alpine plant associations (after Hamann 1972)

Species	Association[1]					
	Emni/Lule	Aruv/Sosp	Arob-Lule	Erau-Lule	Peco-Casp	Phgl/Asal
Empetrum nigrum	5	1	1	1	—	1
Erigeron aureus	5	3	5	5	4	—
Lupinus lepidus	5	3	5	5	5	5
Pedicularis contorta	4	2	2	—	5	2
Arenaria capillaris	3	2	1	1	5	—
Arenaria rubella	3	4	4	—	2	—
Phyllodoce glanduliflora	3	1	1	—	—	5
Vaccinium scoparium	3	3	—	—	—	2
Carex spectabilis	3	—	2	—	4	5
Artemisia trifurcata	3	3	2	1	4	—
Aster alpigenus	3	2	—	1	5	5
Festuca ovina	3	3	5	3	—	—
Luzula spicata	3	3	5	1	—	—
Penstemon procerus	3	5	5	4	4	2
Potentilla diversifolia	3	3	2	—	3	—
Solidago spathulata	3	5	5	3	2	—
Arctostaphylos uva-ursi	1	5	1	—	—	2
Potentilla fruticosa	1	5	4	1	—	—
Carex nigricans	1	5	3	—	2	2
Achillea millefolium	1	5	1	—	—	2
Campanula rotundifolia	1	5	1	—	1	—
Trisetum spicatum	2	5	3	1	2	1
Antennaria alpina	—	3	3	1	1	—
Juniperus communis	1	4	1	—	—	—
Phlox diffusa	2	4	1	3	5	1
Arenaria obtusiloba	3	3	5	4	—	—
Penstemon davidsonii	1	4	5	2	—	—
Carex phaeocephala	2	—	4	5	4	—
Poa gracillima	2	2	3	—	—	—
Silene suksdorfii	—	2	3	—	—	—
Spraguea umbellata	1	—	3	—	1	—
Sibbaldia procumbens	2	1	3	1	2	1
Luetkea pectinata	—	—	—	2	—	3
Cassiope mertensiana	—	—	—	—	—	4
Vaccinium deliciosum	—	—	—	—	—	3
Kalmia polifolia	—	—	—	—	—	3
Antennaria lanata	2	—	1	—	4	4
Potentilla flabellifolia	1	—	—	—	2	4
Veronica cusickii	2	—	—	—	4	3

[1] Emni/Lule = *Empetrum nigrum/Lupinus lepidus* association
  Aruv/Sosp = *Arctostaphylos uva-ursi/Solidago spathulata* association
  Arob-Lule = *Arenaria obtusiloba-Lupinus lepidus* association
  Erau-Lule = *Erigeron aureus – Lupinus lepidus* association
  Peco-Casp = *Pedicularis contorta-Carex spectabilis* association
  Phgl/Asal = *Phyllodoce glanduliflora/Aster alpigenus* association.

288

alpine soils are associated with this community. The *Phlox/Arenaria* community is considered to be a pioneer community which evolves into the *Pedicularis-Carex* type. Flat mats of *Phlox diffusa* cover 25 to 75 percent of the ground in this association. *Lupinus lepidus, Arenaria capillaris, Trisetum spicatum, Carex phaeocephala*, and *C. spectabilis* also have significant coverage.

Bliss (1969) provides some data on alpine community composition in the northeastern Olympic Mountains and notes that most alpine tundra is found at 1,800 to 1,850 meters with upper limits of about 2,250 meters due to rock and permanent snow and ice. South- and west-facing slopes where the snow melts in May and early June are characterized by *Phlox diffusa, Lupinus lepidus, Oxytropis campestris, Antennaria racemosa*, and *Erigeron compositus* var. *discoideus*. Where striping occurs on scree slopes 20 to 75 centimeters wide, vegetation stripes alternate with 50- to 120-centimeter stone stripes; important species in the vegetation stripes include *Phlox diffusa, Arenaria obtusiloba, Draba lonchocarpa, Synthyris pinnatifida* var. *lanuginosa, Carex spectabilis*, and *Potentilla diversifolia*. *Lupinus lepidus* is more common in the stone stripes. A closed herb meadow characterizes less exposed sites with greater winter snow cover. Dominants there are *Carex phaeocephala, Potentilla diversifolia, Phlox diffusa, Draba lonchocarpa*, and *Solidago spathulata*. Bliss describes communities around several different types of snowbanks, since slope aspect as well as time of melt are important: (1) steep, north slopes—scattered *Douglasia laevigata, Lomatium angustatum*, and *Ranunculus eschscholtzii*; (2) gentle slopes—*Antennaria lanata, Carex spectabilis*, and *Polygonum bistortoides* (mid-June snowmelt),

Figure 206.—Cushion plants are major components of the sparse pioneer vegetation on this flat alpine area covered by recent pumice falls (Burroughs Mountain, Mount Rainier National Park, Washington).

Carex nigricans (August snowmelt), and *Hedysarum occidentale* (warm, relatively dry microsites); and (3) low-elevation snowbank sites—mostly subalpine meadow species such as *Hedysarum occidentale, Artemisia trifurcata, Elmera racemosa,* and *Arnica rydbergii.* Rock outcrops low in the alpine regions which are released in mid-June are dominated by *Douglasia laevigata, Phlox diffusa, Draba lonchocarpa, Saxifraga bronchialis, Sedum divergens,* and *Penstemon procerus* var. *tolmiei.*

Van Ryswyk's (1969) and McLean's (1970) study area is actually in extreme southern British Columbia, immediately north of the border in the Okanogan Highlands Province. The alpine communities here resemble much more closely the types described from the Rocky Mountains. The landscape is generally much gentler than that encountered in the Cascade Range and Olympic Mountains. On stable, well-drained tracts, a closed alpine tundra or turf in which Carices are prominent is typical: *Carex scirpoidea, C. nardina, C. pyrenaica, C. albonigra, Kobresia bellardii, Festuca ovina, Luzula spicata, Trisetum spicatum, Antennaria alpina, Silene acaulis, Potentilla nivea, Poa alpina, Juncus drummondii, J. parryi, Penstemon procerus,* and *Sibbaldia procumbens.* On dry, shallow soils an *Antennaria lanata-Vaccinium scoparium* community is conspicuous; it includes abundant mosses and lichens and common occurrences of *Arenaria formosa, Luzula glabrata, Veronica wormskjoldii, Juncus parryi,* and *J. drummondii.* In moist, snowy depressions *Carex nigricans, C. pyrenaica,* and *Polytrichum* spp. are constant with *Veronica wormskjoldii, Juncus drummondii, Potentilla diversifolia, Deschampsia atropurpurea, Luzula glabrata, Juncus mertensianus,* and *Caltha leptosepala* varying from site to site. *Salix* spp./*Carex macrochaeta* communities (up to 1 meter and 3 decimeters tall, respectively) may also occur in poorly drained sites at about the tree line along with *Sphagnum* spp. and some of the aforementioned vascular plants. As Van Ryswyk (1969) points out, there is a distinctive lack of ericaceous shrub communities so common in the Cascade Range; they occur here but only as minor elements of the mosaic. Tall herbaceous communities are also lacking.

Obviously, except east of the Cascade Range proper, the alpine turf communities so characteristic of the Rocky Mountains are essentially absent. Just as in the case of forested and subalpine parkland zones, the wet, mild coastal climate has tended to evolve communities which are distinct in composition and structure, certainly from those typical of continental mountain regions in North America, and probably the temperate zone in general.

CHAPTER XI. VEGETATION OF SOME UNIQUE HABITATS

There are many habitats supporting unusual floras or communities which are of particular interest to the geneticist, ecologist, or plant geographer. These are typically "azonal" types of habitats which are the result of unusual substrates (e.g., serpentine), unique topographic features, or other unusual environmental factors (e.g., salt spray). Sufficient data are available to consider four such areas: (1) the ocean-front community complex including sand dune and strand communities and tidelands; (2) areas of recent vulcanism such as mudflows and lava beds; (3) serpentine tracts; and (4) other botanically interesting areas—the Columbia Gorge, upper Skagit River-Ross Lake area, and the San Juan Islands.

Ocean-Front Communities

A rich variety of specialized community types are found along the edge of the Pacific Ocean reflecting a mosaic of distinctive habitat types—estuaries, tidal flats, sand dunes, headlands, etc. (fig. 207). Some features of shoreline vegetation are immediately obvious— e.g., communities of sand colonizers and stabilizers, the often impenetrable, bordering belts of shrubs and forests of *Pinus contorta* and *Picea sitchensis*. Strong, seashore winds greatly influence composition and form of the vegetation by desiccating foliage, transporting salt spray, and abrading the plants with sand. As a consequence, the stands of *Pinus contorta* and *Picea sitchensis* are frequently deformed on the oceanside and increase in height to the lee (fig. 208). We will consider three major categories of ocean-front communities: (1) sand dune and strand communities, (2) tideland communities, and (3) herb- and shrub-dominated communities.

Sand Dune and Strand Communities

Sand dunes are the major locale where ocean-facing vegetation types have been studied by ecologists. In Oregon and Washington, there are extensive areas of such dunes— on 225 kilometers of 500 kilometers of shoreline in Oregon alone (Wiedemann 1966, Wiedemann et al. 1969) (fig. 209). The greatest development is the Coos Bay dune sheet covering 86 kilometers of continuous coastline and a major site of Cooper's (1958) monumental study of dune origin and form.

Dune and strand vegetation of the Oregon coast has been studied by many scientists, including Egler (1934), Green (1965), Byrd (1950), Hanneson (1962), Kumler (1963), Wiedemann (1966), and Wiedemann et al. (1969). Any discussion should probably begin with the early plant colonizers which begin sand stabilization. Some of the more important species are *Glehnia leiocarpa*, *Carex macrocephala*, *Franseria chamissonis*, *Abronia latifolia*, *Convolvulus soldanella*, *Lupinus littoralis*, *Poa macrantha*, *Polygonum paronychia*, *Juncus lesueurii* and *falcatus*, *Potentilla anserina*, *Calamagrostis nutkaensis*, *Elymus mollis*, *Plantago maritima*, and *Cotula coronopifolia* (Heusser 1960, Kumler 1963). *Ammophila arenaria* is an important pioneer deserving special mention. It was first introduced in the late 1800's for use in dune control planting on the Oregon coast (Green 1965). Since that time, it has become naturalized and occurs widely, especially along the immediate shoreline (next to the high tide line). In fact, *Ammophila arenaria* is responsible

Figure 207.—Mosaic of ocean-front habitats; visible are tidal marshes in an estuary, coastal *Picea sitchensis* and *Pinus contorta* on uplands, a series of sand beaches, and in the foreground, a grassy, headland area (Cascade Head, Oregon).

Figure 208.—Ocean winds and spray strongly influence form and composition of beach-bordering communities; *Picea sitchensis* at Cape Perpetua, Oregon (*photo courtesy Siuslaw National Forest*).

Figure 209.—Sand dunes are very extensive along the Oregon coast, occupying nearly half the 500 kilometers of shoreline; pictured is the north end of the Coos Bay dune sheet, the most extensive single area of dune development in the Northwest *(photo courtesy Siuslaw National Forest).*

for development of foredunes along portions of the coast, a dune type relatively uncommon prior to 1930 (Wiedemann 1966, Wiedemann et al. 1969).

The shrub communities are often extremely dense and 1 to 3 meters in height. The most constant element is apparently *Gaultheria shallon* (Heusser 1960, Kumler 1963, Wiedemann 1966, Wiedemann et al. 1969). Other species typically present are *Vaccinium ovatum, Myrica californica, Rhododendron macrophyllum, Arctostaphylos uva-ursi, Rubus spectabilis,* and *Arctostaphylos columbiana.* Shrub stands on wetter sites (e.g., deflation plains) may include *Salix hookeriana, Alnus rubra, Ledum glandulosum* var. *columbianum, Spiraea douglasii,* and *Lonicera involucrata* (Wiedemann 1966). *Pinus contorta* and *Picea sitchensis* seedlings are frequently present.

The shrub communities develop rapidly into forests typically composed of *Pinus contorta* or *Picea sitchensis* or both (Wiedemann 1966, Wiedemann et al. 1969). Other, less common constituents are *Pseudotsuga menziesii* in places well protected from winds and *Tsuga heterophylla* and *Thuja plicata* in low, moist areas of older forests. In southern Oregon, *Chamaecyparis lawsoniana* can be a pioneer forest species. The understory in these forests is a thick tangle of shrubs of which *Rhododendron macrophyllum* and *Vaccinium ovatum* are most conspicuous. *Gaultheria shallon* is often present in small amounts but develops best on the forest edges where *Arctostaphylos columbiana* and *A. uva-ursi* are also found.

Various plant communities, including the forests, may be engulfed by moving sand dunes (fig. 210). If there is no new sand activity, *Picea sitchensis* and *Pseudotsuga menziesii* will become dominant over the *Pinus contorta* which appears to be a seral species on most (but not all) sand dune sites. Both are longer lived and grow to larger size than the *Pinus.* The major climax tree species on most stabilized dunes would appear to be *Tsuga heterophylla,* with *Thuja plicata* and *Picea sitchensis* as possible associates.

Wiedemann (1966, Wiedemann et al. 1969) conducted the most comprehensive study of succession on Oregon sand dunes. He concentrated on deflation plains, areas where

293

Figure 210.—Active or moving sand dunes are common along parts of the Oregon coast (Eel Creek, Oregon Dunes National Recreation Area, Siuslaw National Forest). Left: Dune burying a stand of *Pseudotsuga menziesii* and *Pinus contorta*. Right: Previously buried forest being exposed by dune migration.

moist sand near the water table has been exposed, effectively stopping sand movement. Earliest stages in succession are herbaceous communities which differ compositionally, depending on moisture conditions: dry meadow, meadow, rush meadow, and marsh. Major species in these communities are:

Community	Species
Dry Meadow	*Lupinus littoralis, Ammophila arenaria, Poa macrantha, Festuca rubra*
Meadow	*Festuca rubra, Lupinus littoralis, Aira praecox, Hypochaeris radicata, Fragaria chiloensis*
Rush Meadow	*Trifolium willdenovii, Juncus nevadensis* and *falcatus, Aster subspicatus, Sisyrinchium californicum*
Marsh	*Carex obnupta, Potentilla anserina, Hypericum anagalloides, Ranunculus flammula, Lilaeopsis occidentalis*

A wet shrub community can develop directly from any of the last three communities and is followed, in turn, by forests of *Pinus contorta* and *Picea sitchensis*. Some exceptions to this general outline include: open *Pinus contorta/Carex obnupta* forests and forest bogs of *Pinus contorta, Thuja plicata, Ledum groenlandicum, Darlingtonia californica, Lysichitum americanum,* and *Blechnum spicant.* Successional sequences on deflation plains and areas of dry active and inactive sand are schematically illustrated in fig. 211.

Tideland Communities

There are a variety of tideland communities associated with estuaries along both the Oregon and Washington coasts but their characteristics are little known. Peck (1961) lists

294

BARE SAND

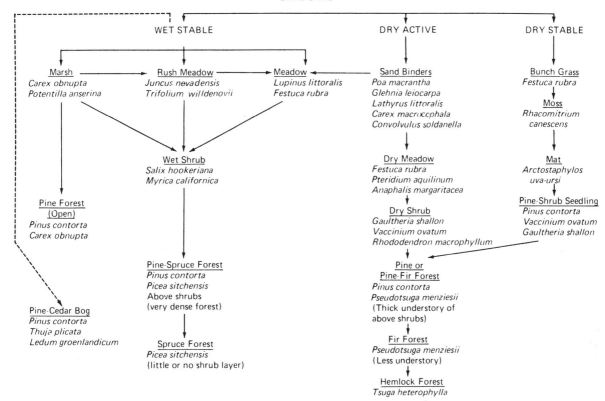

Figure 211.—Plant succession on Oregon coastal sand dunes (from Wiedemann (1966) with later modifications by Wiedemann).

the following species as those most characteristic of marshes on tidal flats along Oregon's northern coast: *Triglochin maritimum, Distichlis spicata, Puccinellia pumila, Scirpus americanus, S. pacificus, Juncus effusus, J. balticus, Salicornia virginica, Glaux maritima, Cuscuta salina, Grindelia stricta,* and *Jaumea carnosa.* Heusser (1960) mentions that *Distichlis spicata* and *Salicornia virginica* are codominants on tidal flats on the east side of the Olympic Peninsula. Johannessen (1964) studied prograding tidal marsh at Nehalem Bay in Oregon. He found *Triglochin maritimum* was typical of the lowest habitats, near deep river channels. Higher and broader expanses of mudflats were dominated by *Scirpus pacificus* and *Carex lyngbyei* and those farther inland, by *Deschampsia caespitosa* and *Juncus balticus.* Muenscher (1941) has provided species lists for salt marshes and tidal flats along Puget Sound in Whatcom County, Washington.

In the lower Columbia River, there are a large number of islands subject to major tidal influences (fig. 212). The composition of these communities is presently unknown as well as the degree to which each is influenced by salt water. The islands closer to the mouth of the river are mainly marshlands (fig. 212), and those farther upriver are characterized by dense, tall shrub communities with scattered *Picea sitchensis* (fig. 213) and, less often, other trees such as *Alnus rubra, Populus trichocarpa,* and *Salix* spp. The coastal climatic influence, as indicated by the presence of *Picea sitchensis,* seems to disappear east of Cathlamet, Washington. Above that point, river island communities are the more typical riparian forests of *Populus trichocarpa, Salix* spp., and *Fraxinus latifolia.*

Picea sitchensis is, of course, the characteristic tree species of tideland areas and has been referred to as "tideland spruce" almost since its discovery. Sprawling, open-grown *P. sitchensis* (fig. 213) border tidal flats and channels all along the Oregon and Washington coasts.

Figure 212.—Marshy, tideland islands are abundant in the lower Columbia River; their composition and environmental characteristics are presently unreported (Lower Columbia National Wildlife Refuge, Oregon).

Figure 213.—Communities of scattered, sprawling *Picea sitchensis* and dense shrubs are characteristic of borders of tidal flats and channels all along the coast; this community in the lower Columbia River has *Populus trichocarpa* as an associated tree and *Cornus stolonifera* and several *Salix* spp. as tall shrub dominants (Columbia White-Tail National Wildlife Refuge, Washington).

296

Herb- and Shrub-Dominated Communities

Northern Oregon Headlands

Forests dominate the ocean-front vegetation along most of the northern Oregon and Washington coasts. Herb- or shrub-dominated communities often occur on exposed portions of headlands, however (fig. 214). Unfortunately, only one study of a grassy headland prairie has been reported (Davidson 1967).

Davidson (1967) recognized five communities on a 10-acre headland prairie. The *Polystichum munitum-Rubus parviflorus*, *Artemisia suksdorfii-Solidago canadensis*, and *Solidago canadensis* communities were most important. *Pteridium aquilinum* was present in most stands but was not a dominant. Notable species in one or more communities besides the aforementioned were: *Equisetum telmateia*, *Ranunculus occidentalis*, *Heracleum lanatum*, *Stachys mexicana*, *Galium aparine*, *Marah oreganus*, *Rubus spectabilis*, *Angelica lucida*, *Carex obnupta*, *Lupinus littoralis*, *Achillea millefolium*, *Plantago lanceolata*, *Holcus lanatus*, *Agrostis idahoensis*, *Elymus glaucus*, *Anthoxanthum odoratum*, *Festuca rubra*, *Dactylis glomerata*, *Bromus sitchensis*, and *Rosa nutkana*. Gradual invasion of the prairie by *Picea sitchensis* appears to be taking place, primarily along trees which have fallen into the prairie and become nurse logs (fig. 215).

Figure 214.—Herbaceous communities occupy exposed portions of many headlands along the Oregon coast as shown here at Heceta Head (*photo courtesy Siuslaw National Forest*).

Figure 215.—*Picea sitchensis* **invading a headland grassland by reproducing atop rotten, down logs (Cascade Head, Siuslaw National Forest, Oregon).**

Southern Oregon Coast

On the southern Oregon coast (from about Port Orford south), the coastline vegetation differs significantly from that found farther north (fig. 216). The shrub and forest stands of *Gaultheria shallon, Pinus contorta,* and *Picea sitchensis* are absent. In their place (above the beach) are communities of herbs and low shrubs grading into a zone of taller shrubs. Typical species on the exposed seaward slopes are *Juniperus communis, Zigadenus fremontii, Mesembryanthemum chilense, Eriogonum latifolium, Sidalcea malvaeflora, Plantago subnuda, Agoseris hirsuta, Scirpus setaceus, Brodiaea coronaria* var. *macropoda,* and *Iris douglasiana* var. *oregonensis.* The shrub zone (fig. 217) is dominated by *Ceanothus integerrimus, Garrya elliptica, Rhododendron occidentale, Alnus sinuata, Ribes menziesii,* and *Arctostaphylos columbiana* (Heusser 1960, Peck 1961, Gratkowski 1961a). *Ulex europaeus,* an introduced shrub, is an important weed species along the southern coast where it creates impenetrable thickets and a serious fire hazard (Wiedemann 1966, Gratkowski 1961a).

This shift in the ocean-front landscape from a domination by forest to herbaceous and shrubby communities is a distinctive change which reaches its culmination in the non-forested ocean-front communities so characteristic of the central and northern California coast. The environmental factors responsible have not been analyzed. Certainly the warmer, drier climate along the southern Oregon coast, as compared with northern

Figure 216.—High bluffs and mountain slopes are characteristically adjacent to the ocean along the southern Oregon coast; herbaceous (pictured here) or shrubby communities typically dominate these ocean-front landscapes rather than forest or dunes as is characteristic of the northern and central Oregon sections of coast, respectively (between Brookings and Gold Beach, Oregon).

Figure 217.—Characteristic shrub-dominated community found on steep slopes and cliffs facing the Pacific Ocean in southern Oregon; visible are *Arctostaphylos columbiana, Ceanothus integerrimus, Vaccinium ovatum,* and *Pteridium aquilinum,* with *Picea sitchensis* and *Pinus contorta* present in small numbers (between Brookings and Gold Beach, Oregon).

299

Oregon, particularly in the summer, must be one factor; the greater frequency of fogs on the southern Oregon coast and predominance of high bluffs and steep mountain slopes adjacent to the ocean may be other factors (Peck 1961).

Areas of Recent Vulcanism

The eruption and deposition of volcanic materials has been taking place in the Northwest up into recent times, particularly in the Cascade Range. Mount Lassen (in California) erupted early in this century; Mount Rainier, between 110 and 150 years ago (Crandell 1969a, 1969b, 1971); and Mount St. Helens was last active about the middle of the 19th century. This activity has resulted in frequent occurrence of cinder cones, pumice and ash fields, lava flows, and laharic deposits[1] which support communities quite unlike those found on adjacent habitats. Some of these have been discussed in preceding chapters, but at this time we would like to consider lava flow and mudflow communities in detail.

Lava Flows and Lava-Dam Lakes

Pleistocene-Recent lava flows are conspicuous at many places in the Cascade Range; e.g., in central Oregon (Peterson and Groh 1966), the Santiam and McKenzie Pass areas (Taylor 1968) in the Oregon Cascade Range, and in the vicinity of Mount St. Helens and Mount Adams, Washington Cascade Range (Wise 1970). These lava fields provide a rugged environment for pioneer plant communities, yet they are often surprisingly rich floristically, particularly in cryptogams.

The most extensive study of lava flow communities was conducted by Roach (1952) on the Nash Crater lava flows near Santiam Pass, Oregon. He recognized three associations which represented a series in development of substrate and community density and organization: the *Aceretum circinati lavosum*, *Pseudotsugeto-abietum lasiocarpae*, and *Pseudotsugetum-abietum grandis* associations. The major species found in these associations are listed in table 42. All these associations are edaphic climaxes in which *Acer circinatum*, *Pseudotsuga menziesii* and *Abies lasiocarpa*, and *Pseudotsuga menziesii* are the respective dominants.

Some studies have also been made of the basaltic flows found in southern Washington (Franklin 1966, Franklin and Mitchell 1967). On one flow (near Wind River), a *Pseudotsuga menziesii-Abies lasiocarpa-Acer circinatum* community, very similar to the one described by Roach (1952), is present. The Wind River and Nash Crater flows lie almost entirely within the *Abies amabilis* Zone, however. Near Mount Adams is a large flow (Big Lava Beds) which extends from about 520 to 1,100 meters in elevation (Wise 1970). Communities vary from *Pseudotsuga menziesii-Quercus garryana* at the lowest through *Pinus contorta/ Arctostaphylos uva-ursi* to *Abies lasiocarpa/Xerophyllum tenax* (fig. 218) at the highest elevations (Franklin 1966). All these communities are very rich in mosses and lichens, and there are many vascular species present in niches and crevices, surprising in view of the xeric, sterile appearance of the substrate. We will not discuss the *Pinus contorta* community here since it is similar to those communities occurring on lahars and glacial outwash which are described later in this chapter.

Lava flows frequently blocked drainages of many small streams, and seasonal lakes or lakes with widely fluctuating water levels are not uncommon. Kienholz (1931) studied one of the latter type adjacent to Big Lava Beds (southern Washington). He recognized

---

[1] "The term, lahar, includes all of the broad textural range of debris flows and mudflows of volcanic origin . . . any unsorted or poorly sorted deposit of volcanic debris that moved and was deposited as a mass and owed its mobility to water." (Mullineaux and Crandell 1962).

Table 42. — Representative species in three plant associations on lava flows near Santiam Pass, Oregon

Item	Aceretum circinati lavosum	Pseudotsugeto-abietum lasiocarpae	Pseudotsugetum-abietum grandis
Substrate	Block basalt	Block basalt with smaller size scoria in broken crust	Grayer, rounded block basalts with high proportion sand and ash
Trees	None	Pseudotsuga menziesii, Abies lasiocarpa	Pseudotsuga menziesii, Abies grandis, Pinus contorta, P. monticola
Shrubs	Acer circinatum, Rhamnus purshiana, Holodiscus glabrescens, Arctostaphylos columbiana	Acer circinatum, Arctostaphylos nevadensis, A. columbiana, Castanopsis chrysophylla, Pachistima myrsinites	Castanopsis chrysophylla, Pachistima myrsinites, Acer circinatum, Rubus parviflorus, Ceanothus velutinus
Herbs	Cryptogramma crispa, Penstemon davidsonii, Sedum oregonense, Juncus parryi	Sedum oregonense, Penstemon davidsonii, Chimaphila umbellata occidentalis	Chimaphila umbellata occidentalis, Linnaea borealis, Festuca occidentalis, Xerophyllum tenax, Penstemon cardwellii
Mosses	Rhacomitrium patens, R. lanuginosum, Dicranum scoparium, Hypnum fertile	Rhacomitrium patens, R. lanuginosum, Dicranum scoparium, Hypnum fertile	Rhacomitrium patens, R. lanuginosum, Dicranum scoparium, Hypnum fertile, Aulacomnium androgynum, Rhytidiadelphus triquetrus, Polytrichum juniperinum

Source: Roach (1952).

Figure 218. — An *Abies lasio-carpa/Xerophyllum tenax* **community is found at about 1,200-meter elevation on the Big Lava Beds near Mount Adams (Gifford Pinchot National Forest, Washington).**

five zones around Goose Lake (fig. 219): Willow-Alder (*Salix sitchensis-Alnus rubra*); Sedge (*Carex* spp.); Fontinalis (*Fontinalis antipyretica gigantea*); Cottonwood (*Populus trichocarpa*); and Weed (*Artemisia tilesii-Stachys cooleyae-Scrophularia lanceolata*).

More frequently, lava-dammed lakes are filled during the winter months when underground drainage systems are inadequate and drain completely during the summer. Roach (1952) describes one of these which is covered primarily by a bog-type vegetation dominated by *Carex sitchensis* and *Vaccinium occidentale*. In nearby Fish Lake where the substrate is thin and rocky, a dense, nearly pure community of *Carex aperta* and *C. rostrata* 1 meter or more in height develops during the short growing season.

Mudflows

Lahars, including mud and debris flows as defined earlier, are common near the major volcanoes in the Cascade Range (Mullineaux and Crandell 1962, Tidball 1965, Crandell and Mullineaux 1967, Crandell 1969a, 1971). Some of these are occupied by communities relatively normal in composition for the zone in which they occur (Franklin 1966). Others are sufficiently recent that they are obvious sites for studies of vegetation succession and soil development, such as some at Mount Rainier (Frehner 1957, Ballard 1963), Mount Lassen (Bailey 1963), and Mount Shasta (Dickson and Crocker 1953a, 1953b, 1954). Still others in the *Tsuga heterophylla* and *Abies amabilis* Zones are occupied by climax stands of *Pinus contorta* (Franklin 1966). The status of the vegetation appears to be primarily a function of age and nature of the substrate.

Frehner's (1957) study of succession on the 1947 Kautz Creek mudflow (fig. 220) at Mount Rainier has shown that many conifers, typical of adjacent zonal forests, will invade recent mudflows. *Populus trichocarpa*, *Salix* spp., and *Alnus rubra* have been major pioneer species, however, and *Alnus rubra* had special importance as a consequence of its nitrogen-fixing abilities. Occurrence of standing snags appeared to be an important factor in the composition of the pioneering vegetation.

Figure 219.—Lava flows have created lakes with widely fluctuating water levels by blocking stream drainages as at Goose Lake (Gifford Pinchot National Forest, Washington), an impoundment of this type (Kienholz 1931). Top: General view of lake showing vegetational mosaic. Middle: Close view of Sedge, Willow-Alder, and Cottonwood Zones. Lower: *Fontinalis* and Sedge Zones occupied much of the lake area. (Photographs from collection of R. Kienholz, taken about 1930, prior to major vegetative alteration following plugging of lake outlet by USDA Forest Service.)

Figure 220.—Mudflows are common features near the major volcanoes in the Cascade Range; the Kautz Creek mudflow occurred in 1947, burying and killing the original forest *(photo courtesy University of Washington).*

303

A *Pinus contorta/Arctostaphylos* community is found on some xeric laharic deposits (fig. 221), lava flows, and coarse glacial outwash in southwestern Washington and northwestern Oregon (Franklin 1966, Stephens 1966). This appears to be a stable climax type on such habitats in many areas, although occasional specimens of more tolerant tree species such as *Pseudotsuga menziesii, Tsuga heterophylla,* or *Abies lasiocarpa* may be present; droughty years can eliminate many of these. In other cases, it is clearly successional to a *Pseudotsuga menziesii, P. menziesii-Pinus contorta,* or *Tsuga heterophylla* community. Although details of composition vary, the dominance of *Pinus contorta* and either *Arctostaphylos nevadensis* or *A. uva-ursi* in their respective layers (fig. 221) and a rich cryptogammic ground cover are distinguishing. Typical cryptogams are: *Rhacomitrium canescens* var. *ericoides, Cladonia grayii, Polytrichum juniperinum, Cladonia bellidiflora, C. ecmocyna* vars. *ecmocyna* and *intermedia, Stereocaulon paschale, Cladonia phyllophora, C. rangiferina, Polytrichum piliferum, Cladonia coniocraea, Lecidea quadricolor, Cladonia macrophyllodes, Rhacomitrium heterostichum, R. sudeticum,* and *Aulacomnium androgynum.*

Figure 221.— *Pinus contorta/ Arctostaphylos* **communities are widespread on mudflows, glacial outwash, and lava flows in the Cascade Range of southern Washington and northern Oregon; this community is distinguished by a depauperate understory of** *Arctostaphylos uva-ursi* **or** *A. nevadensis* **and many species of lichens and mosses (Kalama River lahar, Gifford Pinchot National Forest, Washington).**

Serpentine Areas

Serpentine areas are characterized by unusual plant communities and floras. Vegetation is invariably stunted on serpentine sites in comparison with that on adjacent nonserpentine soils. Sharp contrasts in physiognomy, composition, and productivity of communities are typical at margins of serpentine outcrops (fig. 222). The floras are unusual, including endemics restricted to serpentine species not usually found in adjacent communities and "bodenvag" species which appear edaphically indifferent but may, in fact, develop special races tolerant to serpentine (Kruckeberg 1954, 1964, 1967, 1969a; Whittaker 1954b, 1960; Waring 1969; White 1971).

Serpentine areas in this discussion refer to habitats with soils derived from ultramafic rocks either as peridotite and dunite (igneous forms) or as serpentinite (the metamorphic derivative) (Kruckeberg 1967). Such soils are typically low in total and adsorbed calcium and high in magnesium, chromium, and nickel (Walker 1954). Walker (1954) has analyzed the factors affecting plant growth on such sites and concluded the plants growing there must be tolerant of low calcium levels and one or more additional conditions; e.g., high nickel, chromium, or magnesium and physically unfavorable shallow soils. White (1971), working in southwestern Oregon, found high levels of calcium in some serpentine soils and concluded that neither calcium supply nor calcium/magnesium ratios were critical in controlling plant growth on these soils. Through a combination of soil and plant analyses, he found that distributions of individual plant species were largely controlled by high levels of magnesium, nickel, and chromium, in decreasing order of importance. The limiting effects of these elements were ameliorated to some extent by factors influencing water availability such as elevation, aspect, and areas of localized seepage.

Figure 222. — **Vegetations on serpentine sites and adjacent nonserpentine soils often contrast sharply; in this Siskiyou Mountain area, open** *Pinus jeffreyi* **stands on serpentine (right) contrast with those of** *Pseudotsuga menziesii, Pinus lambertiana,* **and** *Libocedrus decurrens* **on nonserpentine soils (left).**

There are two major serpentine areas in the Pacific Northwest (Whittaker 1954a; Kruckeberg 1964, 1967): (1) a large area in the Siskiyou Mountains of Oregon and (2) about 100 square miles in the Wenatchee Mountains, an eastern outlier of the Cascade Range. In addition, there are small, scattered outcroppings of ultramafic rocks in northwestern Washington, including parts of the San Juan Islands, and a small area in Grant County, Oregon. Serpentine communities have been most thoroughly analyzed in the Siskiyou Mountains (Whittaker 1954b, 1960; Waring 1969; White 1971), but Kruckeberg (1964, 1967, 1969a, 1969b) has provided general descriptions for the Washington serpentines.

Siskiyou Mountains

Perhaps the outstanding feature of the Siskiyou serpentines is the *Pinus jeffreyi*/grass woodland (fig. 223) which occupies the most xeric serpentine sites from 300- to 2,000-meter elevation (Whittaker 1960, Waring 1969). *Pinus jeffreyi* is typically the only tree species present in these open woodlands (fig. 223) along with a rather sparse growth of grasses (e.g., *Stipa lemmonii, Sitanion jubatum, Melica geyeri, Elymus glaucus,* and *Festuca ovina*) and occasional *Arctostaphylos viscida*. Forests, intermediate in elevation and moisture regime, are typified by a sparse and xerophytic appearance and dominated by a mixture of several conifers—*Pseudotsuga menziesii, Libocedrus decurrens, Pinus jeffreyi, P. monticola, P. lambertiana,* and *P. attenuata* (Whittaker 1960). Associated with them is a dense layer of sclerophyllous shrubs such as *Quercus vaccinifolia, Lithocarpus densiflorus, Vaccinium parvifolium, Garrya buxifolia,* and *Umbellularia californica*. Herb coverage is generally low but rich in species. Whittaker (1954b, 1960) has commented at length on the "two-phase" or patchwise distribution of the shrub cover in these forests with essentially closed shrub patches alternating with herbaceous openings. On more mesic sites the shrubs form the matrix, but on more xeric sites the herbaceous openings are dominant.

Other community types described on Siskiyou serpentines include: *Chamaecyparis lawsoniana-Pinus monticola-Pseudotsuga menziesii* stands, with a dense shrubby understory in ravines and draws, and higher elevation forests dominated by *Abies concolor* and *Pseudotsuga menziesii* and *Pinus monticola*, singly or collectively, over a *Xerophyllum tenax* and *Arctostaphylos nevadensis* understory.

Whittaker (1960) has provided a list of more useful serpentine indicator plants in the Siskiyou area. These include: (1) some dominants—*Pinus jeffreyi, P. monticola, P. attenuata, Quercus chrysolepis, Ceanothus cuneatus,* and *Arctostaphylos nevadensis*; and (2) species which are of more frequent occurrence on smaller serpentine areas—*Galium ambiguum, Pyrola dentata, Lomatium macrocarpum, Cheilanthes siliquosa,*[2] *Rhododendron macrophyllum,* and *Darlingtonia californica*. *Xerophyllum tenax* is considered the most useful single indicator of small serpentine outcrops.

White (1971) studied the vegetation on 21 serpentine outcrops and adjacent nonserpentine areas occupying a wide range of elevations in southwestern Oregon. He found a larger number of plant taxa growing on serpentine soils than were present on adjacent nonserpentine sites. He suggested that the plant species occurring within his study areas could be grouped into four main groupings: (1) species restricted to nonserpentine soils (typified by leguminous species, *Quercus kelloggii, Pseudotsuga menziesii,* and *Rhus diversiloba*); (2) species restricted to serpentine areas (*Cheilanthes siliquosa, Ceanothus pumilus, Eriophyllum lanatum,* and *Epilobium minutum* were some of the most common); (3) species largely restricted to ecotonal areas (exemplified by *Dodecatheon hendersonii* and *Berberis pumila*); and (4) those plants which were apparently indifferent to soil; i.e., "bodenvag" species (*Galium ambiguum* and *Polygala californica* were two of the most important species within this group).

According to White, his vegetation data indicated the existence of five principal plant communities on serpentine sites: open, grassy vegetation on gentle, stony slopes situated above 1,200 meters in elevation; a more floristically diverse community with scattered shrubs at elevations of 860 to 980 meters; and two successional stages and a climax savanna community at elevations below 860 meters.

The serpentine community found above 1,200-meter elevation is characterized by grasses in the *Festuca idahoensis-oregana* complex and in the *Poa sandbergii* complex and by

---

[2] The name for *Cheilanthes siliquosa* has recently been changed to *Aspidotis densa*. However, since the new epithet has not yet found its way into regional floras or manuals, we continue to use *Cheilanthes siliquosa* in the text.

Figure 223.—*Pinus jeffreyi*/grass woodland is the climax community on dry serpentine sites in the Siskiyou Mountains of southwestern Oregon.

serpentine indicator species such as *Allium falcifolium, Cheilanthes siliquosa*, and *Ceanothus pumilus*. Other species present are largely those which thrive on shallow, stony soils at high elevations—for example, *Sedum laxum, Claytonia lanceolata*, and *Eriogonum ternatum*. The community occupying serpentine outcrops at 860 to 980 meters in elevation includes many of the same species found at higher elevations, plus a considerable number of additional ones. Important herbaceous species include *Gilia capitata, Eriophyllum lanatum, Sitanion jubatum*, and *Arenaria howellii*. In addition, shrub species, such as *Ceanothus cuneatus, Chrysothamnus nauseosus* var. *albicaulis*, and *Eriodictyon californicum*, are much more abundant within this community than on high elevational sites.

The earliest successional stage occupying low elevational (below 860 meters) serpentine sites is largely made up of species which are annuals and are indicators of open mineral soil. The most commonly occurring species are *Epilobium minutum, Festuca microstachys, Aira caryophyllea, Agoseris heterophylla, Githopsis specularioides*, and *Madia exigua*. The intermediate successional stage is typified by reduced importance of open ground indicators and increased abundance of *Stipa lemmonii* and *Eriophyllum lanatum* and other perennial grasses and shrubs.

White (1971) describes the climax community on serpentine at low elevations as "an open grassland savannah with scattered *Pinus jeffreyi* and occasionally *Libocedrus decurrens*." The ground cover is principally made up of a dense assemblage of perennial grasses, specifically *Stipa lemmonii, Danthonia californica, Koeleria cristata*, and *Festuca* spp. He viewed the presence of *Polygala californica, Ranunculus occidentalis, Monardella villosa*, and the absence of *Cheilanthes siliquosa* as also indicative of climax vegetation.

Wenatchee Mountains

Vegetation characteristic of serpentine outcrops in the Wenatchee Mountains of Washington has been described in some detail by Kruckeberg (1964, 1967, 1969a). His studies have indicated the existence of three main groupings of plant species in the

Figure 224.—Ultramafic rock outcrops in Washington typically consist of open woodlands and barren slopes; upper, landscape mosaic on ultramafic slopes and, lower, local outcrop of serpentine showing the extremely barren openings (Teanaway River drainage, Wenatchee National Forest; *photos courtesy A. R. Kruckeberg).*

area: (1) serpentine indicator species, (2) species which tend to avoid serpentine areas, and (3) species apparently indifferent to edaphic factors ("bodenvag" species).

The most depauperate flora on ultramafic outcrops occurs in areas of steep talus or on exposed ridgetops (fig. 224). Trees do not occur on these sites, and the vegetation is largely made up of perennial herbs which, for the most part, are the most faithful serpentine indicator species. They are *Cheilanthes siliquosa*, *Polystichum mohrioides* var. *lemmoni*, *Eriogonum pyrolaefolium* var. *coryphaeum*, *Lomatium cuspidatum*, *Polygonum newberryi*, *Poa curtifolia*, *Chaenactis thompsonii*, *Cryptantha thompsonii*, *Douglasia nivalis*, and *Anemone drummondii*.

Less severe sites, such as moist swales, gentle slopes, and the base of screes and talus slopes, support a greater diversity of species and life-forms. Here occur not only serpentine indicators, but also more representatives of the regional flora. Species present here which apparently prefer ultramafic sites but are not restricted to them include *Lomatium brandegei*, *Thlaspi alpestre*, *Adiantum pedatum* var. *aleuticum*, *Claytonia megarhiza* var. *nivalis*, *Erysimum arenicola* var. *torulosum*, *Ivesia tweedyi*, *Castilleja elmeri*, *Salix brachycarpa*, *Ledum glandulosum*, and *Arenaria obtusiloba*. "Bodenvag" species which extend their range onto those sites include many of regionally important coniferous tree species. Examples are *Pseudotsuga menziesii*, *Abies lasiocarpa*, *Tsuga mertensiana*, *Pinus monticola*, *P. albicaulis*, *P. ponderosa*, *P. contorta*, *Taxus brevifolia*, and *Picea engelmannii*. The most important understory shrub species on these ultramafic sites are *Arctostaphylos nevadensis* and *Juniperus communis*. Important herbaceous "bodenvag" species include *Agropyron spicatum*, *Sitanion jubatum*, *Festuca viridula*, *Achillea millefolium*, *Fragaria virginiana*, *Eriophyllum lanatum*, *Senecio pauperculus*, and *Erysimum arenicola* var. *torulosum*.

Small, Scattered Outcrops of Serpentine Parent Materials

Northwestern Washington

Scattered outcrops of ultramafic rock occur in Whatcom, Skagit, and Snohomish Counties of northwestern Washington (Kruckeberg 1964, 1967, 1969a). The largest area encompasses approximately 30 square miles of dunite outcrop in the vicinity of Twin Sisters Mountain which is located just southwest of Mount Baker in Whatcom County. An unusual feature of the vegetation at this location is the occurrence of *Pinus contorta* as the timberline tree species on steep talus slopes. The understory beneath the "krumholz"-formed *P. contorta* is dominated by *Phyllodoce glanduliflora*. The herb layer contains the serpentine indicators, *Cheilanthes siliquosa* and *Polystichum mohrioides* var. *lemmoni*, plus *Phlox diffusa* and *Erigeron aureus*. In the southern portion of the area, localized swales or benches are occupied by thick mats of the moss *Rhacomitrium canescens* var. *ericoides* overtopped by *Juniperus communis* and *Arctostaphylos uva-ursi*. These areas also support scattered, stunted specimens of *Pseudotsuga menziesii*, *Taxus brevifolia*, *Pinus contorta*, and *Pinus monticola*. Adjacent steep slopes are devoid of woody cover and are principally occupied by scattered herbaceous perennials such as *Cerastium arvense*, *Arenaria rubella*, and *Cheilanthes siliquosa*.

Serpentine outcrops on Fidalgo and Cypress Islands within the San Juan Archipelago are marked only by the occurrence of the faithful serpentine indicator *Cheilanthes siliquosa* (Kruckeberg 1969a). The remaining species are largely those characteristic of all grassy balds on exposed headlands of the islands.

Peridotite bedrock outcrops in eastern Snohomish County along the South Fork of the Stillaguamish River. For the most part, the vegetation consists of a dense cover of grasses and herbs which are also common to soils from nonultramafic parent materials. However,

Phlox diffusa, typical of the east slopes of the Cascades, is restricted here to ultramafic sites. Other singular elements of the flora include three ferns—*Adiantum pedatum*, *Cheilanthes siliquosa*, and *Polystichum mohrioides* var. *lemmoni*.

Grant County in Central Oregon

Ultramafic outcrops are common in the Aldrich Mountains-Strawberry Range of Grant County, Oregon, with the largest single exposure in the vicinity of Baldy Mountain (Kruckeberg 1969a). The elevational range of ultramafic outcrops is considerable, stretching from the *Juniperus occidentalis* type at low elevations to above timberline at the summit of Baldy Mountain. At lowest exposures, the vegetation is largely marked by very small amounts of plant cover, and even *Cheilanthes siliquosa* is absent on these xeric sites. At higher elevations, within the area of *Pseudotsuga menziesii-Pinus ponderosa* forests, the ultramafic areas are singularly lacking in trees and are dominated by scattered *Cercocarpus ledifolius* and a rich grass-herb cover. Outcrops of peridotite near the summit of Baldy Mountain (2,316 meters) support a variety of perennial herbs plus two faithful ultramafic ferns—*Cheilanthes siliquosa* and *Polystichum mohrioides* var. *lemmoni*.

Other Botanically Interesting Areas

The Columbia Gorge

The Columbia Gorge is an unusual physiographic feature which is of considerable importance to biologists (fig. 225). The Columbia River cut this nearly sea level route directly across the axis of the Cascade Range. As Detling (1966) pointed out, this is the only point between the Fraser River (in British Columbia) and Klamath River (in California) where a feature of this type is found. It has provided a major route for both plant and animal migration between the western and eastern halves of Washington and Oregon (Detling 1961, 1966, 1968). At the same time, it contains many species which are endemics or constitute relict populations.

The weather of the Columbia Gorge is as unique as its topography (Lynott 1966), since it provides a sea level transition from marine to continental climate in a region where they are otherwise separated by major mountain barriers. Strong winds are a dominant feature. During the winter, low-pressure systems move through the gorge on westerly winds, bringing heavy rains as a consequence of streamline convergence. Strong high-pressure systems east of the Cascade Range can bring gale-force easterly winds through the gorge, resulting in extremely hot dry weather during summer and fall and cold continental air during the winter. Marine low-pressure systems and this cold air may collide, particularly in the west end of the gorge, with blizzards, ice storms, and freezing rain the result.

These climatic features have profound effects upon the vegetation as Lawrence (1939) and Troll (1955) have pointed out. Tree crowns are markedly deformed or one-sided. In the east end of the gorge, they are flagged toward the east by the dominantly westerly winds. This is the result of wind training, not breakage. At the west end of the gorge, trees typically lack branches on the easterly sides of the stem. This deformation is the result of branch breakage during the severe ice storms and destruction of buds and branches by dry east winds.

Another interesting ecological feature of the gorge is the opportunity it provides to study the transition from xerophytic *Pinus ponderosa-Quercus garryana* forests on the east to the mesophytic *Pseudotsuga menziesii-Tsuga heterophylla* types on the west. Troll (1955) has provided an interesting vegetation profile (fig. 226) indicating the distribution and interdigitation of the various zones through this area.

Figure 225.—The Columbia Gorge, the only nearly sea level break in the Cascade Range in Oregon and Washington and a major route for plant and animal migration between the western and eastern halves of these States; Beacon Rock, a large monolith, is visible (left center) in this east-facing view through the center of the gorge.

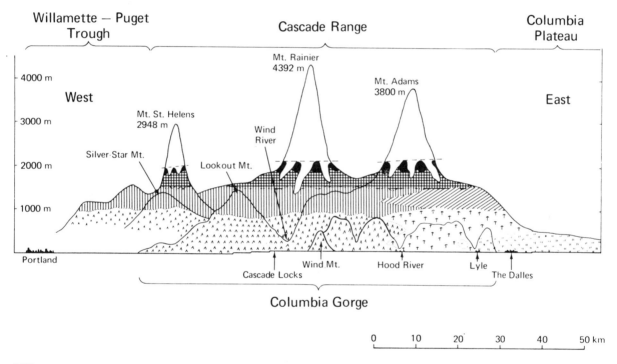

Willamette – Puget Trough   Cascade Range   Columbia Plateau

West

East

4000 m

Mt. Rainier
4392 m

Mt. Adams
3800 m

3000 m

Mt. St. Helens
2948 m

Wind River

Silver-Star Mt.

2000 m

Lookout Mt.

1000 m

Portland

Cascade Locks   Wind Mt.   Hood River   Lyle   The Dalles

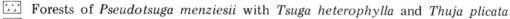

Columbia Gorge

0   10   20   30   40   50 km

⌈∧∧⌉ Forests of *Pseudotsuga menziesii* with *Tsuga heterophylla* and *Thuja plicata*

⌈↑↑⌉ Forests of *Pinus ponderosa* and *Quercus garryana*

⌈∴⌉ Prairie (bunchgrass steppe of *Agropyron spicatum*)

▒ Forests of *Abies amabilis, A. procera, Pinus monticola, Tsuga heterophylla,* and *Chamaecyparis nootkatensis*

▨ Forests of *Pinus contorta, P. ponderosa, P. monticola, Larix occidentalis, Pseudotsuga menziesii,* and *Abies grandis*

▦ Subalpine forests of *Abies lasiocarpa, Tsuga mertensiana,* and *Pinus albicaulis*

■ Alpine communities

☐ Snowfields and glaciers

Figure 226.—Vegetation profile through the Columbia Gorge and Cascade Range (from Troll 1955).

311

Upper Skagit River-Ross Lake Area

The upper Skagit River-Ross Lake area is located west of the crest of the northern Washington Cascade Range and, for that reason, portions located at low to moderate elevations might logically be considered part of the *Tsuga heterophylla* Zone. The region is an anomaly, however, and actually represents a transitional area between the moist coastal and dry interior forests. The major reason is the great breadth of the Cascade Range at this point; the Ross Lake area is actually in the rain shadow of the extensive mountain massifs which lie to the west. Mean annual precipitation around the lake is only about 1,250 millimeters, and it declines to less than 1,000 millimeters near the Canadian border (Larson 1972).

There are numerous species and communities along the upper Skagit River that are more characteristic of interior (east of the Cascade Range) than of coastal forest regions; e.g., *Pinus ponderosa-P. contorta* and *Abies lasiocarpa-Picea engelmannii* forests, as well as *Juniperus scopulorum* and *Larix lyallii*. In general, the slopes and valleys on the west side of Ross Lake tend to have more coastal characteristics with some continental elements, and those east of the lake have more continental characteristics with some coastal influences.

The only portion of this landscape mosaic which has been examined in detail (other than valley-bottom *Thuja plicata* stands discussed in Chapter IV) are the *Pinus contorta* forests (Larson 1972). Fire has been an important factor in developing the existing mosaic of communities. Larson (1972) considers in some detail the successional relationships of *Pinus contorta* and other species, particularly its most constant associate in the Ross Lake area, *Pseudotsuga menziesii*.

The most coastal of Larson's (1972) *Pinus contorta* community groups is the *Pseudotsuga menziesii-Pinus contorta/Gaultheria shallon* found mostly below Ross Dam. On more mesic habitats, *Pinus contorta* is successional to *Pseudotsuga* and, perhaps, to *Tsuga heterophylla* and *Thuja plicata*. Associated with *Gaultheria* in the understory are *Vaccinium membranaceum*, *V. parvifolium*, *Berberis nervosa*, *Pachistima myrsinites*, *Linnaea borealis*, and *Pteridium aquilinum*, among other species. On drier sites, *Pseudotsuga menziesii* or *Pinus contorta* (on the most xeric habitats) or both are climax and *Arctostaphylos uva-ursi*, *Apocynum androsaemifolium*, and *Spiraea betulifolia* are distinctive understory components.

Above Ross Dam, the *Pseudotsuga menziesii-Pinus contorta/Berberis nervosa-Spiraea betulifolia* community group characterizes most *Pinus contorta* stands at middle elevations. *Pseudotsuga* is the major climax species except on the most mesic and xeric habitats where *Tsuga heterophylla-Thuja plicata'* and *Pinus contorta* may be the respective climax species. Besides the *Spiraea* and *Berberis*, some characteristic understory species are *Pachistima myrsinites*, *Salix* sp., *Chimaphila umbellata*, *Linnaea borealis*, *Pteridium aquilinum*, *Amelanchier alnifolia*, *Ceanothus sanguineus*, *Trientalis latifolia*, *Berberis aquifolium*, *Lonicera ciliosa*, *Corylus cornuta* var. *californica*, and *Holodiscus discolor*. On drier sites, *Arctostaphylos uva-ursi* replaces the *Berberis* as the major *Spiraea* associate, and grasses, *Apocynum androsaemifolium*, and *Ceanothus velutinus* are among the minor understory components.

The *Pinus contorta* stands at higher elevations were not studied in detail, but they exhibit considerable variation in climax tree species and total community composition.

In all, Larson (1972) recognizes eight subgroups of these three major groups and relates them to location, elevation, and topography. Individual type groups have obviously strong relationships with one or several communities already known from other adjacent regions including the *Pinus contorta/Gaultheria shallon* of the Puget trough, *Pseudotsuga menziesii/Holodiscus discolor* of the *Tsuga heterophylla* Zone, *Pinus contorta/Arctostaphylos* types found on mudflows, glacial outwash, and lava flows, and pioneer *Pinus contorta* forests found in eastern Washington.

San Juan Islands

The San Juan Islands are situated in the rain shadow of the Olympic Mountains and, as a result, include some of the driest sites encountered in western Washington. However, because of highly variable topography and aspect, most islands possess a surprisingly diverse assemblage of plant communities, ranging from very dry to moist types. The following discussion is based on the work of Fonda and Bernardi,[3] who have recently described the vegetation on Sucia Island located at the north end of the San Juan Islands.

Exposed, south-facing slopes which receive the full force of prevailing dry summer winds are occupied by grassland vegetation frequently accompanied by scattered trees. The open grasslands are dominated by *Festuca idahoensis* and include smaller amounts of *Agropyron dasystachyum* and *Poa pratensis*. Open woodlands, characterized by *Pseudotsuga menziesii* and *Arbutus menziesii* in a *Festuca* matrix, are also common. Other tree species which may be found on such sites include *Juniperus scopulorum*, *Pinus contorta*, and *Quercus garryana*.

Sheltered south slopes, which are not exposed to the drying effects of winds, support closed forest largely made up of *Pseudotsuga menziesii* and *Arbutus menziesii*. The understory is variable, but *Gaultheria shallon* or *Lonicera hispidula* are the most common dominants.

The more mesic north-facing slopes and level areas in the interior of the island are generally occupied by such forest communities as the *Pseudotsuga menziesii/Gaultheria shallon* and *Thuja plicata-Abies grandis/Pachistima myrsinites*. Other understory species include *Symphoricarpos albus* and *Rosa nutkana*.

The moistest sites in interior valleys are occupied by a *Thuja plicata-Abies grandis/ Polystichum munitum* community in undisturbed locations. Areas disturbed by logging commonly support an *Alnus rubra-Acer macrophyllum/Rubus spectabilis* community. Other species abundant on disturbed sites include *Sambucus racemosa* and *Urtica dioica*.

---

[3] R. W. Fonda and J. A. Bernardi. Preliminary report on the vegetation of the San Juan Islands, Washington. 4 p., 1973. Unpublished report on file at the Forestry Sciences Laboratory, Corvallis, Oregon.

REFERENCES

Alban, David H.
   1967. The influence of western hemlock and western redcedar on soil properties. 167 p., illus. (Unpublished Ph.D. thesis on file at Wash. State Univ., Pullman.)

Aldous, A. E., and H. L. Shantz
   1924. Types of vegetation in the semiarid portion of the United States and their economic significance. J. Agric. Res. 28(2): 99-128, illus.

Aldrich, Frank Thatcher
   1972. A chorological analysis of the grass balds in the Oregon Coast Range. 156 p., illus. (Unpublished Ph.D. thesis on file at Oreg. State Univ., Corvallis.)

Aller, Alvin R.
   1956. A taxonomic and ecologic study of the flora of Monument Peak, Oregon. Am. Midl. Nat. 56:454-472, illus.

Anderson, E. William
   1956. Some soil-plant relationships in eastern Oregon. J. Range Manage. 9(4): 171-175, illus.

Anderson, H. G.
   1969. Growth form and distribution of vine maple (*Acer circinatum*) on Mary's Peak, western Oregon. Ecology 50(1): 127-130, illus.

Anderson, Howard George
   1967. The phytosociology of some vine maple communities in the Mary's Peak watershed. 118 p., illus. (Unpublished M.S. thesis on file at Oreg. State Univ., Corvallis.)

Archer, Anthony Clifford
   1963. Some synecological problems in the Alpine Zone of Garibaldi Park. 108 p., illus. (Unpublished M.S. thesis on file at Univ. B.C., Vancouver.)

Arno, Stephen F.
   1966. Interpreting the timberline: an aid to help park naturalists to acquaint visitors with the subalpine-alpine ecotone of western North America. 206 p., illus. (Unpublished M.F. thesis on file at Univ. Mont., Missoula.)

————
   1970. Ecology of alpine larch (*Larix lyallii* Parl.) in the Pacific Northwest. 264 p., illus. (Unpublished Ph.D. thesis on file at Univ. Mont., Missoula.)
———— and James R. Habeck
   1972. Ecology of alpine larch (*Larix lyallii* Parl.) in the Pacific Northwest. Ecol. Monogr. 42(4): 417-450.

Atzet, Thomas, and R. H. Waring
   1970. Selective filtering of light by coniferous forests and minimum light energy requirements for regeneration. Can. J. Bot. 48(12): 2163-2167.

Axelrod, Daniel I.
   1958. Evolution of the Madro-Tertiary geoflora. Bot. Rev. 24(7): 433-509, illus.

Bailey, Arthur W., and William W. Hines
   1971. A vegetation-soil survey of a wildlife-forestry research area and its application to management in northwestern Oregon. Oreg. State Game Comm. Res. Div. Game Rep. No. 2, 36 p., illus.
———— and Charles E. Poulton
   1968. Plant communities and environmental interrelationships in a portion of the Tillamook Burn, northwestern Oregon. Ecology 49: 1-13, illus.

Bailey, Arthur Wesley
  1966. Forest associations and secondary plant succession in the southern Oregon Coast Range. 164 p., illus. (Unpublished Ph.D. thesis on file at Oreg. State Univ., Corvallis. Abstr. published in Diss. Abstr., Sect. B27(8): 2605B-2606B, 1967.)
Bailey, Vernon
  1936. The mammals and life zones of Oregon. U.S. Dep. Agric. North Am. Fauna 55, 416 p., illus.
Bailey, Warren Hutchinson
  1963. Revegetation in the 1914-1915 devastated area of Lassen Volcanic National Park. 195 p., illus. (Unpublished Ph.D. thesis on file at Oreg. State Univ., Corvallis. Abstr. published in Diss. Abstr. 24(8), 1964.)
Baldwin, Ewart M.
  1964. Geology of Oregon. Ed. 2, 165 p., illus. Eugene: Univ. Oreg. Coop. Bookstore.
Ballard, Timothy Mangan
  1963. Some chemical changes accompanying soil development at the Kautz Creek mudflows, Mount Rainier National Park. 75 p., illus. (Unpublished M.F. thesis on file at Univ. Wash., Seattle.)
Barrett, John W. (ed.)
  1962. Regional silviculture of the United States. 610 p., illus. New York: Ronald Press Co.
Beaulieu, John D.
  1971. Geologic formations of western Oregon (west of longitude 121°30'). Oreg. State Dep. Geol. & Miner. Ind. Bull. 70, 72 p., illus.
Becking, R. W.
  1956. Die natürlichen Douglasien-Waldgeschellschaften Washingtons und Oregons. Allg. Forstl. u Jagdztg. 127: 42-56.
Becking, Rudolf Willem
  1954. Site indicators and forest types of the Douglas fir region of western Washington and Oregon. 133 p., illus. (Unpublished Ph.D. thesis on file at Univ. Wash., Seattle.)
Bell, M. A. M.
  1965. The dry subzone of the interior Western Hemlock Zone. Part 1. In V. J. Krajina (ed.), Ecology of western North America. Vol. 1, p. 42-56, illus. Univ. B. C. Dep. Bot.
Berntsen, Carl Martin
  1967. Relative low temperature tolerance of lodgepole and ponderosa pine seedlings. 158 p., illus. (Unpublished Ph.D. thesis on file at Oreg. State Univ., Corvallis.)
Berry, Dick Wallace
  1963. An ecological study of a disjunct ponderosa pine forest in the northern Great Basin in Oregon. 291 p., illus. (Unpublished Ph.D. thesis on file at Oreg. State Univ., Corvallis.)
Billings, W. D.
  1949. The shadscale vegetation zone of Nevada and eastern California in relation to climate and soil. Am. Midl. Nat. 42: 87-109.

  1951. Vegetational zonation in the Great Basin of western North America. In Les bases ecologiques de la regeneration de la vegetation des zones arides. Union Int. Sci. Biol. Ser. B, No. 9: 101-122, illus.
Blaisdell, James P.
  1953. Ecological effects of planned burning of sagebrush-grass range on the upper Snake River Plains. U.S. Dep. Agric. Tech. Bull. 1075, 39 p., illus.

_____ and Walter F. Mueggler

1956. Sprouting of bitterbrush (*Purshia tridentata*) following burning or top removal. Ecology 37: 365-370, illus.

Bliss, L. C.

1969. Alpine community pattern in relation to environmental parameters. *In* K. N. H. Greenidge (ed.), Essays in plant geography and ecology, p. 167-184, illus. N.S. Mus., Halifax, N.S., Can.

Bockheim, James Gregory

1972. Effects of alpine and subalpine vegetation on soil development, Mount Baker, Washington. 171 p., illus. (Unpublished Ph.D. thesis on file at Univ. Wash., Seattle.)

Brayshaw, T. C.

1965. The dry forest of southern British Columbia. *In* V. J. Krajina (ed.), Ecology of western North America. Vol. 1, p. 65-75, illus. Univ. B.C. Dep. Bot.

Bretz, J. Harlen

1959. Washington's Channeled Scabland. Wash. Div. Mines & Geol. Bull. 45, 57 p., illus.

Brink, V. C.

1959. A directional change in the subalpine forest-heath ecotone in Garibaldi Park, British Columbia. Ecology 40: 10-16, illus.

Brockman, C. Frank

1949. Trees of Mount Rainier National Park. 49 p., illus. Seattle: Univ. Wash. Press.

Brooke, Robert C.

1965. The Subalpine Mountain Hemlock Zone. Part II. Ecotypes and biogeocoenotic units. *In* V. J. Krajina (ed.), Ecology of western North America. Vol. 1, p. 79-101, illus. Univ. B.C. Dep. Bot.

_____, E. B. Peterson, and V. J. Krajina

1970. The Subalpine Mountain Hemlock Zone. *In* V. J. Krajina (ed.), Ecology of western North America. Vol. 2, p. 147-349, illus. Univ. B.C. Dep. Bot.

Brooks, William H.

1969. Some quantitative aspects of the present shrub community in Ginkgo State Park, Washington. Northwest Sci. 43(4): 185-190, illus.

Brown, Eugene Allen

1963. Early stages in plant succession on Douglas-fir clearcuts on the Mary's Peak watershed near Corvallis, Oregon. 61 p., illus. (Unpublished M.S. thesis on file at Oreg. State Univ., Corvallis.)

Byrd, Nathan Lemuel

1950. Vegetation zones of coastal dunes near Waldport, Oregon. 44 p., illus. (Unpublished M.S. thesis on file at Oreg. State Univ., Corvallis.)

Campbell, Alcetta Gilbert

1973. Vegetative ecology of Hunts Cove, Mt. Jefferson, Oregon. 89 p., illus. (Unpublished M.S. thesis on file at Oreg. State Univ., Corvallis.)

Campbell, C. D.

1953. Washington geology and resources. State Coll. Wash. Res. Stud. 21: 114-153, illus.

Carpenter, J. Richard

1956. An ecological glossary. 306 p., illus. New York: Hafner Publ. Co.

Chaney, R. W.

1938. Ancient forests of Oregon: A study of earth history in western America. Carnegie Inst. Wash. Publ. 501: 631-648, illus.

_____

1948. The ancient forests of Oregon. 56 p., illus. Oreg. State Syst. Higher Educ., Eugene.

Chohlis, G. John
   1952. Range conditions in eastern Washington fifty years ago and now. J. Range Manage. 5(3): 129-134, illus.
Clausen, Jens
   1963. Treelines and germ plasm—a study in evolutionary limitations. Proc. Natl. Acad. Sci. 50(5): 860-868.

_____
   1965. Population studies of alpine and subalpine races of conifers and willows in the California High Sierra Nevada. Evolution 19(1): 56-68.
Cleary, B. D., and R. H. Waring
   1969. Temperature: collection of data and its analysis for the interpretation of plant growth and distribution. Can. J. Bot. 47: 167-173, illus.
Cochran, P. H.
   1969. Thermal properties and surface temperatures of seedbeds, a guide for foresters. USDA Forest Serv. Pac. Northwest Forest & Range Exp. Stn., 19 p., illus. Portland, Oreg.

_____
   1971. Pumice particle bridging and nutrient levels affect lodgepole and ponderosa pine seedling development. USDA Forest Serv. Res. Note PNW-150, 10 p. Pac. Northwest Forest & Range Exp. Stn., Portland, Oreg.

_____
   1972. Tolerance of lodgepole and ponderosa pine seeds and seedlings to high water tables. Northwest Sci. 46(4): 322-331, illus.
Coleman, Babette Brown, Walter C. Muenscher, and Donald R. Charles
   1956. A distributional study of the epiphytic plants of the Olympic Peninsula, Washington. Am. Midl. Nat. 56: 54-87, illus.
Collins, Duane Francis
   1947. Potential timber values as compared to grazing values of the Oregon State College foothill pasture lands. 82 p., illus. (Unpublished M.S. thesis on file at Oreg. State Univ., Corvallis.)
Colville, F. V.
   1896. The sage plains of Oregon. Natl. Geogr. Mag. 7: 395-404, illus.
Cooke, Wm. Bridge
   1955. Fungi, lichens and mosses in relation to vascular plant communities in eastern Washington and adjacent Idaho. Ecol. Monogr. 25: 119-180, illus.
Cooper, William S.
   1957. Vegetation of the Northwest-American Province. Proc. Eighth Pac. Sci. Congr. 8(4): 133-138.

_____
   1958. Coastal sand dunes of Oregon and Washington. Geol. Soc. Am. Mem. 72, 169 p., illus.
Corliss, J. F., and C. T. Dyrness
   1965. A detailed soil-vegetation survey of the Alsea area in the Oregon Coast Range. In C. T. Youngberg (ed.), Forest-soil relationships in North America, p. 457-483, illus. Corvallis: Oreg. State Univ. Press.
Cotton, J. S.
   1904. A report on the range conditions of central Washington. Wash. Agric. Exp. Stn. Bull. 60, 45 p., illus.
Countryman, C. M., and D. R. Cornelius
   1957. Some effects of fire on a perennial range type. J. Range Manage. 10: 39-41.

Crandell, Dwight R.
    1969a. Surficial geology of Mount Rainier National Park, Washington. U.S. Geol. Surv. Bull. 1288, 41 p., illus.

———
    1969b. The geologic story of Mount Rainier. U.S. Geol. Surv. Bull. 1292, 43 p., illus.

———
    1971. Postglacial lahars from Mount Rainier volcano, Washington. U.S. Geol. Surv. Prof. Pap. 677, 75 p., illus.
——— and Donal R. Mullineaux
    1967. Volcanic hazards at Mount Rainier, Washington. U.S. Geol. Surv. Bull. 1238, 26 p., illus.
Crouch, Glenn L.
    1971. Susceptibility of ponderosa, Jeffrey, and lodgepole pines to pocket gophers. Northwest Sci. 45(4): 252-256, illus.
Culver, Roger Norman
    1964. An ecological reconnaissance of the *Artemisia* steppe on the east central Owyhee uplands of Oregon. 99 p., illus. (Unpublished M.S. thesis on file at Oreg. State Univ., Corvallis.)
Dalquest, W. W., and V. B. Scheffer
    1942. The origin of the Mima mounds of western Washington. J. Geol. 50: 68-84, illus.
Daniels, Jess Donald
    1969. Variation and intergradation in the grand fir-white fir complex. 235 p., illus. (Unpublished Ph.D. thesis on file at Univ. Idaho, Moscow.)
Danner, Wilbert R.
    1955. Geology of Olympic National Park. 68 p., illus. Seattle: Univ. Wash. Press.
Daubenmire, R.
    1952. Forest vegetation of northern Idaho and adjacent Washington, and its bearing on concepts of vegetation classification. Ecol. Monogr. 22: 301-330, illus.

———
    1953. Classification of the conifer forests of eastern Washington and northern Idaho. Northwest Sci. 27(1): 17-24.

———
    1954. Alpine timberlines in the Americas and their interpretation. Butler Univ. Bot. Stud. 11: 119-136, illus.

———
    1956. Climate as a determinant of vegetation distribution in eastern Washington and northern Idaho. Ecol. Monogr. 26: 131-154, illus.

———
    1961. Vegetative indicators of rate of height growth in ponderosa pine. Forest Sci. 7: 24-34, illus.

———
    1966. Vegetation: Identification of typal communities. Science 151: 291-298, illus.

———
    1968a. Plant communities. 300 p., illus. New York, Evanston [etc.]: Harper & Row.

———
    1968b. Soil moisture in relation to vegetation distribution in the mountains of northern Idaho. Ecology 49: 431-438, illus.

———
    1969a. Ecologic plant geography of the Pacific Northwest. Madroño 20(3): 111-128, illus.

———
    1970. Steppe vegetation of Washington. Wash. Agric. Exp. Stn. Tech. Bull. 62, 131 p., illus.

1972. Annual cycles of soil moisture and temperature as related to grass development in the steppe of eastern Washington. Ecology 53(3): 419-424, illus.

_____ and Jean B. Daubenmire

1968. Forest vegetation of eastern Washington and northern Idaho. Wash. Agric. Exp. Stn. Tech. Bull. 60, 104 p., illus.

Daubenmire, R. F.

1940. Plant succession due to overgrazing in the *Agropyron* bunchgrass prairie of southeastern Washington. Ecology 21: 55-64, illus.

_____

1946. The life zone problem in the northern intermountain region. Northwest Sci. 20(2): 28-38, illus.

_____

1947. Origin and distribution of North American plant formations. *In* Biogeography, p. 17-22. Corvallis: Oreg. State Coll.

Daubenmire, Rexford

1969b. Structure and ecology of coniferous forests of the northern Rocky Mountains. *In* Center for Natural Resources (ed.), Coniferous forests of the northern Rocky Mountains, p. 25-41. Univ. Mont. Found., Missoula.

Daubenmire, Rexford F.

1942. An ecological study of the vegetation of southeastern Washington and adjacent Idaho. Ecol. Monogr. 12: 53-79, illus.

_____ and W. E. Colwell

1942. Some edaphic changes due to overgrazing in the *Agropyron-Poa* prairie of southeastern Washington. Ecology 23: 32-40.

Davidson, Eric Duncan

1967. Synecological features of a natural headland prairie on the Oregon coast. 79 p., illus. (Unpublished M.S. thesis on file at Oreg. State Univ., Corvallis.)

Dealy, J. Edward

1971. Habitat characteristics of the Silver Lake mule deer range. USDA Forest Serv. Res. Pap. PNW-125, 99 p., illus. Pac. Northwest Forest & Range Exp. Stn., Portland, Oreg.

Dean, Milton Lee

1960. A taxonomic and ecologic study of the vascular plants of a section of the Owyhee River Canyon in Oregon. 147 p., illus. (Unpublished M.S. thesis on file at Oreg. State Univ., Corvallis.)

Del Moral, Roger

1973. The vegetation of Findley Lake Basin. Am. Midl. Nat. 89(1): 26-40.

Dennis, La Rae June

1959. A taxonomic study of the vascular flora of Ashland Peak, Jackson County, Oregon. 144 p., illus. (Unpublished M.A. thesis on file at Oreg. State Univ., Corvallis.)

Detling, LeRoy E.

1954. Significant features of the flora of Saddle Mountain, Clatsop County, Oregon. Northwest Sci 28(2): 52-60.

_____

1961. The Chaparral Formation of southwestern Oregon, with considerations of its postglacial history. Ecology 42: 348-357, illus.

_____

1966. The flora of the Columbia River Gorge. Northwest Sci. 40: 133-137.

_____

1968. Historical background of the flora of the Pacific Northwest. Univ. Oreg. Mus. Nat. Hist. Bull. 13, 57 p., illus.

Dick-Peddie, W. A., and W. H. Moir
  1970. Vegetation of the Organ Mountains, New Mexico. Range Sci. Dep. Sci. Ser. No. 4, 28 p., illus. Colo. State Univ., Ft. Collins.
Dickson, B. A., and R. L. Crocker
  1953a. A chronosequence of soils and vegetation near Mt. Shasta, California. I. Definition of the ecosystem investigated and features of the plant succession. J. Soil Sci. 4: 123-141, illus.
———— and R. L. Crocker
  1953b. A chronosequence of soils and vegetation near Mt. Shasta, California. II. The development of the forest floors and the carbon and nitrogen profiles of the soils. J. Soil Sci. 4: 142-154, illus.
———— and R. L. Crocker
  1954. A chronosequence of soils and vegetation near Mount Shasta, California. III. Some properties of the mineral soils. J. Soil Sci. 5: 173-191, illus.
Dirks-Edmunds, Jane C.
  1947. A comparison of biotic communities of the cedar-hemlock and oak-hickory associations. Ecol. Monogr. 17: 235-260, illus.
Dott, R. H., Jr.
  1971. Geology of the southwestern Oregon coast west of the 124th meridian. Oreg. State Dep. Geol. & Miner. Ind. Bull. 69, 63 p., illus.
Douglas, George W.
  1969. Subalpine tree groups in the western North Cascades. (Abstr.) Northwest Sci. 43: 34-35.

————
  1971. The alpine-subalpine flora of the North Cascade Range, Washington. Wasmann J. Biol. 29: 129-168.

————
  1972. Subalpine plant communities of the western North Cascades, Washington. Arctic & Alpine Res. 4: 147-166, illus.
———— and T. M. Ballard
  1971. Effects of fire on alpine plant communities in the North Cascades, Washington. Ecology 52: 1058-1064, illus.
Douglas, George Wayne
  1970. A vegetation study in the subalpine zone of the western North Cascades, Washington. 293 p., illus. (Unpublished M.S. thesis on file at Univ. Wash., Seattle.)
Driscoll, Richard S.
  1962. Characteristics of some ecosystems in the juniper zone in central Oregon. J. Range Manage. 15: 347.

————
  1963. Sprouting bitterbrush in central Oregon. Ecology 44: 820-821, illus.

————
  1964a. A relict area in the central Oregon juniper zone. Ecology 45: 345-353, illus.

————
  1964b. Vegetation-soil units in the central Oregon juniper zone. USDA Forest Serv. Res. Pap. PNW-19, 60 p., illus. Pac. Northwest Forest & Range Exp. Stn., Portland, Oreg.
Dyrness, C. T.
  1965. The effect of logging and slash burning on understory vegetation in the H. J. Andrews Experimental Forest. USDA Forest Serv. Res. Note PNW-31, 13 p., illus. Pac. Northwest Forest & Range Exp. Stn., Portland, Oreg.

1973. Early stages of plant succession following logging and burning in the western Cascades of Oregon. Ecology 54(1): 57-69, illus.

———— and C. T. Youngberg

1958. Soil-vegetation relationships in the central Oregon pumice region. *In* First North American Forest Soils Conference, p. 57-66. Mich. State Univ. Agric. Exp. Stn., East Lansing.

———— and C. T. Youngberg

1966. Soil-vegetation relationships within the ponderosa pine type in the central Oregon pumice region. Ecology 47: 122-138, illus.

————, Jerry F. Franklin, and Chris Maser

1973. Wheeler Creek Research Natural Area. Supplement No. 1 to Federal Research Natural Areas in Oregon and Washington: A guidebook for scientists and educators, 16 p., illus. USDA Forest Serv., Pac. Northwest Forest & Range Exp. Stn., Portland, Oreg.

Dyrness, Christen Theodore

1960. Soil-vegetation relationships within the ponderosa pine type in the central Oregon pumice region. 217 p., illus. (Unpublished Ph.D. thesis on file at Oreg. State Univ., Corvallis.)

Easterbrook, Don J., and David A. Rahm

1970. Landforms of Washington. 156 p., illus. Bellingham, Wash.: Union Printing Co.

Eckert, Richard Edgar, Jr.

1957. Vegetation-soil relationships in some *Artemisia* types in northern Harney and Lake Counties, Oregon. 208 p., illus. (Unpublished Ph.D. thesis on file at Oreg. State Univ., Corvallis.)

Egler, Frank E.

1934. Communities and successional trends in the vegetation of the Coos Bay sand dunes, Oregon. 49 p., illus. (Unpublished M.S. thesis on file at Univ. Minn., St. Paul.)

Eis, Slavoj

1962. Statistical analysis of several methods for estimation of forest habitats and tree growth near Vancouver, B.C. Univ. B.C. Fac. Forest., Forest. Bull. 4, 76 p., illus.

Faegri, Knut

1966. A botanical excursion to Steens Mountain, SE Oregon, U.S.A. Saertr. av Blyttia 24: 173-181, illus.

Fenneman, Nevin M.

1931. Physiography of Western United States. 534 p., illus. New York and London: McGraw-Hill Book Co.

Fiske, Richard S., Clifford A. Hopson, and Aaron C. Waters

1963. Geology of Mount Rainier National Park, Washington. U.S. Geol. Surv. Prof. Pap. 444, 93 p., illus.

Fitzgerald, Betty Jo

1966. The microenvironment in a Pacific Northwest bog and its implications for establishment of conifer seedlings. 164 p., illus. (Unpublished M.S. thesis on file at Univ. Wash., Seattle.)

Folsom, Michael M.

1970. The glacial geomorphology of the Puget lowland, Washington and British Columbia: comments and selected references. Northwest Sci. 44(2): 143-146.

Fonda, R. W., and L. C. Bliss

1969. Forest vegetation of the montane and subalpine zones, Olympic Mountains, Washington. Ecol. Monogr. 39(3): 271-301, illus.

Fonda, Richard Weston
  1967. Ecology of montane and subalpine forests Olympic Mountains, Washington. 145 p., illus. (Unpublished Ph.D. thesis on file at Univ. Ill., Urbana.)
Forest Soils Committee of the Douglas-fir Region
  1957. An introduction to the forest soils of the Douglas-fir region of the Pacific Northwest. Various paging, illus. Seattle: Univ. Wash.
Foster, Robert J.
  1960. Tertiary geology of a portion of the central Cascade Mountains, Washington. Bull. Geol. Soc. Am. 71: 99-126, illus.
Fowells, H. A. (comp.)
  1965. Silvics of forest trees of the United States. U.S. Dep. Agric. Handb. 271, 762 p., illus.
_____ and B. M. Kirk
  1945. Availability of soil moisture to ponderosa pine. J. For. 43: 601-604.
Franklin, Jerry F.
  1964. Some notes on the distribution and ecology of noble fir. Northwest Sci. 38(1): 1-13, illus.

_____
  1965a. Ecology and silviculture of the true fir-hemlock forests of the Pacific Northwest. Soc. Am. For. Proc. 1964: 28-32, illus.

_____
  1965b. Tentative ecological provinces within the true fir-hemlock forest areas of the Pacific Northwest. USDA Forest Serv. Res. Pap. PNW-22, 31 p., illus. Pac. Northwest Forest & Range Exp. Stn., Portland, Oreg.
_____ and Norman A. Bishop
  1969. Notes on the natural history of Mount Rainier National Park. 24 p., illus. Mount Rainier Nat. Hist. Assoc., Longmire, Wash.
_____ and Dean S. DeBell
  1973. Effects of various harvesting methods on forest regeneration. *In* Richard K. Hermann and Denis P. Lavender (eds.), Even-age management. Sch. For. Pap. 848, p. 29-57, illus. Oreg. State Univ., Corvallis.
_____ and C. T. Dyrness
  1969. Vegetation of Oregon and Washington. USDA Forest Serv. Res. Pap. PNW-80, 216 p., illus. Pac. Northwest Forest & Range Exp. Stn., Portland, Oreg.
_____, C. T. Dyrness, and W. H. Moir
  1970. A reconnaissance method for forest site classification. Shinrin Richi XII(1): 1-14.
_____, C. T. Dyrness, Duane G. Moore, and Robert F. Tarrant
  1968. Chemical soil properties under coastal Oregon stands of alder and conifers. *In* J. M. Trappe, J. F. Franklin, R. F. Tarrant, and G. M. Hansen (eds.), Biology of alder. Northwest Sci. Assoc. Fortieth Annu. Meet., Symp. Proc. 1967: 157-172. Pac. Northwest Forest & Range Exp. Stn., Portland, Oreg.
_____, Frederick C. Hall, C. T. Dyrness, and Chris Maser
  1972. Federal Research Natural Areas in Oregon and Washington: A guidebook for scientists and educators. 498 p., illus. USDA Forest Serv. Pac. Northwest Forest & Range Exp. Stn., Portland, Oreg.
_____ and Russel G. Mitchell
  1967. Successional status of subalpine fir in the Cascade Range. USDA Forest Serv. Res. Pap. PNW-46, 16 p., illus. Pac. Northwest Forest & Range Exp. Stn., Portland, Oreg.

————, William H. Moir, George W. Douglas, and Curt Wiberg

1971. Invasion of subalpine meadows by trees in the Cascade Range, Washington and Oregon. Arctic & Alpine Res. 3: 215-224, illus.

———— and Anna A. Pechanec

1968. Comparison of vegetation in adjacent alder, conifer, and mixed alder-conifer communities. I. Understory vegetation and stand structure. *In* J. M. Trappe, J. F. Franklin, R. F. Tarrant, and G. M. Hansen (eds.), Biology of alder. Northwest Sci. Assoc. Fortieth Annu. Meet. Symp. Proc. 1967: 37-43, illus. Pac. Northwest Forest & Range Exp. Stn., Portland, Oreg.

———— and James M. Trappe

1963. Plant communities of the northern Cascade Range: A reconnaissance. (Abstr.) Northwest Sci. 37(4): 163-164.

Franklin, Jerry Forest

1966. Vegetation and soils in the subalpine forests of the southern Washington Cascade Range. 132 p., illus. (Unpublished Ph.D. thesis on file at Wash. State Univ., Pullman.)

Frehner, Hans K.

1957. Development of soil and vegetation on the Kautz Creek flood deposit in Mount Rainier National Park. 83 p., illus. (Unpublished M.F. thesis on file at Univ. Wash., Seattle.)

Fujimori, Takao

1971. Primary productivity of a young *Tsuga heterophylla* stand and some speculations about biomass of forest communities on the Oregon coast. USDA Forest Serv. Res. Pap. PNW-123, 11 p., illus. Pac. Northwest Forest & Range Exp. Stn., Portland, Oreg.

————

1972. Discussion about the large forest biomasses on the Pacific Northwest in U.S.A. J. Jap. Forest. Soc. 54: 230-233, illus.

Furman, Thomas E., and James M. Trappe

1971. Phylogeny and ecology of mycotrophic achlorophyllous angiosperms. Q. Rev. Biol. 46(3): 219-225.

Galbraith, William A., and E. William Anderson

1971. Grazing history of the Northwest. J. Range Manage. 24(1): 6-12, illus.

Garrison, G. A.

1961. Recovery of ponderosa pine range in eastern Oregon and eastern Washington by seventh year after logging. Soc. Am. For. Proc. 1960: 137-139, illus.

————

1965. Changing conditions in Northwest forests and their relation to range use. Soc. Am. For. Proc. 1964: 67-68.

Garrison, George A., and Robert S. Rummell

1951. First-year effects of logging on ponderosa pine forest range lands of Oregon and Washington. J. For. 49: 708-713, illus.

————, Jon M. Skovlin, and Charles E. Poulton

1967. Northwest range-plant symbols. USDA Forest Serv. Res. Pap. PNW-40, 121 p. Pac. Northwest Forest & Range Exp. Stn., Portland, Oreg.

Gashwiler, Jay S.

1970. Plant and mammal changes on a clearcut in west-central Oregon. Ecology 51: 1018-1026, illus.

Gilkeson, R. H., W. A. Starr, and E. C. Steinbrenner

1961. Soil survey of the Snoqualmie Falls Tree Farm. Weyerhaeuser Co., Centralia, Wash. 10 p. + maps.

Graham, Alan (ed.)

    1972. Floristics and paleofloristics of Asia and eastern North America. 278 p. New York: Elsevier Publ. Co.

Gratkowski, H.

    1961a. Brush problems in southwestern Oregon. USDA Forest Serv. Pac. Northwest Forest & Range Exp. Stn., 53 p., illus. Portland, Oreg.

Gratkowski, H. J.

    1961b. Use of herbicides on forest lands in southwestern Oregon. USDA Forest Serv. Res. Note 217, 18 p. Pac. Northwest Forest & Range Exp. Stn., Portland, Oreg.

Gratkowski, Henry John

    1962. Heat as a factor in germination of seeds of *Ceanothus velutinus* var. *laevigatus* T. & G. 122 p., illus. (Unpublished Ph.D. thesis on file at Oreg. State Univ., Corvallis.)

Green, Diantha Louise

    1965. Developmental history of European beachgrass [*Ammophila arenaria* (L.) Link] plantings on the Oregon coastal sand dunes. 64 p., illus. (Unpublished M.S. thesis on file at Oreg. State Univ., Corvallis.)

Griffin, James R.

    1967. Soil moisture and vegetation patterns in northern California forests. USDA Forest Serv. Res. Pap. PSW-46, 22 p., illus. Pac. Southwest Forest & Range Exp. Stn., Berkeley, Calif.

Griffiths, D.

    1902. Forage conditions on the northern border of the Great Basin, being a report upon investigations made during July and August, 1901, in the region between Winnemucca, Nevada, and Ontario, Oregon. U.S. Bur. Plant Ind. Bull. 15, 60 p., illus.

    ———

    1903. Forage conditions and problems in eastern Washington, eastern Oregon, northeastern California, and northwestern Nevada. U.S. Bur. Plant Ind. Bull. 38, 52 p., illus.

Habeck, James R.

    1961. The original vegetation of the mid-Willamette Valley, Oregon. Northwest Sci. 35(2): 65-77, illus.

    ———

    1962. Forest succession in Monmouth Township, Polk County, Oregon since 1850. Mont. Acad. Sci. Proc. 21: 7-17, illus.

    ———

    1967. Mountain hemlock communities in western Montana. Northwest Sci. 41: 169-177, illus.

    ——— and Ernest Hartley

    1968. A glossary of alpine terminology. 35 p. Univ. Mont. Dep. Bot., Missoula.

Hall, Fred C.

    1968. The influences of variations in elevation on range vegetation. Range Manage. Workshop, p. 30-34. Wash. State Univ., Pullman.

Hall, Frederick C.

    1971. Some uses and limitations of mathematical analysis in plant ecology and land management. *In* G. P. Patil, E. C. Pielou, and W. E. Waters (eds.), Statistical ecology. Vol. 3, Many species populations, ecosystems, and systems analysis, p. 377-395. University Park: Pa. State Univ. Press.

Hall, Frederick Columbus
   1956. Use of oak woodlands (*Quercus garryana*) for farm forestry and grazing in the Willamette valley. 80 p., illus. (Unpublished M.S. thesis on file at Oreg. State Univ., Corvallis.)

   1967. Vegetation-soil relations as a basis for resource management on the Ochoco National Forest of central Oregon. 207 p., illus. (Unpublished Ph.D. thesis on file at Oreg. State Univ., Corvallis.)
Hamann, M. J.
   1972. Vegetation of alpine and subalpine meadows of Mount Rainier National Park, Washington. 120 p., illus. (Unpublished M.S. thesis on file at Wash. State Univ., Pullman.)
Hamrick, J. L., and W. J. Libby
   1972. Variation and selection in western U.S. montane species. I. White fir. Silvae Genet. 21: 29-35, illus.
Hamrick, James Lewis, III
   1966. Geographic variation in white fir. 143 p., illus. (Unpublished M.S. thesis on file at Univ. Calif., Berkeley.)
Hanneson, Bill
   1962. Changes in the vegetation on coastal dunes in Oregon. 103 p., illus. (Unpublished M.S. thesis on file at Univ. Oreg., Eugene.)
Hansen, Charles Goodman
   1956. An ecological survey of the vertebrate animals on Steen's Mountain, Harney County, Oregon. 200 p., illus. (Unpublished Ph.D. thesis on file at Oreg. State Univ., Corvallis.)
Hansen, Henry P.
   1946. Post glacial forest succession and climate in the Oregon Cascades. Am. J. Sci. 244: 710-734.

   1947. Postglacial forest succession, climate, and chronology in the Pacific Northwest. Am. Philos. Soc. Trans. New Ser., vol. 37, part 1, 130 p., illus.
Hanson, Herbert C.
   1962. Dictionary of ecology. 382 p. New York: Philos. Libr.
Harris, Grant A., and A. M. Wilson
   1970. Competition for moisture among seedlings of annual and perennial grasses as influenced by root elongation at low temperature. Ecology 51: 530-534, illus.
Harris, S. W.
   1954. An ecological study of the waterfowl of the potholes area, Grant County, Washington. Am. Midl. Nat. 52: 403-432, illus.
Harthill, Marion Paul
   1964. Mosses of the Olympic Peninsula. 151 p., illus. (Unpublished M.S. thesis on file at Univ. Utah, Salt Lake City.)
Hawk, Glenn Martin
   1973. Forest vegetation and soils of terraces and floodplains along the McKenzie River, Oregon. 188 p., illus. (Unpublished M.S. thesis on file at Oreg. State Univ., Corvallis.)
Hayes, G. L.
   1959. Forest and forest-land problems of southwestern Oregon. USDA Forest Serv. Pac. Northwest Forest & Range Exp. Stn., 54 p., illus. Portland, Oregon.

_____ and William E. Hallin
    1962. Tree species to grow in the South Umpqua drainage. USDA Forest Serv. Pac.
        Northwest Forest & Range Exp. Stn. Res. Note 221, 7 p., illus. Portland, Oreg.
Head, Serge Conrade
    1959. Plant taxonomy and ecology of the East Eagle Creek drainage of the Wallowa
        Mountains northeastern Oregon. 249 p., illus. (Unpublished Ph.D. thesis on
        file at Oreg. State Univ., Corvallis.)
Heady, Harold F.
    1968. Grassland response to changing animal species. J. Soil & Water Conserv. 23(5):
        173-176.
Henderson, Jan Alan
    1970. Biomass and composition of the understory vegetation in some *Alnus rubra*
        stands in western Oregon. 64 p., illus. (Unpublished M.S. thesis on file at Oreg.
        State Univ., Corvallis.)

_____
    1973. Composition, distribution and succession of subalpine meadows in Mount Rainier
        National Park, Washington. 150 p., illus. (Unpublished Ph.D. thesis on file at
        Oreg. State Univ., Corvallis.)
Hermann, Richard K., and Roger G. Petersen
    1969. Root development and height increment of ponderosa pines in pumice soils of
        central Oregon. Forest Sci. 15: 226-237, illus.
Herring, H. G.
    1968. Soil-moisture depletion by a central Washington lodgepole pine stand. Northwest
        Sci. 42(1): 1-4, illus.
Heusser, Calvin J.
    1960. Late-Pleistocene environments of North Pacific North America. Am. Geogr.
        Soc. Spec. Publ. 35, 308 p., illus.
Hickman, James Craig
    1968. Disjunction and endemism in the flora of the central western Cascades of Oregon:
        An historical and ecological approach to plant distributions. 335 p., illus. (Un-
        published Ph.D. thesis on file at Univ. Oreg., Eugene.)
Higinbotham, N., and Betty Wilson Higinbotham
    1954. Quantitative relationships of terrestrial mosses with some coniferous forests at
        Mt. Rainier National Park. Butler Univ. Bot. Stud. 11: 149-168.
Hines, William W.
    1971. Plant communities in the old-growth forests of north coastal Oregon. 146 p.,
        illus. (Unpublished M.S. thesis on file at Oreg. State Univ., Corvallis.)
Hitchcock, C. Leo, Arthur Cronquist, Marion Ownbey, and J. W. Thompson
    1955. Vascular plants of the Pacific Northwest. Part 5. Compositae. 343 p., illus.
        Seattle: Univ. Wash. Press.
_____, Arthur Cronquist, Marion Ownbey, and J. W. Thompson
    1959. Vascular plants of the Pacific Northwest. Part 4. Ericaceae through Campanu-
        laceae. 510 p., illus. Seattle: Univ. Wash. Press.
_____, Arthur Cronquist, Marion Ownbey, and J. W. Thompson
    1961. Vascular plants of the Pacific Northwest. Part 3. Saxifragaceae to Ericaceae.
        614 p., illus. Seattle: Univ. Wash. Press.
_____, Arthur Cronquist, Marion Ownbey, and J. W. Thompson
    1964. Vascular plants of the Pacific Northwest. Part 2. Salicaceae to Saxifragaceae.
        597 p., illus. Seattle: Univ. Wash. Press.
_____, Arthur Cronquist, Marion Ownbey, and J. W. Thompson
    1969. Vascular plants of the Pacific Northwest. Part 1. Vascular cryptogams, gymno-
        sperms, and monocotyledons. 914 p., illus. Seattle: Univ. Wash. Press.

Horn, Elizabeth Mueller
  1968. Ecology of the Pumice Desert, Crater Lake National Park. Northwest Sci. 42: 141-149, illus.
Howard, Grace E.
  1950. Lichens of the State of Washington. 191 p., illus. Seattle: Univ. Wash. Press.
Howell, Joseph, Jr.
  1932. The development of seedlings of ponderosa pine in relation to soil types. J. For. 30: 944-947.
Humphrey, R. R.
  1945. Common range forage types of the inland Pacific. Northwest Sci. 19: 3-11.

Huntting, Marshall T., W. A. G. Bennett, Vaughan E. Livingstone, Jr., and Wayne S. Moen
  1961. Geologic map of Washington. Wash. Dep. Conserv., Div. Mines & Geol.
Illingworth, K., and J. W. C. Arlidge
  1960. Interim report on some forest site types in lodgepole pine and spruce-alpine fir stands. B.C. Forest Serv. Res. Note 35, 44 p., illus.
Ingram, Douglas C.
  1931. Vegetative changes and grazing use on Douglas-fir cutover land. J. Agric. Res. 43: 387-417, illus.
Irwin, W. P.
  1966. Geology of Klamath Mountains Province. In E. H. Bailey (ed.), Geology of northern California. Calif. Div. Mines & Geol. Bull. 190: 19-38, illus.
Isaac, Leo A.
  1940. Vegetative succession following logging in the Douglas-fir region with special reference to fire. J. For. 38: 716-721.
Johannessen, Carl L.
  1964. Marshes prograding in Oregon: aerial photographs. Science 146: 1575-1578, illus.
  _____, William A. Davenport, Artimus Millet, and Steven McWilliams
  1971. The vegetation of the Willamette valley. Ann. Assoc. Am. Geogr. 61(2): 286-302, illus.
Johnsgard, G. A.
  1963. Temperature and the water balance of Oregon weather stations. Oreg. State Univ. Agric. Exp. Stn. Spec. Rep. 150, 127 p. Corvallis, Oreg.
Johnson, John Morris
  1961. Taxonomy and ecology of the vascular plants of Black Butte, Oregon. 193 p., illus. (Unpublished M.S. thesis on file at Oreg. State Univ., Corvallis.)
Johnson, Walter Van-Gale
  1959. Forage utilization estimates in relation to ecological units in the Wallowa Mountains of northeastern Oregon. 138 p., illus. (Unpublished M.S. thesis on file at Oreg. State Univ., Corvallis.)
Jones, George Neville
  1936. A botanical survey of the Olympic Peninsula, Washington. Univ. Wash. Publ. Biol. vol. 5, 286 p., illus.

  _____
  1938. The flowering plants and ferns of Mount Rainier. Univ. Wash. Publ. Biol. vol. 7, 192 p., illus.
Kellman, M. C.
  1969. Plant species interrelationships in a secondary succession in coastal British Columbia. Syesis 2: 201-212, illus.
Kerr, H. S.
  1913. Notes on the distribution of lodgepole pine and yellow pine in the Walker Basin. For. Q. 11: 509-515.

Kienholz, Raymond
  1929. Revegetation after logging and burning in the Douglas-fir region of western Washington. Ill. State Acad. Sci. Trans. 21: 94-108, illus.

———— 1931. The vegetation of a lava-formed lake in the Cascade Mountains. Am. J. Bot. 18: 641-648, illus.

Kirk, Ruth
  1966. The Olympic rain forest. 86 p., illus. Seattle: Univ. Wash. Press.

Kirkwood, J. E.
  1902. The vegetation of northwestern Oregon. Torreya 2(9): 129-134.

Knapp, Rüdiger
  1965. Die Vegetation von Nord- und Mittelamerika und der Hawaii-Inseln. [The vegetation of North and Central America and of the Hawaiian Islands.] 373 p., illus. Stuttgart: Gustav Fischer Verlag.

Knox, Ellis G.
  1962. Soils. *In* Richard M. Highsmith, Jr. (ed.), Atlas of the Pacific Northwest. Ed. 3. p. 43-46, illus. Corvallis: Oreg. State Univ. Press.

Kotor, John
  1972. Ecology of *Abies amabilis* in relation to its altitudinal distribution and in contrast to its common associate *Tsuga heterophylla*. 171 p., illus. (Unpublished Ph.D. thesis on file at Univ. Wash., Seattle.)

Krajina, V. J.
  1969. Ecology of forest trees in British Columbia. *In* V. J. Krajina (ed.), Ecology of western North America. Vol. 2, p. 1-146, illus. Univ. B. C. Dep. Bot.

Krajina, Vladimir J.
  1965. Biogeoclimatic zones and classification of British Columbia. *In* V. J. Krajina (ed.), Ecology of western North America. Vol. 1, p. 1-17, illus. Univ. B.C. Dep. Bot.

Kruckeberg, Arthur R.
  1954. The ecology of serpentine soils. III. Plant species in relation to serpentine soils. Ecology 35: 267-274, illus.

———— 1964. Ferns associated with ultramafic rocks in the Pacific Northwest. Am. Fern J. 54: 113-126, illus.

———— 1967. Ecotypic response to ultramafic soils by some plants species of northwestern United States. Brittonia 19: 133-151, illus.

———— 1969a. Plant life on serpentinite and other ferromagnesian rocks in northwestern North America. Syesis 2: 15-114, illus.

———— 1969b. Soil diversity and the distribution of plants, with examples from western North America. Madroño 20: 129-154, illus.

Küchler, A. W.
  1946. The broadleaf deciduous forests of the Pacific Northwest. Ann. Assoc. Am. Geogr. 36: 122-147.

———— 1964. Manual to accompany the map of potential natural vegetation of the conterminous United States. Am. Geogr. Soc. Spec. Publ. 36, various paging, illus.

Kumler, Marion Lawrence
1963. Succession and certain adaptative features of plants native to the sand dunes of the Oregon coast. 149 p., illus. (Unpublished Ph.D. thesis on file at Oreg. State Univ., Corvallis.)

Kuramoto, R. T., and L. C. Bliss
1970. Ecology of subalpine meadows in the Olympic Mountains, Washington. Ecol. Monogr. 40: 317-347, illus.

Kuramoto, Richard Tatsuo
1968. Ecology of subalpine meadows in the Olympic Mountains, Washington. 150 p., illus. (Unpublished Ph.D. thesis on file at Univ. Ill., Urbana.)

Lang, Frank Alexander
1961. A study of vegetation change in the gravelly prairies of Pierce and Thurston Counties, western Washington. 109 p., illus. (Unpublished M.S. thesis on file at Univ. Wash., Seattle.)

Larson, James Wesley
1972. Ecological role of lodgepole pine in the Upper Skagit River valley, Washington. 77 p., illus. (Unpublished M.S. thesis on file at Univ. Wash., Seattle.)

Lawrence, Donald B.
1938. Trees on the march. Mazama 20(12): 49-54, illus.

_____
1939. Some features of the vegetation of the Columbia River Gorge with special reference to asymmetry in forest trees. Ecol. Monogr. 9: 217-257, illus.

Lawton, Elva
1965. Keys for the identification of the mosses of Washington and Oregon. The Bryologist 68: 141-184.

_____
1971. Moss flora of the Pacific Northwest. 362 p., illus. Nichinan, Miyazaki, Japan: The Hattori Bot. Lab.

Little, Elbert L., Jr.
1953. Check list of native and naturalized trees of the United States (including Alaska). U.S. Dep. Agric. Agric. Handb. 41, 472 p.

Livingston, Kenneth Hubert
1953. Composition of and production from an improved and an unimproved hill pasture. 112 p., illus. (Unpublished M.S. thesis on file at Oreg. State Univ., Corvallis.)

Lotspeich, Frederick B., Jack B. Secor, Rose Okazaki, and Henry W. Smith
1961. Vegetation as a soil-forming factor on the Quillayute physiographic unit in western Clallam County, Washington. Ecology 42: 53-68, illus.

Lowery, Robert Franklin
1972. Ecology of subalpine zone tree clumps in the North Cascades Mountains of Washington. 137 p., illus. (Unpublished Ph.D. thesis on file at Univ. Wash., Seattle.)

Lu, K. C., C. S. Chen, and W. B. Bollen
1968. Comparison of microbial populations between red alder and conifer soils. *In* J. M. Trappe, J. F. Franklin, R. F. Tarrant, and G. M. Hansen (eds.), Biology of alder. Northwest Sci. Assoc. Fortieth Annu. Meet. Symp. Proc. 1967: 173-178, illus. Pac. Northwest Forest & Range Exp. Stn., Portland, Oreg.

Lynott, Robert E.
1966. Weather and climate of the Columbia Gorge. Northwest Sci. 40(4): 129-132.

McConnell, Burt R., and Justin G. Smith
    1971. Effect of ponderosa pine needle litter on grass seedling survival. USDA Forest Serv.Res. Note PNW-155, 6 p., illus. Pac. Northwest Forest & Range Exp. Stn., Portland, Oreg.

McKell, Cyrus Milo
    1956. Some characteristics contributing to the establishment of rabbitbrush, *Chrysothamnus* spp. 130 p., illus. (Unpublished Ph.D. thesis on file at Oreg. State Univ., Corvallis.)

McLean, A., and W. D. Holland
    1958. Vegetation zones and their relationship to the soils and climate of the upper Columbia valley. Can. J. Plant Sci. 38: 328-345, illus.

McLean, Alastair
    1970. Plant communities of the Similkameen valley, British Columbia and their relationships to soils. Ecol. Monogr. 40(4): 403-424, illus.

McMinn, R. G.
    1952. The role of soil drought in the distribution of vegetation in the northern Rocky Mountains. Ecology 33: 1-15, illus.

_____
    1960. Water relations and forest distribution in the Douglas-fir region on Vancouver Island. Can. Dep. Agric. Publ. 1091, 71 p., illus.

Mackin, J. Hoover, and Allen S. Cary
    1965. Origin of Cascade landscapes. Wash. Dep. Conserv., Div. Mines & Geol. Inf. Circ. 41, 35 p., illus.

Macnab, James A.
    1958. Biotic aspection in the Coast Range mountains of northwestern Oregon. Ecol. Monogr. 28: 21-54, illus.

Mark, David M., and Peter M. Ojamaa
    1972. The glacial geomorphology of the Puget lowland, Washington and British Columbia: Further comments and references. Northwest Sci. 46: 336-338.

Mason, H. L.
    1947. Evolution of certain floristic associations in western North America. Ecol. Monogr. 17:201-210, illus.

Merkle, John
    1951. An analysis of the plant communities of Mary's Peak, western Oregon. Ecology 32: 618-640, illus.

Meurisse, Robert T., and Chester T. Youngberg
    1971. Soil-vegetation survey and site classification report for Tillamook and Munson Falls Tree Farms. Oreg. State Univ. Dep. Soils, report to Publishers Paper Co., 116 p.

Meurisse, Robert Thomas
    1972. Site quality of western hemlock and chemical characteristics of some western Oregon Andic soils. 164 p., illus. (Unpublished Ph.D. thesis on file at Oreg. State Univ., Corvallis.)

Minore, Don
    1972a. A classification of forest environments in the South Umpqua basin. USDA Forest Serv. Res. Pap. PNW-129, 28 p., illus. Pac. Northwest Forest & Range Exp. Stn., Portland, Oreg.

_____
    1972b. The wild huckleberries of Oregon and Washington—a dwindling resource. USDA Forest Serv. Res. Pap. PNW-143, 20 p., illus. Pac. Northwest Forest & Range Exp. Stn., Portland, Oreg.

_____ and Clark E. Smith

1971. Occurrence and growth of four northwestern tree species over shallow water tables. USDA Forest Serv. Res. Note PNW-160, 9 p., illus. Pac. Northwest Forest & Range Exp. Stn., Portland, Oreg.

Mitchell, Roderic Jamie

1972. An analysis of the vegetation of the Abbott Creek Natural Area, Oregon. 131 p., illus. (Unpublished Ph.D. thesis on file at Oreg. State Univ., Corvallis.)

Moir, William H.

1966. Influence of ponderosa pine on herbaceous vegetation. Ecology 47: 1045-1048, illus.

_____

1969. The Lodgepole Pine Zone in Colorado. Am. Midl. Nat. 81:87-98, illus.

Morris, W. G.

1934. Forest fires in western Oregon and western Washington. Oreg. Hist. Q. 35: 313-339.

Morris, William G.

1958. Influence of slash burning on regeneration, other plant cover, and fire hazard in the Douglas-fir region. USDA Forest Serv. Pac. Northwest Forest & Range Exp. Stn. Res. Pap. 29, 49 p., illus. Portland, Oreg.

Mowat, Edwin L.

1960. No serotinous cones on central Oregon lodgepole pine. J. For. 58: 118-119.

Mueggler, Walter F.

1965. Ecology of seral shrub communities in the cedar-hemlock zone of northern Idaho. Ecol. Monogr. 35: 165-185, illus.

Mueller-Dombois, Dieter

1965. Initial stages of secondary succession in the Coastal Douglas-fir and Western Hemlock Zones. In V. J. Krajina (ed.), Ecology of western North America. Vol. 1, p. 38-41. Univ. B. C. Dep. Bot.

Muenscher, W. C.

1941. The flora of Whatcom County, State of Washington. 134 p., illus. Ithaca, N.Y.

Mullineaux, D. R., and D. R. Crandell

1962. Recent lahars from Mount St. Helens, Washington. Geol. Soc. Am. Bull. 73: 855-870, illus.

Munger, T. T.

1914. Replacement of western yellow pine by lodgepole pine on the pumice soils of central Oregon. Soc. Am. For. Proc. 9: 396-406.

_____

1930. Ecological aspects of the transition from old forests to new. Science 72: 327-332.

_____

1940. The cycle from Douglas-fir to hemlock. Ecology 21:451-459, illus.

Neiland, Bonita J.

1958. Forest and adjacent burn in the Tillamook Burn area of northwestern Oregon. Ecology 39: 660-671, illus.

Newcomb, R. C.

1952. Origin of the Mima mounds, Thurston County region, Washington. J. Geol. 60: 461-472, illus.

Newton, Michael, B. A. El Hassan, and Jaroslav Zavitkovski

1968. Role of red alder in western Oregon forest succession. In J. M. Trappe, J. F. Franklin, R. F. Tarrant, and G. M. Hansen (eds.), Biology of alder. Northwest Sci. Assoc. Fortieth Annu. Meet. Symp. Proc. 1967: 73-84, illus. Pac. Northwest Forest & Range Exp. Stn., Portland, Oreg.

Oosting, H. J., and W. D. Billings
   1943. The red fir forest of the Sierra Nevada: *Abietum magnificae*. Ecol. Monogr. 13: 259-274, illus.
_____ and J. F. Reed
   1952. Virgin spruce-fir of the Medicine Bow Mountains, Wyoming. Ecol. Monogr. 22: 69-91, illus.
Oosting, Henry J.
   1956. The study of plant communities. Ed. 2, 440 p., illus. San Francisco: W. H. Freeman & Co.
Orloci, Laszlo
   1965. The Coastal Western Hemlock Zone on the south-western British Columbia mainland. *In* V. J. Krajina (ed.), Ecology of western North America. Vol. 1, p. 18-34. Univ. B. C. Dep. Bot.
Owen, Herbert Elmer, Jr.
   1953. Certain factors affecting establishment of the Douglas-fir [*Pseudotsuga taxifolia* (Lamb.) Britt.] seedlings. 71 p., illus. (Unpublished M.S. thesis on file at Oreg. State Univ., Corvallis.)
Pacific Northwest River Basins Commission
   1969. Columbia-North Pacific Region comprehensive framework study. Appendix IV: Vol. 1, p. 1-286; vol. 2, p. 287-545. *In* Land & mineral resources. Vancouver, Wash.
Pase, Charles P.
   1958. Herbage production and composition under immature ponderosa pine stands in the Black Hills. J. Range Manage. 11(5): 238-243, illus.
Pearson, G. A.
   1923. Natural reproduction of western yellow pine in the Southwest. U.S. Dep. Agric. Bull. 1105, 143 p., illus.
Pechanec, Anna A., and Jerry F. Franklin
   1968. Comparison of vegetation in adjacent alder, conifer, and mixed alder-conifer communities. II. Epiphytic, epixylic, and epilithic crypogams. *In* J. M. Trappe, J. F. Franklin, R. F. Tarrant, and G. M. Hansen (eds.), Biology of alder. Northwest Sci. Assoc. Fortieth Annu. Meet. Symp. Proc. 1967: 85-98, illus. Pac. Northwest Forest & Range Exp. Stn., Portland, Oreg.
Pechanec, Anna Alice
   1961. Some aspects of the ecology of the bryophytes in the Three Sisters Primitive Area. 187 p., illus. (Unpublished Ph.D. thesis on file at Oreg. State Univ., Corvallis.)
Pechanec, Joseph F., and George Stewart
   1944. Sagebrush burning—good and bad. U.S. Dep. Agric. Farmer's Bull. 1948, 32 p., illus.
Peck, Dallas L., Allen B. Griggs, Herbert G. Schlicker, and others
   1964. Geology of the central and northern parts of the western Cascade Range in Oregon. U.S. Geol. Surv. Prof. Pap. 449, 56 p., illus.
Peck, Morton Eaton
   1961. A manual of the higher plants of Oregon. Ed. 2, 936 p., illus. Portland, Oreg.: Binfords & Mort.
Peterson, Everett B.
   1965. The Subalpine Mountain Hemlock Zone. Part I. Phytocoenoses. *In* V. J. Krajina (ed.), Ecology of western North America. Vol. 1, p. 76-78. Univ. B. C. Dep. Bot.
_____
   1969. Radiosonde data for characterization of a mountain environment in British Columbia. Ecology 50(2): 200-205, illus.

Peterson, N. V., and E. A. Groh

    1966. State of Oregon Lunar Geological Field Conference guide book. 51 p., illus. Portland, Oreg.: Schultz-Wack-Weir, Inc.

Pettit, Russell Dean

    1968. Effects of seeding and grazing on a clearcut-burn in a mixed-coniferous forest stand of the Wallowa Mountain foothills. 133 p., illus. (Unpublished M.S. thesis on file at Oreg. State Univ., Corvallis.)

Pharis, Richard P.

    1967. Comparative drought resistance of five conifers and foliage moisture content as a viability index. Ecology 47: 211-221, illus.

Pickford, G. D., and Elbert H. Reid

    1942. Basis for judging subalpine grassland ranges of Oregon and Washington. U.S. Dep. Agric. Circ. 655, 38 p., illus.

Piper, Charles V.

    1906. Flora of the State of Washington. Contrib. U.S. Natl. Herb. vol. 11, 637 p., illus.

Post, Austin, Don Richardson, Wendell V. Tangborn, and F. L. Rosselot

    1971. Inventory of glaciers in the North Cascades, Washington. U.S. Geol. Surv. Prof. Pap. 705-A, 26 p., illus.

Poulton, C. E.

    1962. Range types. *In* Richard M. Highsmith, Jr. (ed.), Atlas of the Pacific Northwest. Ed. 3, p. 61-64, illus. Corvallis: Oreg. State Univ. Press.

Poulton, Charles Edgar

    1955. Ecology of the non-forested vegetation in Umatilla and Morrow County, Oregon. 166 p., illus. (Unpublished Ph.D. thesis on file at Wash. State Univ., Pullman.)

Reid, Elbert H., and G. D. Pickford

    1946. Judging mountain meadow range condition in eastern Oregon and eastern Washington. U.S. Dep. Agric. Circ. 748, 31 p., illus.

Rigg, G. B.

    1913. Forest distribution in the San Juan Islands: A preliminary note. Plant World 16(6): 177-182.

————

    1917. Forest succession and rate of growth in sphagnum bogs. J. For. 15: 726-739, illus.

————

    1919. Early stages of bog succession. Puget Sound Biol. Stn. Publ. 2: 195-210.

————

    1922a. Birch succession in sphagnum bogs. J. For. 20: 848-850.

————

    1922b. The sphagnum bogs of Mazama Dome. Ecology 3: 321-324, illus.

————

    1922c. A bog forest. Ecology 3: 207-213, illus.

Rigg, George B.

    1958. Peat resources of Washington. Wash. State Div. Mines & Geol. Bull. 44, 272 p., illus.

Roach, Archibald W.

    1952. Phytosociology of the Nash Crater lava flows, Linn County, Oregon. Ecol. Monogr. 22: 169-193, illus.

Robinson, Andrew Frederick, Jr.

    1967. The influence of tree cover and shade pattern upon the distribution of understory plants in ponderosa pine stands of central Oregon. 77 p., illus. (Unpublished M.A. thesis on file at Oreg. State Univ., Corvallis.)

Roe, Arthur L.
  1967. Productivity indicators in western larch forests. USDA Forest Serv. Res. Note INT-59, 4 p. Intermt. Forest & Range Exp. Stn., Ogden, Utah.
Roemer, Hans Ludwig
  1972. Forest vegetation and environments on the Saanich Peninsula, Vancouver Island. 405 p., illus. (Unpublished Ph.D. thesis on file at Univ. Victoria, Victoria, B.C.)
Rothacher, Jack, C. T. Dyrness, and Richard L. Fredriksen
  1967. Hydrologic and related characteristics of three small watersheds in the Oregon Cascades. USDA Forest Serv. Pac. Northwest Forest & Range Exp. Stn., 54 p., illus. Portland, Oreg.
Rummell, R. S.
  1951. Some effects of livestock grazing on ponderosa pine forest and range in central Washington. Ecology 32: 594-607, illus.
Ruth, Robert H.
  1954. Cascade Head climatological data 1936 to 1952. USDA Forest Serv. Pac. Northwest Forest & Range Exp. Stn., 29 p. Portland, Oreg.
Sabhasri, Sanga, and William K. Ferrell
  1960. Invasion of brush species into small stand openings in the Douglas-fir forests of the Willamette foothills. Northwest Sci. 34(3): 77-88.
Schmidt, R. L.
  1957. The silvics and plant geography of the genus *Abies* in the coastal forests of British Columbia. B. C. Forest Serv. Tech. Publ. T.46, 31 p., illus.
Schofield, W. B.
  1969. Phytogeography of northwestern North America: Bryophytes and vascular plants. Madroño 20(3): 155-207.
Scott, David R. M.
  1962. Plant associations of western Washington. Univ. Wash. Arbor. Bull. 25(1): 11-14, 26.
Secor, J. B.
  1960. Volcanic ash content of the soil of a *Festuca viridula* "bald" on a mountain in northern Idaho. Ecology 41: 390-391, illus.
Shantz, H. L., and Raphael Zon
  1924. Natural vegetation. Atlas Am. Agric. Part 1, Sect. E, 29 p., illus.
Sharpe, Grant William
  1956. A taxonomical-ecological study of the vegetation by habitats in eight forest types of the Olympic rain forest, Olympic National Park, Washington. 335 p., illus. (Unpublished Ph. D. thesis on file at Univ. Wash., Seattle.)
Shelford, Victor E.
  1963. The ecology of North America. 610 p., illus. Urbana: Univ. Ill. Press.
Sherman, Robert J., and William W. Chilcote
  1972. Spatial and chronological patterns of *Purshia tridentata* as influenced by *Pinus ponderosa*. Ecology 53: 294-298, illus.
Sherman, Robert James
  1966. Spatial and chronological patterns of *Purshia tridentata* as influenced by *Pinus ponderosa* overstory. 81 p., illus. (Unpublished M.S. thesis on file at Oreg. State Univ., Corvallis.)
————
  1969. Spatial and developmental patterns of the vegetation of Black Butte, Oregon. 80 p., illus. (Unpublished Ph.D. thesis on file at Oreg. State Univ., Corvallis.)
Silen, Roy R.
  1962. A discussion of forest trees introduced into the Pacific Northwest. J. For. 60: 407-408.

Smith, J. E.
  1949. Natural vegetation in the Willamette valley, Oregon. Science 109: 41-42.
Smith, Richard B.
  1965. The dry subzone of the interior Western Hemlock Zone. Part II. Edaphotopes. *In* V. J. Krajina (ed.), Ecology of western North America. Vol. 1, p. 57-64. Univ. B. C. Dep. Bot.
Snavely, P. D., Jr., and H. C. Wagner
  1963. Tertiary geologic history of western Oregon and Washington. Wash. Div. Mines and Geol. Rep. Invest. No. 22, 25 p., illus.
Snyder, Robert V., and John M. Wade
  1970. Mt. Baker National Forest soil resource inventory. 267 p., plus atlas of maps and interpretive tables. Northwest Reg., USDA Forest Serv., Portland, Oreg.
Soeriaatmadja, Roehajat Emon
  1966. Fire history of the ponderosa pine forests of the Warm Springs Indian Reservation, Oregon. 123 p., illus. (Unpublished Ph.D. thesis on file at Oreg. State Univ., Corvallis.)
Soil Conservation Service (comp.)
  1970. Distribution of principal kinds of soils: Orders, suborders, and great groups. *In* National atlas of the USA: Soils. Sheet No. 86-87. U.S. Geol. Surv., Washington, D.C.
Spilsbury, R. H., and D. S. Smith
  1947. Forest site types of the Pacific Northwest. B. C. Forest Serv. Tech. Publ. T.30, 46 p., illus.
_____ and E. W. Tisdale
  1944. Soil-plant relationships and vertical zonation in the southern interior of British Columbia. Sci. Agric. 24: 395-436, illus.
Sprague, F. LeRoy, and Henry P. Hansen
  1946. Forest succession in the McDonald Forest, Willamette valley, Oregon. Northwest Sci. 20: 89-98, illus.
Steen, Harold K.
  1966. Vegetation following slash fires in one western Oregon locality. Northwest Sci. 40(3): 113-120.
Stein, William I.
  1963. Comparative juvenile growth of five western conifers. 194 p., illus. (Unpublished Ph.D. thesis on file at Yale Univ., New Haven.)
Stephens, F. R.
  1965. Ponderosa pine thrives on wet soils in southwestern Oregon. J. For. 63: 122-123, illus.
_____
  1966. Lodgepole pine—soil relations in the northwest Oregon Cascade Mountains. J. For. 64: 184-186, illus.
Stone, E. C., and H. A. Fowells
  1955. Survival value of dew under laboratory conditions with *Pinus ponderosa*. Forest Sci. 1: 183-188, illus.
Stone, Edward C., Rudolf F. Grah, and Paul J. Zinke
  1972. Preservation of the primeval redwoods in the Redwood National Park, Part I. Am. Forests 78(4): 50-55, illus.
Strickler, Gerald S.
  1961. Vegetation and soil condition changes on a subalpine grassland in eastern Oregon. USDA Forest Serv. Pac. Northwest Forest & Range Exp. Stn. Res. Pap. 40, 46 p., illus. Portland, Oreg.

1966. Soil and vegetation on the Starkey Experimental Forest and Range. Soc. Am. For. Proc. 1965: 27-30.

Swedberg, Kenneth C.
1973. A transition coniferous forest in the Cascade Mountains of northern Oregon. Am. Midl. Nat. 89(1): 1-25, illus.

Swedberg, Kenneth Charles
1961. The coniferous ecotone of the east slope of the northern Oregon Cascades. 118 p., illus. (Unpublished Ph.D. thesis on file at Oreg. State Univ., Corvallis.)

Tarrant, R. F.
1953. Soil moisture and the distribution of lodgepole and ponderosa pine; a review of the literature. USDA Forest Serv. Pac. Northwest Forest & Range Exp. Stn. Res. Note 8, 10 p. Portland, Oreg.

1964. Forest soil improvement through growing red alder (*Alnus rubra* Bong.) in Pacific Northwestern United States. 8th Int. Congr. Soil Sci. Trans., vol. 5, p. 1029-1043. Bucharest, Romania.

Tarrant, Robert F., and James M. Trappe
1971. The role of Alnus in improving the forest environment. Plant & Soil Spec. Vol., p. 335-348.

Taylor, Edward M.
1968. Roadside geology Santiam and McKenzie Pass highways, Oregon. *In* Hollis M. Dole (ed.), Andesite conference guidebook. Oreg. Dep. Geol. & Miner. Ind. Bull. 62: 3-33, illus.

Thilenius, John F.
1968. The *Quercus garryana* forests of the Willamette valley, Oregon. Ecology 49: 1124-1133, illus.

Thilenius, John Fredrick
1964. Synecology of the white-oak (*Quercus garryana* Douglas) woodlands of the Willamette valley, Oregon. 151 p., illus. (Unpublished Ph.D. thesis on file at Oreg. State Univ., Corvallis.)

Thornburgh, Dale Alden
1969. Dynamics of the true fir-hemlock forests of the west slope of the Washington Cascade Range. 210 p., illus. (Ph.D. thesis on file at Univ. Wash., Seattle.)

Tidball, Ronald
1965. A study of soil development on dated pumice deposits from Mount Mazama, Oregon. 235 p., illus. (Unpublished Ph.D. thesis on file at Univ. Calif., Berkeley.)

Tiedemann, Arthur R.
1972. Soil properties and nutrient availability in tarweed communities of central Washington. J. Range Manage. 25(6): 438-443, illus.

Tisdale, E. W.
1947. The grasslands of the southern interior of British Columbia. Ecology 28: 346-382, illus.

1968. Principal habitat types of forest ranges in the Pacific Northwest. Range Manage. Workshop, p. 106-109. Wash. State Univ., Pullman.

_____ and A. McLean
1957. The Douglas-fir Zone of southern interior British Columbia. Ecol. Monogr. 27: 247-266.

Trappe, James M., and Robert W. Harris
1958. Lodgepole pine in the Blue Mountains of northeastern Oregon. USDA Forest Serv. Pac. Northwest Forest & Range Exp. Stn. Res. Pap. 30, 22 p., illus. Portland, Oreg.

Troll, C.
1955. Der Mount Rainier und das mittlere Cascaden-gebirge. Erdkunde 9: 264-274, illus.

Tueller, Paul Teuscher
1962. Plant succession on two *Artemisia* habitat types in southeastern Oregon. 249 p., illus. (Unpublished Ph.D. thesis on file at Oreg. State Univ., Corvallis.)

Turner, Robert B., Charles E. Poulton, and Walter L. Gould
1963. Medusahead—a threat to Oregon rangeland. Oreg. State Univ. Agric. Exp. Stn. Spec. Rep. 149, 22 p., illus. Corvallis.

Turner, Robert Bruce
1969. Vegetation changes of communities containing medusahead [*Taeniatherum asperum* (Sim.) *Nevski*] following herbicide, grazing, and mowing treatments. 199 p., illus. (Unpublished Ph.D. thesis on file at Oreg. State Univ., Corvallis.)

U.S. Soil Conservation Service, Soil Survey Staff
1951. Soil survey manual. U.S. Dep. Agric. Handb. 18, 503 p., illus.

_____
1962. Identification and nomenclature of soil horizons. Suppl. to U.S. Dep. Agric. Handb. 18, p. 173-188, illus.

U.S. Weather Bureau
1956. Climatic summary of the United States—supplement for 1931 through 1952, Washington. Climatography of the United States 11-39, 79 p., illus.

_____
1960a. Climates of the States, Oregon. Climatography of the United States 60-35, 20 p., illus.

_____
1960b. Climates of the States, Washington. Climatography of the United States 60-45, 23 p., illus.

_____
1965a. Climatic summary of the United States—supplement for 1951 through 1960, Oregon. Climatography of the United States 86-31, 96 p., illus.

_____
1965b. Climatic summary of the United States—supplement for 1951 through 1960, Washington. Climatography of the United States 86-39, 92 p., illus.

USDA Forest Service
1973. Silvicultural systems for the major forest types of the United States. U.S. Dep. Agric. Agric. Handb. 445, 114 p., illus.

Valassis, Vlassios Thomas
1955. Some factors affecting the establishment and growth of improved forage species on Laughlin-like soils in western Oregon. 181 p., illus. (Unpublished Ph.D. thesis on file at Oreg. State Univ., Corvallis.)

Van Ryswyk, Albert Leonard
1969. Forest and alpine soils of south-central British Columbia. 178 p., illus. (Unpublished Ph.D. thesis on file at Wash. State Univ., Pullman.)

VanVechten, George Wendell, III
1960. The ecology of the timberline and alpine vegetation of the Three Sisters, Oregon. 111 p., illus. (Unpublished Ph.D. thesis on file at Oreg. State Univ., Corvallis.)

Volland, Leonard Allen
    1963. Phytosociology of the ponderosa pine type on pumice soils in the upper Williamson River Basin, Klamath County, Oregon. 166 p., illus. (Unpublished M.S. thesis on file at Oreg. State Univ., Corvallis.)

Walker, Richard B.
    1954. The ecology of serpentine soils. II. Factors affecting plant growth on serpentine soils. Ecology 35: 259-266, illus.

Waring, R. H.
    1969. Forest plants of the eastern Siskiyous: Their environmental and vegetational distribution. Northwest Sci. 43: 1-17, illus.

————, K. L. Reed, and W. H. Emmingham
    1972. An environmental grid for classifying conferous forest ecosystems. *In* Jerry F. Franklin, L. J. Dempster, and Richard H. Waring (eds.), Proceedings—Research on coniferous forest ecosystems—a symposium, p. 79-91, illus. Pac. Northwest Forest & Range Exp. Stn., Portland, Oreg.

Weaver, Harold
    1943. Fire as an ecological and silvicultural factor in the ponderosa pine region of the Pacific slope. J. For. 41: 7-15, illus.

————
    1955. Fire as an enemy, friend, and tool in forest management. J. For. 53: 499-504, illus.

————
    1959. Ecological changes in the ponderosa pine forest of the Warm Springs Indian Reservation in Oregon. J. For. 57: 15-20, illus.

————
    1961. Ecological changes in the ponderosa pine forest of Cedar valley in southern Washington. Ecology 42: 416-420, illus.

————
    1964. Fire and management problems in ponderosa pine. Third Annu. Tall Timbers Fire Ecol. Conf. Proc., p. 61-79, illus.

————
    1968. Fire and its relationship to ponderosa pine. Calif. Tall Timbers Fire Ecol. Conf. Proc., p. 127-149, illus.

Weaver, J. E.
    1917. A study of the vegetation of southeastern Washington and adjacent Idaho. Univ. Nebr., Univ. Stud. 17: 1-114, illus.

Weaver, John E., and Frederic E. Clements
    1938. Plant ecology. Ed. 2, 601 p., illus. New York and London: McGraw-Hill Book Co., Inc.

Wells, Francis G., and Dallas L. Peck
    1961. Geologic map of Oregon west of the 121st Meridian. U.S. Geol. Surv., Misc. Geol. Invest. Map I-325.

West, Neil E.
    1968. Rodent-influenced establishment of ponderosa pine and bitterbrush seedlings in central Oregon. Ecology 49: 1009-1011, illus.

————
    1969a. Successional changes in the montane forest of the central Oregon Cascades. Am. Midl. Nat. 81: 265-271, illus.

————
    1969b. Tree patterns in central Oregon ponderosa pine forests. Am. Midl. Nat. 81: 584-590, illus.

_____ and William W. Chilcote

    1968. *Senecio sylvaticus* in relation to Douglas-fir clear-cut succession in the Oregon Coast Range. Ecology 49: 1101-1107, illus.

West, Neil Elliott

    1964. An analysis of montane forest vegetation on the east flank of the central Oregon Cascades. 272 p., illus. (Unpublished Ph.D. thesis on file at Oreg. State Univ., Corvallis.)

Western Land Grant Universities and Colleges and U.S. Soil Conservation Service

    1964. Soils of the Western United States (exclusive of Hawaii and Alaska). 69 p., illus., map. Pullman: Wash. State Univ.

White, Charles David

    1971. Vegetation-soil chemistry correlations in serpentine ecosystems. 274 p., illus. (Unpublished Ph.D. thesis on file at Univ. Oreg., Eugene.)

Whittaker, R. H.

    1954a. The ecology of serpentine soils. I. Introduction. Ecology 35: 258-259.

    1954b. The ecology of serpentine soils. IV. The vegetational response to serpentine soils. Ecology 35: 275-288, illus.

    1960. Vegetation of the Siskiyou Mountains, Oregon and California. Ecol. Monogr. 30: 279-338, illus.

    1961. Vegetation history of the Pacific coast States and the "central" significance of the Klamath region. Madroño 16(1): 5-23.

Wiedemann, Alfred M., LaRea J. Dennis, and Frank H. Smith

    1969. Plants of the Oregon coastal dunes. 117 p., illus. Corvallis, Oreg.: O.S.U. Book-stores, Inc.

Wiedemann, Alfred Max

    1966. Contributions to the plant ecology of the Oregon coastal sand dunes. 255 p., illus. (Unpublished Ph.D. thesis on file at Oreg. State Univ., Corvallis.)

Williams, Carroll B., Jr., and C. T. Dyrness

    1967. Some characteristics of forest floors and soils under true fir-hemlock stands in the Cascade Range. USDA Forest Serv. Res. Pap. PNW-37, 19 p., illus. Pac. Northwest Forest & Range Exp. Stn., Portland, Oreg.

Williams, Howell

    1942. The geology of Crater Lake National Park. Carnegie Inst. Wash. Publ. 540, 162 p., illus.

Winward, Alma H.

    1970. Taxonomic and ecological relationships of the big sagebrush complex in Idaho. 90 p., illus. (Unpublished Ph.D. thesis on file at Univ. Idaho, Moscow.)

Wise, William S.

    1970. Cenozoic volcanism in the Cascade Mountains of southern Washington. Wash. State Dep. Nat. Resour. Div. Mines & Geol. Bull. 60, 45 p., illus.

Wolfe, Jack A.

    1969. Neogene floristic and vegetational history of the Pacific Northwest. Madroño 20(3): 83-110, illus.

Wollum, A. G., II, C. T. Youngberg, and F. W. Chichester

    1968. Relation of previous timber stand age to nodulation of *Ceanothus velutinus*. Forest Sci. 14: 114-118, illus.

Wollum, Arthur George, II
    1962. The role of certain non-leguminous woody species in the nitrogen nutrition of some conifer seedlings. 53 p., illus. (Unpublished M.S. thesis on file at Oreg. State Univ., Corvallis.)

———
    1965. Symbiotic nitrogen fixation by *Ceanothus* species. 67 p., illus. (Unpublished Ph.D. thesis on file at Oreg. State Univ., Corvallis.)
Yerkes, Vern P.
    1960. Occurrence of shrubs and herbaceous vegetation after clearcutting old-growth Douglas-fir in the Oregon Cascades. USDA Forest Serv. Pac. Northwest Forest & Range Exp. Stn. Res. Pap. 34, 12 p., illus. Portland, Oreg.
Youngberg, C. T.
    1963. Forest soils—their characteristics in the Pacific Northwest. *In* Symposium of forest watershed management, p. 21-32, illus. Corvallis: Oreg. State Univ.
——— and W. G. Dahms
    1970. Productivity indices for lodgepole pine on pumice soils. J. For. 68: 90-94, illus.
——— and C. T. Dyrness
    1959. The influence of soils and topography on the occurrence of lodgepole pine in central Oregon. Northwest Sci. 33: 111-120, illus.
——— and C. T. Dyrness
    1965. Biological assay of pumice soil fertility. Soil Sci. Soc. Am. Proc. 29: 182-187, illus.
Zavitkovski, J., and M. Newton
    1968. Ecological importance of snowbrush *Ceanothus velutinus* in the Oregon Cascades. Ecology 49: 1134-1145, illus.
——— and M. Newton
    1971. Litterfall and litter accumulation in red alder stands in western Oregon. Plant & Soil 35: 257-268, illus.
——— and R. D. Stevens
    1972. Primary productivity of red alder ecosystems. Ecology 53: 235-242, illus.
Zavitkovski, Jaroslav
    1966. Snowbrush, *Ceanothus velutinus* Dougl., its ecology and role in forest regeneration in the Oregon Cascades. 102 p., illus. (Unpublished Ph.D. thesis on file at Oreg. State Univ., Corvallis.)

APPENDIX I

Brief Description of Soil Great Groups
(1967 Classification System)

Descriptions follow those contained on the soils sheet (No. 86-87) within the National Atlas (Soil Conservation Service 1970). To describe the soil great groups, it is necessary to describe characteristics of the soil orders (names ending with the suffix "sol") and suborders (two syllable words which are suffixes for great group names).

Since great groups defined in the 1967 and 1938 soil classification systems overlap, it is not possible to give equivalents that would apply in every case. However, the 1938 great soil groups shown in parentheses are probably the most commonly occurring equivalents to the great groups as defined in the new classification system.

ALFISOLS

Soils that are medium to high in bases with gray to brown surface horizons and sub-surface horizons of clay accumulation. Soils are usually moist but may be dry for a portion of the warm season.

Aqualfs

Seasonally wet Alfisols with mottles, iron-manganese concretions, or gray colors; used for general crops where drained, and pasture and woodland where undrained.

Albaqualfs

Aqualfs with a bleached (white) upper horizon and an abrupt change in texture into an underlying horizon of clay accumulation (formerly Planosols).

Udalfs

Alfisols situated in temperate to tropical regions. Soils are usually moist, but during the warm season may be intermittently dry in some horizons for short periods; used for row crops, small grain, and pasture.

Hapludalfs

Udalfs with a subsurface horizon of clay accumulation that is relatively thin or brownish (formerly Gray-Brown Podzolic soils without fragipans).

Xeralfs

Alfisols that are in climates with rainy winters and dry summers; during the summer, these soils are continuously dry for a long period; used for range, small grain, and irrigated crops.

Haploxeralfs

Xeralfs with a subsurface horizon of clay accumulation that is relatively thin or brownish (formerly Noncalcic Brown soils).

ARIDISOLS

Soils with pedogenic horizons low in organic matter and never moist for as long as 3 consecutive months.

Argids

Aridisols with a horizon in which clay has accumulated with or without alkali (sodium); mostly used for range with some irrigated crops.

Durargids

Argids with a hardpan (duripan) which is cemented with silica (formerly Desert, Red Desert, Sierozem, and some Brown soils, all with hardpan).

Haplargids

Argids with a loamy horizon of clay accumulation with or without alkali (sodium) (formerly Sierozem, Desert, Red Desert, and some Brown soils).

Natrargids

Argids with a horizon of clay and alkali (sodium) accumulation (formerly Solonetz soils).

Orthids

Aridisols with accumulations of calcium carbonate, gypsum, or other salts but no horizon of clay accumulation. They may have horizons from which some materials have been removed or altered; used mostly for range with some irrigated crops.

Camborthids

Orthids with horizons from which some materials have been removed or altered but which have no large accumulations of calcium carbonate or gypsum (formerly Sierozem, Desert, and Red Desert soils).

Durorthids

Orthids with a hardpan (duripan) cemented with silica (formerly Regosols, Calcisols, or Alluvial soils, all with hardpan).

ENTISOLS

Soils that have no pedogenic horizons.

Fluvents

Entisols with organic matter content that decreases irregularly with depth; formed in loamy or clayey alluvial deposits; used for range or irrigated crops in dry regions and for general farming in humid regions.

Udifluvents

Fluvents which are usually moist (formerly Alluvial soils).

Orthents

Loamy or clayey Entisols that have a regular decrease in organic matter content with depth; used for range or irrigated crops in dry regions and for general farming in humid regions.

Haplorthents

Orthents which lack diagnostic horizons due to geologically recent deposition or erosion (formerly Regosols, Lithosols, and Alluvial soils).

Xerorthents

Orthents developed in climates with rainy winters but dry summers; they are continuously dry for a long period during the summer (formerly Regosols, Brown, and Alluvial soils).

Xerorthents (shallow)

Orthents that are shallower than 20 inches to bedrock (formerly Lithosols).

Psamments

Entisols that have textures of loamy fine sand or coarser; used for range, woodland, small grains, and irrigated crops.

Torripsamments

Psamments that contain easily weatherable minerals; they are never moist for as long as 3 consecutive months (formerly Regosols).

Xeropsamments

Psamments in climates with rainy winters but dry summers; they are continuously dry for a long period during the summer (formerly Regosols).

INCEPTISOLS

Soils which have weakly differentiated horizons; materials in the soil have been altered or removed but have not accumulated. These soils are usually moist, but some are dry part of the time during the warm season.

Andepts

Inceptisols that either have formed in volcanic ash materials, or have low bulk density and large amounts of amorphous materials, or both; used for woodland and range or pasture.

Cryandepts

Andepts of cold regions (formerly Brown Podzolic or Gray-Brown Podzolic soils).

Dystrandepts

Andepts with a thick dark-colored surface horizon that is low in bases, or with a light colored surface horizon (formerly Ando soils).

Vitrandepts
Andepts mostly formed in pumice or slightly weathered volcanic ash (formerly Regosols).

Aquepts
Seasonally wet Inceptisols with an organic surface horizon, sodium saturation, and mottles or gray colors; used for woodland and pasture.

Haplaquepts
Aquepts with either a light-colored or a thin black surface horizon (formerly Low-Humic Gley soils).

Ochrepts
Inceptisols formed in materials with crystalline clay minerals, with light-colored surface horizons and altered subsurface horizons that have lost mineral materials; used for woodland and range.

Dystrochrepts
Ochrepts that are usually moist and low in bases and have no free carbonates in the subsurface horizons (formerly Sols Bruns Acides and some Brown Podzolic and Gray-Brown Podzolic soils).

Xerochrepts
Ochrepts that are in climates with rainy winters but dry summers; the soils are continuously dry for a long period during the summer (formerly Regosols).

Umbrepts
Inceptisols with crystalline clay minerals, thick dark-colored surface horizons, and altered subsurface horizons that have lost mineral materials and are low in bases; used for woodland and range.

Cryumbrepts
Umbrepts of cold regions (formerly Tundra soils).

Haplumbrepts
Umbrepts of temperate to warm regions (formerly Western Brown Forest soils).

Xerumbrepts
Umbrepts formed in climates with rainy winters but dry summers; the soils are continuously dry for a long period during the summer (formerly Regosols).

MOLLISOLS
Soils with nearly black, friable, organic-rich surface horizons high in bases; formed mostly in subhumid and semiarid warm to cold climates.

Aquolls
Seasonally wet mollisols with a thick, nearly black surface horizon and gray subsurface horizons; used for pasture.

Haplaquolls

Aquolls with horizons in which materials have been altered or removed, but no clay or calcium carbonate has accumulated (formerly Humic Gley soils).

Udolls

Mollisols of temperate climates. Udolls are usually moist and have no horizon in which calcium carbonate or gypsum has accumulated.

Argiudolls

Udolls with a subsurface horizon in which clay has accumulated (formerly Prairie soils).

Xerolls

Mollisols formed in climates with rainy winters and dry summers; these soils are continuously dry for a long period during the summer; used for wheat, range, and irrigated crops.

Argixerolls

Xerolls with a subsurface horizon of clay accumulation that is relatively thin or brownish (formerly Prairie and Chernozem soils).

Durixerolls

Xerolls with a hardpan (duripan) cemented with silica (formerly Prairie soils with hardpan).

Haploxerolls

Xerolls with a subsurface horizon high in bases but lacking large accumulations of clay, calcium carbonate, or gypsum (formerly Prairie, Chernozem, Chestnut, and Brown soils).

Palexerolls

Xerolls with a hardpan cemented with carbonates or a horizon of clay accumulation that is thick and reddish or is clayey in the upper part and changes abruptly in texture into an overlying horizon (formerly Prairie soils).

SPODOSOLS

Soils with low base supply having in subsurface horizons an accumulation of amorphous materials consisting of organic matter plus compounds of aluminum and usually iron; formed in acid, coarse-textured materials in humid and mostly cool or temperate climates.

Orthods

Spodosols with a horizon in which organic matter plus compounds of iron and aluminum have accumulated; used for woodland, hay, and pasture.

Cryorthods

Orthods of cold regions (formerly Podzols).

Fragiorthods
Orthods with a dense, brittle, but not indurated horizon (fragipan) below a horizon that has an accumulation of organic matter and compounds of iron and aluminum (formerly Podzols and Brown Podzolic soils, both with fragipans).

Haplorthods
Orthods of cool regions with a horizon in which organic matter plus compounds of iron and aluminum have accumulated; they have no dense, brittle, or indurated horizon (fragipan) (formerly Podzols and Brown Podzolic soils).

ULTISOLS

Soils which are low in bases and have subsurface horizons of clay accumulation; they are generally moist, but during the warm season some are dry part of the time.

Humults
Ultisols with a high content of organic matter; formed in temperate or tropical climates typified by a large amount of rainfall throughout the year; used for woodland, pasture, and small grain, truck, and seed crops.

Haplohumults
Humults with either a subsurface horizon of clay accumulation that is relatively thin, or a subsurface horizon with appreciable amounts of weatherable minerals, or both; formed in temperate climates (formerly Reddish Brown Lateritic soils).

Xerults
Ultisols that are relatively low in organic matter in the subsurface horizons; formed in climates with rainy winters and dry summers; these soils are continuously dry for a long period during the summer; used for range and woodland.

Haploxerults
Xerults with either a relatively thin subsurface horizon of clay accumulation, or a subsurface horizon with appreciable amounts of weatherable minerals, or both (formerly Reddish Brown Lateritic soils).

VERTISOLS

Clayey soils with wide, deep cracks when dry; most have distinctive wet and dry periods throughout the year.

Xererts
Vertisols with wide, deep cracks that open and close once each year and remain open continuously for more than 2 months; used for irrigated small grains, hay, and pasture.

Chromoxererts
Xererts with a brownish surface horizon (formerly Grumusols).

346

Brief Description of Great Soil Groups
(1938 Classification System)

Descriptions follow those contained in "Soils of the Western United States" (Western Land Grant Universities et al. 1964). Soil horizon thickness classes are approximately: (1) very thin, less than 3 centimeters; (2) thin, 3 to 10 centimeters; (3) moderately thick, 10 to 20 centimeters; (4) thick, 20 to 40 centimeters; and (5) very thick, over 40 centimeters. All other classes and designations follow the "Soil Survey Manual" and Supplement (U.S. Soil Conservation Service 1951, 1962).

Azonal Soils

Alluvial

Alluvial soils are formed on recent alluvium and, therefore, exhibit very little profile development. A horizons are thin to moderately thick, light to dark in color, with low to moderate amounts of organic matter accumulation. B horizons are lacking, and the C is made up of stratified alluvium which is often stony or gravelly. Soil reaction ranges from moderately alkaline to medium acid.

Lithosols

Lithosols are well drained, shallow, generally stony soils over bedrock. A horizons are very thin to moderately thick, light to dark in color, with low to moderate amounts of incorporated organic matter. B horizons are lacking; a transitional AC horizon may be present. Soil reaction may vary from moderately alkaline to medium acid.

Regosols

Regosols are well to excessively drained, poorly developed soils formed in deep, unconsolidated materials. A horizons are very thin to moderately thick, light to dark colored, with low to moderate organic matter content. B horizons are lacking, and the C is made up of uniform or stratified material. Reaction ranges from slightly acid to moderately alkaline. In dry areas, the soil may be calcareous.

Zonal Soils

Alpine Turf

Alpine Turf soils are formed under alpine grasses and herbs in high mountain areas having a cold, humid climate. These well to imperfectly drained soils have thin to thick black A horizons of moderate to high organic matter content. B horizons are lighter colored, generally stony, and may have noticeable increase in clay. The stony or gravelly C horizon may be layered by solifluction processes. Soil reaction is strongly acid in the surface and medium acid in the B.

Brown

Brown soils are formed under shrub-steppe in cool, semiarid climates. These well-drained soils have moderately thick, dark-brown A horizons of low organic matter content. B horizons typically have more clay and subangular blocky to prismatic structure. Soil reaction is slightly alkaline in the surface, and alkalinity increases with depth. A zone of calcium carbonate accumulation is generally present as a Bca or Cca horizon.

Brown Podzolic

Brown Podzolic soils are formed under forest in cool, humid climates. Soil drainage varies from well to imperfectly drained. 0.1 and 02 horizons are usually present. The A1 horizon is thin and dark grayish-brown in color. A very thin, intermittent A2 horizon may also be present. The brown-colored B horizon gives evidence of iron and humus accumulation but has no appreciable clay increase. Soil reaction is medium to strongly acid.

Chernozem

Chernozem soils are formed under steppe or shrub-steppe in cool, subhumid climates. These moderately well to well drained soils have moderately thick, very dark-brown to black A horizons of moderate organic matter content. B horizons usually contain more clay than the A but may be distinguished solely on the basis of color and structural changes. Soil reaction becomes more alkaline with depth, and a zone of carbonate accumulation is generally present in the lower part of the B.

Chestnut

Chestnut soils are formed under steppe or shrub-steppe in cool, semiarid climates. These well-drained soils have dark-brown, moderately thick A horizons containing moderate amounts of organic matter. B horizons often contain more clay than the A but may be distinguished largely by color and structural changes. Structure in the B may be blocky, subangular blocky, or prismatic. Soil reaction becomes more alkaline with depth, and a zone of carbonate accumulation usually occurs in or below the B horizon.

Desert

Desert soils are formed under shrub-steppe in warm, arid climates. These well-drained soils have thin, light-colored A horizons of low organic matter content. Structure of the A is platy. B horizons typically show increased clay content and are as dark as or darker colored than the A1. Reaction varies from neutral to strongly alkaline. Horizons of calcium enrichment, sometimes cemented with lime or silica, occur in or below the B horizon.

Gray-Brown Podzolic

Gray-Brown Podzolic soils are formed under forest in cool to cold, subhumid climates. They are well to imperfectly drained. A1 horizons are thin to moderately thick and very dark gray. A thin, light-colored A2 horizon occurs beneath the A1. B horizons contain more clay than the A2 and have darker colors. Structure is blocky, subangular blocky, or prismatic. Reaction is medium acid in the surface and medium to slightly acid in the B horizon.

Gray Wooded

Gray Wooded soils are formed under forest in cold, subhumid climates. They are well to imperfectly drained. Thin to moderately thick 01 and 02 horizons are present at the soil surface. A thin A1 horizon is generally present. The A2 horizon is light colored, low in organic matter content, and has platy structure. The B horizon contains more clay than the A2 and has blocky or subangular blocky structure. Surface reaction is slight to medium acid and may approach neutrality with depth.

Noncalcic Brown

Noncalcic Brown soils are formed under shrub communities (e.g., chaparral) in warm, subhumid climates. These well to moderately well drained soils have moderately thick to very thick, brown or reddish-brown, massive A horizons of low organic matter content. B horizons contain more clay and are redder than the A. They may be massive or have blocky or prismatic structure. Hard layers caused by silica cementation commonly occur in the B. In addition, small amounts of carbonate may be present in the lower part of the B horizon.

Podzol

Podzol soils are formed under forest in cool to cold, subhumid climates. They are well to moderately well drained. Thin to thick (5 to 40 centimeters) 01 horizons overlie thin to thick (5 to 25 centimeters) 02 horizons. A2 (bleicherde) horizons are thin to thick, white to very pale brown, and very low in organic matter content. B horizons are much darker and contain accumulations of iron and humus. Structure ranges from very weak blocky to strong blocky or prismatic. Soil reaction is strongly to very strongly acid.

Prairie

Prairie soils are formed under grassland vegetation in cool, subhumid to humid climates. These well to imperfectly drained soils have thick, very dark A horizons generally containing large amounts of organic matter. B horizons typically have more clay but may be differentiated from the A largely by color and structural changes. Prairie soils exhibit decreasing acidity with depth, and accumulations of calcium carbonate may occur in the C horizon.

Reddish Brown Lateritic

Reddish Brown Lateritic soils are formed under forest in warm, humid climates. They are moderately well to well drained. A horizons are moderately thick, reddish brown, and of granular structure. Shotlike iron-magnesium concretions are commonly present. B horizons are red or reddish brown and have more clay than the A. Structure is typically moderate blocky or subangular blocky. Reaction varies from moderately to very strongly acid.

Sierozem

Sierozem soils are formed under shrub-steppe in cool, arid climates. These well-drained soils have thin, light-colored A horizons containing low amounts of organic matter. Soil structure is typically platy, especially in the upper portion. B horizons contain more clay and are often darker than the A. Calcium carbonate accumulations, often cemented, generally occur in the lower part of, or just below, the B horizon. Soil reaction ranges from mildly to strongly alkaline.

Sols Bruns Acides

Sols Bruns Acides soils are formed under coniferous forest vegetation in cool, humid climates. Soil drainage may vary from well to imperfectly drained. Thin 01 and 02 horizons are usually present at the soil surface. A horizons are thick, dark brown to dark reddish brown, and moderate to high in organic matter content. B horizons show no evidence of illuviated clay and are distinguished by color and structure (subangular blocky). Soil acidity generally increases with depth and ranges from medium to very strongly acid.

Western Brown Forest

Western Brown Forest soils are formed under forest in areas of forest-steppe transition and cool, semiarid to subhumid climates. These well or moderately well drained soils have moderately thick, dark-colored granular A horizons containing low to moderate amounts of organic matter. B horizons may show increased clay content but are usually distinguished by color and structural changes. Typical B horizon structure is subangular blocky. Reaction varies from slight to medium acid, and acidity commonly decreases with depth. A zone of calcium carbonate accumulation occurs in the lower B or upper C horizon.

Intrazonal Soils

Alpine Meadow

Alpine Meadow soils are formed under high-elevation meadows in cold to very cold and humid climates. These soils are imperfectly to poorly drained. The A1 horizon is thick, very dark brown to black, moderate to high in organic matter content, and frequently stony. B horizons are lacking. C horizons are stony, mottled, and gleyed. Soil reaction is strongly or very strongly acid.

Grumusols

Grumusols occur under grassland or grass-shrub vegetation in a variety of climatic zones. They are generally well drained except in depressional areas. A horizons are very thick, dark colored, and contain low to moderate amounts of organic matter. Montmorillonitic clay content is usually high. B horizons are lacking. C horizons have wide vertical cracks which typically extend up into the A horizon. Carbonate accumulation may occur in the lower A or upper C horizon. Reaction varies from slightly acid to slightly alkaline.

Humic Gley

Humic Gley soils occur under meadows in virtually all climatic zones. They are poorly to very poorly drained. The A horizon is very dark colored and contains large amounts of organic matter. The B horizon is gleyed or mottled and is higher in clay content. Soluble salts may be present in weak to moderate concentrations. Soil reaction ranges from moderately alkaline to strongly acid.

Planosol

Planosols may occur under steppe, shrub-steppe, or forest and in a wide variety of climatic zones. They are imperfectly to poorly drained. A1 horizons are thick, dark colored, with moderate to high organic matter content. An A2 horizon of variable thickness underlies the A1; it is massive and mottled. B horizons contain appreciably more clay and have blocky or prismatic structure. A cemented layer may be present in the lower solum. Reaction varies from medium to mildly alkaline.

Solonchak

Solonchak soils occur under shrub-steppe in arid to semiarid areas. These poorly drained soils have thin, light-colored A horizons low in organic matter content. Salt crusts are commonly present on the surface. B horizons are lacking.

Solonetz

Solonetz soils occur under steppe or shrub-steppe in arid to semiarid areas. Soil drainage ranges from well to imperfectly drained. A horizons are thin, platy, light-colored, and contain low amounts of organic matter. B horizons contain more clay and have blocky, prismatic, or columnar structure. Portions of the B horizon contain more than 15 percent of exchangeable sodium. A zone of carbonate enrichment, commonly cemented, occurs in the lower part of or below the B horizon. Soil reaction varies from neutral to very strongly alkaline.

Miscellaneous Land Type

Rockland

Rockland designates areas with only sparse vegetation, dominated by rock outcrops, rock rubble, boulders, or stones. Restricted areas of thin soils may be included, but in general, soil development is severely limited.

APPENDIX II

List and Partial Index to Plant Species

The major sources for the scientific names of these species are Hitchcock et al. (1955, 1959, 1961, 1964, 1969) for most vascular plants; Little (1953) for trees and some shrubs; and Peck (1961) for the remainder. Mosses follow Lawton (1971), and lichens follow Howard (1950) in most cases. Common names are from a variety of sources, the most important being Garrison et al. (1967), Peck (1961), and Little (1953). Mosses, lichens, and liverworts are marked with an asterisk; no common names are given for these species.

Species which are not extensively cited in the text are indexed in this list; those which are broadly cited are indexed in the General Subject Index.

Arctostaphylos hispidula How.	Howell manzanita	
Arctostaphylos nevadensis Gray	pine-mat manzanita	
Arctostaphylos patula Greene	green manzanita	
Arctostaphylos uva-ursi (L.) Spreng.	kinnikinnick	
Arctostaphylos viscida Parry	white-leaved manzanita	114, 119, 124, 306
Arenaria capillaris Poir.	mountain sandwort	262, 263, 287, 288, 289
Arenaria capillaris var. *americana* Davis	fescue sandwort	207
Arenaria congesta Nutt. in T. & G.	dense-flowered sandwort	244
Arenaria formosa (Fisch.) Reg.	slender mountain sandwort	272, 290
Arenaria howellii Wats.	Howell's sandwort	307
Arenaria macrophylla Hook.	bigleaf sandwort	142, 152, 158, 195, 205, 255
Arenaria obtusiloba (Rydb.) Fern.	blunt-leaved sandwort	287, 288, 289, 309
Arenaria pumicola Cov.	Crater Lake sandwort	270
Arenaria rubella (Wahlenb.) J. E. Smith	varying sandwort	288, 309
Aristida longiseta Steud.	red threeawn	228, 229
Armeria maritima (Mill.) Willd.	thrift	89
Arnica chamissonis Less.	leafy arnica	188
Arnica cordifolia Hook.	heartleaf arnica	158, 176, 195, 196, 205, 206
Arnica latifolia Bong.	broadleaf arnica	158, 192
Arnica rydbergii Greene	Rydberg's arnica	290
Arnica sororia Greene	sisters' arnica	214
Arrhenatherum elatius (L.) Pres.	tall oatgrass	91, 121
Artemisia arbuscula Nutt.	low sagebrush	
Artemisia cana Pursh	silver sagebrush	234, 236, 243
Artemisia douglasiana Bess.	Douglas' wormwood	260
Artemisia ludoviciana Nutt.	western wormwood	259
Artemisia norvegica Fries	arctic wormwood	259
Artemisia rigida (Nutt.) Gray	stiff sagebrush	
Artemisia spinescens Eat.	bud sagebrush	245
Artemisia suksdorfii Piper	Suksdorf sagebrush	297
Artemisia tilesii Ledeb.	mountain wormwood	307
Artemisia tridentata Nutt.	big sagebrush	
Artemisia trifurcata Steph. ex Spreng.	three-forked wormwood	288, 290
Artemisia tripartita Rydb.	threetip sagebrush	
Asarum caudatum Lindl.	wild ginger	74, 78, 157, 195
Aster alpigenus (T. & G.) Gray	alpine aster	252, 254, 261, 262, 267, 270, 287, 288
Aster canescens Pursh	hoary aster	184, 235
Aster chilensis Nees	Chilian aster	122
Aster conspicuus Lindl.	showy aster	205
Aster engelmannii (Eat.) Gray	Engelmann aster	255
Aster foliaceus Lindl.	leafy aster	208, 260
Aster ledophyllus Gray	Cascades aster	252, 254, 263, 264, 265, 267
Aster occidentalis (Nutt.) T. & G.	western aster	109, 199, 269

Cerastium arvense L.	field chickweed	309
Ceratophyllum demersum L.	hornwort	231
Cercocarpus betuloides Nutt. in T. & G.	birchleaf mountainmahogany	114, 124
Cercocarpus ledifolius Nutt. in T. & G.	curlleaf mountainmahogany	
Chaenactis douglasii (Hook.) H. & A.	falseyarrow	235
Chaenactis thompsonii Cronq.	Thompson falseyarrow	309
Chamaecyparis lawsoniana (A. Murr.) Parl.	Port-Orford-cedar	
Chamaecyparis nootkatensis (D. Don) Spach	Alaska-cedar	
Cheilanthes gracillima D.C. Eaton	lace-fern	91, 148
Cheilanthes siliquosa Maxon	podfern	306, 307, 309, 310
Chimaphila menziesii (R. Br.) Spreng.	little prince's pine	82, 158, 195
Chimaphila umbellata (L.) Bart.	western prince's pine	
Chimaphila umbellata var. *occidentalis* Blake	western prince's pine	301
Chrysothamnus nauseosus (Pall.) Brit.	tall gray rabbitbrush	
Chrysothamnus nauseosus var. *albicaulis* Hall. & Clem.	whitestem gray rabbitbrush	124, 215, 216, 307
Chrysothamnus viscidiflorus (Hook.) Nutt.	tall green rabbitbrush	
Cicuta douglasii (DC.) Coult. & Rose	western waterhemlock	109, 228
Circaea alpina L.	alpine circaea	157, 228
Cirsium vulgare (Savi) Airy-Shaw	common thistle	84, 197, 228
Cladonia bellidiflora (Ach.) Schaer.*		304
Cladonia coniocraea (Flk.) Spreng.*		304
Cladonia grayii Merr.*		304
Cladonia ecmocyna (Ach.) Nyl.*		304
Cladonia ecmocyna var. *intermedia* (Robb.) Thoms.*		304
Cladonia macrophyllodes Nyl.*		304
Cladonia phyllophora Hoffm.*		304
Cladonia rangiferina (L.) Wigg.*		304
Cladothamnus pyrolaeflorus Bong.	cladothamnus or copper bush	104, 106, 277, 278
Claytonia lanceolata Pursh	lance-leaved springbeauty	259, 264, 307
Claytonia megarhiza (Gray) Parry ex Wats.	alpine springbeauty	
Claytonia megarhiza var. *nivalis* (English) C. L. Hitchc.	alpine springbeauty	309
Clintonia uniflora (Schult.) Kunth	queencup beadlily	82, 95, 96, 97, 145, 152, 195, 202, 205
Collinsia parviflora Lindl.	littleflower collinsia	
Collomia grandiflora Dougl. ex Lindl.	large-flowered collomia	166
Collomia heterophylla Hook.	varied-leaved collomia	74
Collomia linearis Nutt.	narrow-leaved collomia	216
Collomia tenella Gray	diffuse collomia	241
Convolvulus soldanella L.	coast morningglory	295
Coptis asplenifolia Salisb.	boreal goldthread	106
Coptis laciniata Gray	cutleaf goldthread	74, 77
Coptis occidentalis (Nutt.) T. & G.	western goldthread	205
Corallorhiza maculata Raf.	spotted coralroot	89, 158, 195
Corallorhiza mertensiana Bong.	Mertens' coralroot	158

Cornus canadensis L.	bunchberry dogwood	69, 95, 96, 97, 98
Cornus glabrata Benth.	brown dogwood	124
Cornus nuttallii Aud. ex T. & G.	Pacific dogwood	74, 117, 134
Cornus stolonifera Michx.	red-osier dogwood	296
Corylus cornuta Marsh.	western hazel	111, 113
Corylus cornuta Marsh. var. *californica* (DC.) Sharp	California hazel	
Cotula coronopifolia L.	bird brassbuttons	291
Crataegus columbiana How.	Columbia hawthorn	227
Crataegus douglasii Lindl.	blackhawthorn	113, 122, 174, 183, 221, 227
Crepis acuminata Nutt.	long-leaved hawks beard	201
Crepis atribarba Heller ssp. *originalis* Babc. & Stebb.	slender hawks beard	215
Cryptantha affinis (Gray) Greene	slender cryptantha	177, 178, 196
Cryptantha ambigua (Gray) Greene	obscure cryptantha	166, 239
Cryptantha thompsonii Johnst.	Thompson cryptantha	309
Cryptantha torreyana (Gray) Greene	Torrey's cryptantha	208
Cryptogramma crispa (L.) R. Br. ex Hook.	parsley-fern	91, 300
Cryptogramma densa (Brackenr.)	Oregon cliff-brake	148
Cuscuta salina Engelm.	salt-marsh dodder	295
Cynoglossum grande Dougl. ex Lehm.	great houndstongue	144
Cynosurus echinatus L.	hedgehog dogtail	90, 91, 121, 122, 123
Cytisus scoparius (L.) Link	Scotch broom	89
Dactylis glomerata L.	orchardgrass	113, 121, 122, 296
Danthonia californica Boland.	California danthonia	113, 119, 121, 122, 187, 306, 307
Danthonia intermedia Vasey	timber danthonia	121, 188
Danthonia unispicata (Thurb.) Munro ex Macoun	few-flowered wild oatgrass	239, 240, 241, 245
Darlingtonia californica Torr.	California pitcher-plant	294
Daucus carota L.	wild carrot	121, 123
Delphinium glareosum Greene	rockslide larkspur	264
Delphinium glaucum Wats.	pale larkspur	260
Delphinium menziesii DC.	Menzies' larkspur	91
Deschampsia atropurpurea (Wahl.) Scheele	mountain hairgrass	253, 255, 261, 262, 290
Deschampsia caespitosa (L.) Beauv.	tufted hairgrass	122, 123, 190 199, 200, 295
Deschampsia elongata (Hook.) Munro ex Benth.	slender hairgrass	187
Descurainia pinnata (Walt.) Britt.	pinnate tansymustard	216, 228, 230
Dicranella heteromalla (Hedw.) Schimp.*		270
Dicranum fuscescens Turn.*		
Dicranum scoparium Hedw.*		301
Digitalis purpurea L.	foxglove	63

Gaultheria ovatifolia Gray	slender gaultheria	142
Gaultheria shallon Pursh	salal	
Gayophytum nuttallii T. & G.	Nuttall's gayophytum	166, 177, 178, 196
Gayophytum ramosissimum Nutt. ex T. & G.	hairstem groundsmoke	235
Gentiana calycosa Griseb.	mountain bog gentian	254
Geranium dissectum L.	cut-leaved geranium	122
Geranium molle L.	dovefoot geranium	91
Geranium viscosissimum F. & M.	sticky geranium	214, 215, 220, 227
Geum macrophyllum Willd.	largeleaf avens	228
Geum triflorum Pursh	three-flowered avens	215, 262
Geum triflorum Pursh var. *ciliatum* (Pursh) Fassett	long-plumed avens	220
Gilia capitata Sims	globe gilia	114, 148, 307
Gilia nuttallii Gray	Nuttall gilia	272
Githopsis specularioides Nutt.	bluecup	307
Glaux maritima L.	sea milkwort	295
Glehnia leiocarpa Mathias	beach silver-top	291, 295
Glyceria elata (Nash) M. E. Jones	tall mannagrass	109
Goodyera oblongifolia Raf.	rattlesnake plantain	77, 89, 134, 135, 142, 145
Grayia spinosa (Hook.) Moq.	spiny hopsage	165, 216, 217, 228, 230, 245
Grindelia stricta DC.	Oregon gum-plant	295
Grossularia velutina		165
Gymnocarpium dryopteris (L.) Newm.	oakfern	65, 81, 82, 96, 97, 202
Gymnomitrium varians (Lindb.) Schiffn.*		262
Habenaria dilatata (Pursh) Hook.	boreal bogorchid	260
Habenaria unalascensis (Spreng.) Wats.	Alaska reinorchid	144
Hakelia jessicae (McGregor) Brand	Jessica's tickweed	153
Haplopappus bloomeri Gray	Bloomer's haplopappus	158, 184
Haplopappus liatriformis (Greene) St. John	Palouse haplopappus	214, 215
Haplopappus stenophyllus Gray in Torr.	narrow-leaved haplopappus	226, 235
Hedysarum occidentale Greene	western hedysarum	290
Helianthella uniflora (Nutt.) T. & G.	false sunflower	214
Helianthella uniflora (Nutt.) T. & G. var. *douglasii* (T. & G.) Weber	false sunflower	213, 215, 220
Heracleum lanatum Michx.	common cowparsnip	227, 228, 255, 257, 260, 296
Hieracium albertinum Farr	western hawkweed	207, 213, 214, 215, 220
Hieracium albiflorum Hook.	white hawkweed	
Hieracium cynoglossoides Arv.-Touv.	houndstongue hawkweed	177, 178, 211, 224, 244
Hieracium gracile Hook.	slender hawkweed	255, 261, 262, 269, 272
Holcus lanatus L.	common velvetgrass	63, 69, 113, 121, 122, 296

Menziesia ferruginea Smith	rustyleaf	59, 61, 69, 105, 106, 205, 207, 278
Mertensia ciliata (Torr.) G. Don	broad-leaved lungwort	153
Mesembryanthemum chilense Molina	sea fig	298
Microseris alpestris (Gray) Q. Jones	alpine microseris	254
Microseris troximoides Gray	false agoseris	215, 235
Microsteris gracilis (Hook.) Greene	pink annual phlox	175, 213, 216, 230, 241, 244
Mimulus lewisii Pursh	Lewis monkeyflower	252, 257, 260, 261, 267
Mimulus nanus H. & A.	dwarf monkeyflower	235
Mimulus tilingii Reg.	clustered monkeyflower	260
Mitella breweri Gray	feathery mitrewort	255, 257, 260
Mitella pentandra Hook.	fivepoint mitrewort	261
Mitella stauropetala Piper	sideflower mitrewort	195, 205
Moerckia blyttii (Morck) Brockm.*		270
Monardella villosa Benth.	coyote mint	307
Monotropa hypopitys L.	pinesap	195
Montia linearis (Dougl.) Greene	narrow-leaved montia	122, 175, 216
Montia perfoliata (Donn) How.	miner's lettuce	166, 175
Montia sibirica (L.) How.	western springbeauty	59, 63, 65, 145
Muhlenbergia filiformis (Thurb.) Rydb.	pullup muhly	187, 199
Muhlenbergia richardsonis (Trin.) Rydb.	short-leaved muhly	243
Myosotis micrantha Pall. ex Lehm.	smallflower forgetmenot	175, 213, 214
Myosurus aristatus Benth. ex Hook.	bristly mousetail	216
Myrica californica Cham.	waxmyrtle	293, 295
Myriophyllum spicatum L.	western spiked watermilfoil	231
Navarretia divaricata (Torr.) Greene	short-stemmed navarretia	148
Navarretia tagetina Greene	marigold navarretia	241
Oenanthe sarmentosa Presl ex DC.	water parsley	68
Oligotrichum hercynicum (Hedw.) Lam. & DC.*		262
Oplopanax horridum (J. E. Smith.) Miq.	devilsclub	
Opuntia polyacantha Haw.	plains pricklypear	220
Orthocarpus imbricatus Torr. ex Wats.	mountain owlclover	153
Oryzopsis hymenoides (R. & S.) Ricker	Indian ricegrass	230, 231, 245
Osmaronia cerasiformis (T. & G.) Greene	Indian plum	89, 113
Osmorhiza chilensis H. & A.	mountain sweetroot	82, 113, 114, 117, 145, 157, 158, 174
Oxalis oregana Nutt. ex T. & G.	Oregon oxalis	
Oxyria digyna (L.) Hill	alpine mountainsorrel	
Oxytropis campestris (L.) DC.	Cusick's crazyweed	289
Oxytropis campestris (L.) DC. var. *gracilis* (A. Nels.) Barneby	slender crazyweed	289
Pachistima myrsinites (Pursh) Raf.	Oregon boxwood	
Paeonia brownii Dougl. ex Hook.	western peony	177, 178

Sedum oregonense (Wats.) Peck	creamy stonecrop	148, 149, 301
Selaginella oregana DC. Eat.	Oregon selaginella	65, 66
Selaginella wallacei Hieron.	Wallace's selaginella	91
Senecio canus Hook.	woolly groundsel	167, 184
Senecio fremontii T. & G.	Fremont groundsel	270
Senecio integerrimus Nutt.	western groundsel	143, 178, 208, 214, 260
Senecio integerrimus Nutt. var. *exaltatus* (Nutt.) Cronq.	tall western groundsel	148, 213, 215
Senecio pauperculus Michx.	balsam groundsel	309
Senecio serra Hook.	butterweed groundsel	208
Senecio subnudus DC.	fewleaf groundsel	109
Senecio sylvaticus L.	woodland groundsel	83, 84
Senecio triangularis Hook.	arrowleaf groundsel	269
Sequoia sempervirens (D. Don) Endl.	coast redwood	
Sherardia arvensis L.	bluefield madder	121
Sibbaldia procumbens L.	creeping sibbaldia	263, 288, 290
Sidalcea campestris Greene	meadow checkermallow	122
Sidalcea malvaeflora (DC.) Gray	mallow sidalcea	298
Sidalcea oregana (Nutt.) Gray	Oregon checkermallow	214
Silene acaulis L.	moss silene or campion	290
Silene parryi (Wats.) C. L. Hitchc. & Maguire	Parrys silene or campion	
Silene suksdorfii Robins.	Suksdorf's silene or campion	288
Sisymbrium altissimum L.	tumbleweed, tumble mustard, or Jim Hill mustard	218, 232
Sisyrinchium angustifolium Mill.	Idaho blue-eyedgrass	89
Sisyrinchium californicum (Ker) Dryand.	golden blue-eyedgrass	294
Sisyrinchium inflatum (Suksd.) St. John	purple-eyedgrass	174
Sitanion hystrix (Nutt.) J. G. Smith	bottlebrush squirreltail	
Sitanion jubatum J. G. Smith	big squirreltail	119, 306, 307, 309
Smelowskia calycina (Steph.) C. A. Mey.	alpine smelowskia	270
Smelowskia ovalis M. E. Jones	short-fruit smelowskia	
Smilacina stellata (L.) Desf.	starry solomonplume	89, 96, 145, 155, 157, 195
Solidago canadensis L.	Canada goldenrod	119, 259, 297
Solidago missouriensis Nutt.	Missouri goldenrod	214
Solidago spathulata DC.	coast goldenrod	287, 288, 289
Sorbus sitchensis Roemer var. *grayi* (Wenzig) C. L. Hitchc.	Sitka or Pacific mountain-ash	278
Spartina gracilis Trin.	alkali cordgrass	231
Sphagnum magellanicum Brid.*		108
Sphagnum squarrosum Crome*		108, 270
Spiraea betulifolia Pall. var. *lucida* (Dougl.) C. L. Hitchc.	shinyleaf spirea	173, 174, 192, 196, 197, 205, 227, 312
Spiraea douglasii Hook.	Douglas spirea	68, 109, 188, 293
Spiraea douglasii var. *menziesii* (Hook.) Presl	Menzies spirea	69
Sporobolus cryptandrus (Torr.) Gray	sand dropseed	228, 229

Tiarella unifoliata Hook.	western coolwort	65, 74, 77, 82, 96, 152, 202
Tofieldia glutinosa (Michx.) Pers.	western tofieldia	109, 261
Tolmiea menziesii (Pursh) T. & G.	youth-on-age	68
Torilis arvensis (Huds.) Link	field hedge-parsley	113
Torilis nodosa (L.) Gaertn.	knotted hedge-parsley	121
Tortula brevipes (Lesq.) Broth.*		217
Tortula princeps De Not.*		217
Tortula ruralis (Hedw.) Gaertn., Mey. & Schreb.*		236
Tragopogon dubius Scop.	larger yellow goatsbeard or yellow salsify	174, 215
Trautvetteria caroliniensis (Walt.) Vail	false bugbane	81
Trientalis arctica Fisch. ex Hook.	northern starflower	108
Trientalis latifolia Hook.	starflower	
Trifolium gymnocarpon Nutt.	tufted clover	235
Trifolium longipes Nutt.	long-stalked clover	187, 190
Trifolium macrocephalum (Pursh) Poir.	big-headed clover	183, 235
Trifolium willdenovii Lehm.	spring-bank clover	294, 295
Triglochin maritimum L.	seaside arrow-grass	295
Trillium ovatum Pursh	white trillium	74, 96, 143, 145, 152, 195
Trisetum canescens Buckl.	tall trisetum	260
Trisetum cernuum Trin.	nodding trisetum	65, 82, 114
Trisetum spicatum (L.) Richter	downy oatgrass	262, 269, 272, 288, 289, 290
Tsuga heterophylla (Raf.) Sarg.	western hemlock	
Tsuga mertensiana (Bong.) Carr.	mountain hemlock	
Typha latifolia L.	broad-leaved cattail	231
Ulex europaeus L.	common gorse	298
Umbellularia californica (Hook. & Arn.) Nutt.	California laurel	59, 68, 72, 126, 127, 134, 135, 306
Urtica dioica L.	bigsting nettle	228, 313
Usnea (Dill) Adans.*		73
Utricularia spp. L.	bladderwort	109
Vaccinium alaskaense How.	Alaska huckleberry	69, 95, 96, 97, 98, 104, 106, 278
Vaccinium caespitosum Michx.	dwarf huckleberry	188, 206
Vaccinium deliciosum Piper	blueleaf huckleberry	
Vaccinium membranaceum Dougl. ex Hook.	big huckleberry	
Vaccinium occidentale Gray	westernbog huckleberry	109, 302
Vaccinium ovalifolium Smith	ovalleaf huckleberry	
Vaccinium ovatum Pursh	evergreen huckleberry	
Vaccinium parvifolium Smith	red huckleberry	
Vaccinium scoparium Leiberg	grouse huckleberry	
Vaccinium uliginosum L.	bog huckleberry or blueberry	187, 188

Valeriana sitchensis Bong.	Sitka valerian	
Vancouveria hexandra (Hook.) Morr. & Dec.	white inside-out-flower	74, 78, 96, 97, 145, 152, 157
Veratrum californicum Durand	California false hellebore	153
Veratrum viride Ait.	American false hellebore	104, 106, 252, 254, 255, 256, 258, 260, 265
Veronica cusickii Gray	Cusick's speedwell	254, 263, 288
Veronica peregrina L.	purslane speedwell	121
Veronica wormskjoldii Roem. & Schult.	alpine speedwell	290
Viburnum edule (Michx.) Raf.	high-bush cranberry	109
Vicia americana Muhl. ex Willd.	American vetch	82, 145, 174, 260
Vicia americana var. *truncata* Brew.	American vetch	117, 121, 157
Vicia tetrasperma (L.) Moench	slender vetch	121
Viola adunca Smith	western long-spurred violet	89, 90, 91
Viola glabella Nutt.	wood violet	59, 81, 158, 195, 205, 255, 260, 269
Viola nuttallii Pursh	upland yellow violet	89
Viola nuttallii Pursh var. *vallicola* (A. Nels.) St. John	valley violet	188
Viola palustris L.	marsh violet	108
Viola purpurea Kell.	purple-tinged violet	178
Viola sempervirens Greene	evergreen violet	59, 74, 77, 95, 106, 142, 143, 145
Whipplea modesta Torr.	whipple vine	74, 75, 134, 140, 141, 142, 143, 144
Wyethia spp. Nutt.	sunflower	199
Xerophyllum tenax (Pursh) Nutt.	common beargrass	
Zigadenus fremontii Torr.	Fremont deathcamas	298
Zigadenus paniculatus (Nutt.) Wats.	foothill deathcamas	235
Zigadenus venenosus Wats. var. *gramineus* (Rydb.) Walsh	meadow deathcamas	89, 213, 216

APPENDIX III

Index to Plant Communities

384

APPENDIX IV

General Subject Index

Woodrush (see *Luzula* spp.)

COMMENTARY AND BIBLIOGRAPHIC SUPPLEMENT

Jerry F. Franklin
Chief Plant Ecologist, USDA Forest Service, Pacific Northwest Research Station, Forestry Sciences Laboratory, Corvallis, OR, and Bloedel Professor of Ecosystem Analysis, College of Forest Resources, University of Washington, Seattle, WA

Tawny Blinn
Editorial Assistant, USDA Forest Service, Pacific Northwest Research Station, Forestry Sciences Laboratory, Corvallis, OR

Major advances have occurred in our understanding of the vegetation of the Pacific Northwest since *Natural Vegetation of Oregon and Washington* was published in 1973. These advances are reflected in the large number of articles and reports published in the last 15 years; this supplemental bibliography contains more than 500 citations compared to only about 400 in the original book.

We have included all relevant literature citations of which we are aware including some currently in press. We have tried to be comprehensive regarding vegetative communities and their distribution, including successional dynamics. The coverage is exemplary only regarding ecosystem analysis, autecology, and plant population ecology. We will appreciate readers bringing omissions to our attention.

Plant Community Analysis and Classification

The most comprehensive advance during the last 15 years has been in our understanding of plant communities, including their composition, distribution, and relation to the environment. The greatest single contributor to this advance has been the Area Ecology program of the U.S. Department of Agriculture, Forest Service. Trained plant ecologists have now sampled and classified vegetation on every national forest in Oregon and Washington. The program was created by Frederick C. Hall, based on the work and philosophy of R. F. Daubenmire. Vegetation classifications are now available as processed reports for every national forest (see the literature citations under Atzet, Brockway, Hall, Halverson, Hemstrom, Henderson, Hopkins, C. Johnson, Topik, Volland, and Williams) as are large, compatible synecological data sets.

Vegetation analysis has also been conducted by other individuals and institutions. Major examples include classifications in Olympic, Mount Rainier, and North Cascades National Parks (e.g., Agee and Pickford 1985, Franklin and others 1987a, Smith and Henderson 1986), Bureau of Land Management areas in southwestern Oregon (e.g., Wheeler and others 1986a, 1986b), and studies at various universities (e.g., del Moral 1979b, del Moral and Long 1977, Frenkel and Heinitz 1987, Kratz 1975).

Wetland communities, especially salt marshes, have finally begun to receive the attention they deserve. Studies have examined not only composition of salt marsh communities and its relation to environmental gradients (e.g., Disraeli and Fonda 1979, Eilers and others 1983, Ewing 1983), but also the effects of diking and dike removal (e.g., Mitchell 1981). Mountain wetlands, riparian vegetation, and lowland bogs have been studied (e.g., Campbell and Franklin 1979, Frenkel and others 1986, Kovalchik 1987, Padgett 1982).

Subalpine and alpine meadow communities have always been favorite subjects of ecologists. Efforts to describe these communities have continued (e.g., Douglas and Bliss 1977, Henderson 1974, Mairs 1977), with increased attention to the impacts of human and other animal use (e.g., Schreiner 1982). Population studies of these communities have been conducted, especially by del Moral and his associates in the meadows of the Olympic Mountains.

Disturbance Ecology

Research in the Pacific Northwest reflects the increased interest in effects of disturbance on vegetation patterns, processes, and rates of recovery. Catastrophic as well as small-scale disturbances have been examined. Approaches have included modeling (e.g., Dale and others 1986), chronosequences (e.g., Oliver and others 1985), and stand-age reconstructions (e.g., Franklin and Hemstrom 1981).

The most notable disturbance in the region was the catastrophic eruption of Mount St. Helens on May 18, 1980, which precipitated major studies of successional patterns and processes in affected regions (Bilderback 1987; Keller 1982, 1987). Included were studies on the devastated region close to the mountain (e.g., Wood and del Moral 1987), as well as in the ashfall areas in the Cascade Range (e.g., Antos and Zobel 1985, Zobel and Antos 1986) and in eastern Washington (e.g., Mack 1981a). A major surprise was the importance of biological legacies—living organisms and organic materials including woody debris—in determining rates and pathways of recovery (e.g., Franklin and others 1985). Reviews of the Mount St. Helens research have been made covering one (Keller 1982) and five (Keller 1987) years of recovery.

Other disturbances that have received much attention are wildfire, exotic invasions, grazing, and clearcutting. Fire histories have been analyzed for several geographic regions (e.g., Hemstrom and Franklin 1982; Stewart 1986a, 1986b); they suggest that both large- and small-scale fires have been important. Mack, Rickard, and their associates have analyzed the invasion of *Bromus tectorum* L. (cheatgrass brome) in the steppic regions east of the Cascade Range. Succession after clearcutting continues to receive attention and has been the subject of long-term studies on permanent sample plots (Halpern 1987). Research on tree mortality and the role of forest gaps in northwestern forests is just getting underway.

Paleobotanical studies provide a long-term context on climatic and vegetational dynamics. Such research has continued to contribute to our regional knowledge through a variety of techniques including pollen analysis (e.g., Mack and others

1979), tree-ring analysis (e.g., Brubaker 1980), forest aging (e.g., Yamaguchi 1978), and macrofossils (e.g., Dunwiddie 1987).

Ecosystem Processes

In the last 15 years, we have made great strides in our understanding of structure and function in northwestern ecosystems, both forest and steppe. The International Biological Program's (IBP) Coniferous Forest Biome and Grassland Biome projects, which were in progress in 1973, have now been completed and have spawned numerous successor research projects. Forest work was focused in the Cedar River watershed in the Washington Cascade Range and in the H.J. Andrews Experimental Forest in the Oregon Cascade Range; shrub-steppe research was conducted primarily at the Hanford reserve in eastern Washington (e.g., Rickard 1985). These studies expanded understanding of productivity and its relation to the environment (e.g., Fujimori and others 1976, Gholz 1982, Grier and Logan 1977, Grier and Running 1977, Waring and Franklin 1979), belowground processes (e.g., Vogt and others 1981a, 1981b, 1981c), canopy processes (Massman 1982, Nadkarni 1984), the ecological roles of coarse woody debris (standing dead trees and downed boles) (e.g., Harmon and others 1986, Maser and Trappe 1984), and the influence of individual tree species on soil properties (e.g., Turner and Franz 1985a, 1986). Major progress reports have been published on the IBP-related research (Edmonds 1980, Waring 1980).

Ecosystem studies have also been conducted in the Olympic rain forest. Topics include succession (e.g., Luken and Fonda 1983), forest-river interactions (e.g., Starkey and others 1982), nurse-log phenomena (Harmon 1987), and interactions between elk and vegetation.

Composition, structure, and function of old-growth forest ecosystems have been studied intensively (e.g., Franklin and others 1981, Juday 1977), especially west of the Cascade Range where old-growth forests have become a major land-use issue. The role of such forests as habitat for specialized animal species is of concern. Preservation of such forests is increasingly viewed as a landscape issue (e.g., Franklin and Forman 1987, Harris 1984).

Scientific Reserves

Significant progress has been made during the last 15 years in the establishment of scientific reserves to represent the major vegetation types of Oregon and Washington. A multi-institutional regional plan (Dyrness and others 1975) was an important step, which has been followed by more detailed State plans (Oregon State Land Board 1981, Washington State Department of Natural Resources 1987). Research natural areas and scientific reserves have nearly doubled (Greene and others 1986), with areas established by six Federal agencies, the State of Washington, and The Nature Conservancy. This effort has been aided by the development of natural heritage data bases through the cooperative efforts of The Nature Conservancy and the State governments.

Future Research Needs

Information on the vegetation of the Pacific Northwest will probably continue to develop at a similar rate during the next decade as it has during the last. Although some shifts in topical and regional emphasis are occurring, total research appears to be increasing modestly. Clearly, a shift toward more detailed studies of processes and more use of experimentation and modeling has occurred.

Some major research needs that should be addressed include synthesis and collation of plant-community data; analysis of successional patterns, including attention to multiple pathways and detailed stand reconstructions; regional analysis of disturbance patterns including additional paleobotanical studies; and expanded autecological and population studies of important species, including other than the dominant life forms. An absolutely critical need in all of this research will be the development of long-term data sets on successional changes, populations, and ecosystem processes. Without such sustained efforts, information essential to developing and testing ecological theories will not be available.

BIBLIOGRAPHY

Adams, A.B.; Dale, V.H.; Smith, E.P.; Kruckeberg, A.R. 1987. Plant survival, growth form and regeneration following the 18 May 1980 eruption of Mount St. Helens, Washington. Northwest Science. 61(3): 160-170.

Agee, J.K. 1981. Fire effects on Pacific Northwest forests: flora, fuels, and fauna. In: Northwest Fire Council proceedings. Portland, OR: [Publisher unknown]: 54-66.

Agee, J.K.; Dunwiddie, P.W. 1984. Recent forest development on Yellow Island, San Juan County, WA. Canadian Journal of Botany. 62(10): 2074-2080.

Agee, J.K.; Pickford, S.G. 1985. Vegetation and fuel mapping of North Cascades National Park Service Complex. Final report to National Park Service. 64 p. On file at: University of Washington, College of Forest Resources, National Park Service Cooperative Park Studies Unit, Seattle.

Agee, J.K.; Thomas, T.L. 1986. Prescribed fire effects on mixed conifer forest structure at Crater Lake, Oregon. Canadian Journal of Forest Research. 16: 1082-1087.

Agee, James K. 1981. Initial effects of prescribed fire in a climax *Pinus contorta* forest: Crater Lake National Park. CPSU—UW 81-4. Seattle: National Park Service, Cooperative Park Studies Unit, College of Forest Resources, University of Washington. 10 p.

Agee, James K.; Huff, Mark H. 1987. Fuel succession in a western hemlock/Douglas-fir forest. Canadian Journal of Forest Research. 17(7): 697-704.

Agee, James K.; Kertis, Jane. 1987. Forest types of the North Cascades National Park Service Complex. Canadian Journal of Botany. 65(7): 1520-1530.

Agee, James K.; Smith, Larry. 1984. Subalpine tree reestablishment after fire in the Olympic Mountains, Washington. Ecology. 65(3): 810-819.

Aller, Alvin R.; Fosberg, Maynard A.; LaZelle, Monta C.; Falen, Anita L. 1981. Plant communities and soils of north slopes in the Palouse region of eastern Washington and northern Idaho. Northwest Science. 55(4): 248-262.

Alverson, E.; Arnett, J. 1986. From the steppe to the alpine: a botanical reconnaissance of the Lake Chelan-Sawtooth Ridge area, Washington. In: Plant life of the North Cascades. Douglasia Occasional Papers. 2: 1-63.

Andersen, Douglas C.; MacMahon, James A. 1985a. The effects of catastrophic ecosystem disturbance: the residual mammals at Mount St. Helens. Journal of Mammalogy. 66(3): 581-589.

Andersen, Douglas C.; MacMahon, James A. 1985b. Plant succession following the Mount St. Helens volcanic eruption: facilitation by a burrowing rodent, *Thomomys talpoides*. American Midland Naturalist. 114(1): 62-69.

Antos, J.A.; Zobel, D.B. 1982. Snowpack modification of volcanic tephra effects on forest understory plants near Mount St. Helens. Ecology. 63: 1969-1972.

Antos, J.A.; Zobel, D.B. 1986. Seedling establishment in forests affected by tephra from Mount St. Helens. American Journal of Botany. 73: 495-499.

Antos, Joseph A.; Zobel, Donald B. 1984. Ecological implications of belowground morphology of nine coniferous forest herbs. Botanical Gazette. 145(4): 508-517.

Antos, Joseph A.; Zobel, Donald B. 1985. Recovery of forest understories buried by tephra from Mt. St. Helens. Vegetatio. 64: 103-111.

Antos, Joseph A.; Zobel, Donald B. 1986. Habitat relationships of *Chamaecyparis nootkatensis* in southern Washington, Oregon, and California. Canadian Journal of Botany. 64: 1898-1909.

Arno, Stephen F.; Hammerly, Ramona P. 1984. Timberline: mountain and arctic forest frontiers. Seattle: The Mountaineers. 304 p.

Atzet, Thomas. 1979. Description and classification of the forests of the Upper Illinois River drainage of southwestern Oregon. Corvallis, OR: Oregon State University. 211 p. Ph.D. thesis.

Atzet, Thomas; Wheeler, David L. 1984. Preliminary plant associations of the Siskiyou Mountain province. Portland, OR: U.S. Department of Agriculture, Forest Service, Pacific Northwest Region. 315 p.

Atzet, Thomas A.; Wheeler, David L. 1982. Historical and ecological perspectives on fire activity in the Klamath Geological Province of the Rogue River and Siskiyou National Forests. R6-Range-102-1982. Portland, OR: U.S. Department of Agriculture, Forest Service, Pacific Northwest Region. 16 p.

Atzet, Tom; Wheeler, David; Franklin, Jerry; Smith, Brad. 1983. Vegetation classification in southwestern Oregon—a preliminary report. FIR Report. Corvallis, OR: Oregon State University, Extension Service, Forestry Intensified Research; 4(4): 6-8.

Atzet, Tom; Wheeler, David; Riegel, Gregg; Smith, Brad; Franklin, Jerry. 1984a. The Mountain Hemlock and Shasta Red Fir Series of the Siskiyou Region of southwest Oregon. In: Adaptive FIR annual report: October 1, 1983-September 30, 1984. Corvallis, OR: Oregon State University, Forest Research Laboratory, Forestry Intensified Research: 58-61.

Atzet, Tom; Wheeler, David; Smith, Brad; Riegel, Gregg; Franklin, Jerry. 1984b. The Tanoak Series of the Siskiyou Region of southwest Oregon: [part 1]. FIR Report. Corvallis, OR: Oregon State University, Extension Service, Forestry Intensified Research; 6(3): 6-7.

Atzet, Tom; Wheeler, David; Smith, Brad; Riegel, Gregg; Franklin, Jerry. 1985. The Tanoak Series of the Siskiyou Region of southwest Oregon: [part 2]. FIR Report. Corvallis, OR: Oregon State University, Extension Service, Forestry Intensified Research; 6(4): 7-10.

Axelrod, D.I. 1976. History of the coniferous forests, California and Nevada. Publ. in Bot. 70. Berkeley, CA: University of California. 62 p.

Axelrod, D.I.; Raven, P.H. 1985. Origin of the Cordilleran flora. Journal of Biogeography. 12: 21-47.

Bailey, Robert G. 1983. Delineation of ecosystem regions. Environmental Management. 7(4): 365-373.

Baker, W.L. 1984. A working list of the plant associations of the western United States. Version 1: July 10, 1984. 37 p. Unpublished report. On file at: Forestry Sciences Laboratory, Corvallis, OR.

Barbour, M.G.; Major, J., eds. 1977. Terrestrial vegetation of California. New York: John Wiley and Sons. 1002 p.

Barbour, Michael G.; De Jong, Theodore M.; Johnson, Ann F. 1975. Additions and corrections to a review of North American Pacific Coast beach vegetation. Madrono. 23: 130-134.

Barnosky, Cathy W. 1984. Late Pleistocene and early Holocene environmental history of southwestern Washington State, U.S.A. Canadian Journal of Earth Sciences. 21: 619-629.

Beetle, Alan A. 1979. Autecology of selected woody sagebrush species. In: The sagebrush ecosystem: a symposium: Proceedings; 1978 April. Logan, UT: Utah State University, College of Natural Resources: 23-26.

Beget, James E. 1982. Recent volcanic activity at Glacier Peak. Science. 215: 1389-1390.

Bell, Katherine L.; Bliss, Lawrence C. 1973. Alpine disturbance studies: Olympic National Park, USA. Biological Conservation. 5(1): 25-32.

Belsky, Arlene Joy. 1979. Determinants of ecological amplitude in *Festuca idahoensis* and *Festuca ovina.* Seattle: University of Washington. 206 p. Ph.D. dissertation.

Belsky, J.A.; del Moral, R. 1982. Ecology of an alpine-subalpine complex in the Olympic Mountains. Canadian Journal of Botany. 60: 778-788.

Bilderback, David E., ed. 1987. Mount St. Helens 1980: botanical consequences of the explosive eruptions. Berkeley, CA: University of California Press. 360 p.

Binkley, Dan. 1982. Nitrogen fixation and net primary production in a young Sitka alder stand. Canadian Journal of Botany. 60(3): 281-284.

Binkley, Dan; Graham, Robin Lambert. 1981. Biomass, production, and nutrient cycling of mosses in an old-growth Douglas-fir forest. Ecology. 62(5): 1387-1389.

Binney, Elizabeth P. 1987. Drought relations of *Phyllodoce empetriformis* at timberline in the North Cascades. Bellingham, WA: Western Washington University. 37 p. M.S. thesis.

Borgias, Darren D. 1984. Spruce fir forests of the northeastern Cascade Mountains, Washington. Bellingham, WA: Western Washington University. 64 p. M.S. thesis.

Bork, Joyce. 1985. Fire history in three vegetation types on the eastern side of the Oregon Cascades. Corvallis, OR: Oregon State University. 94 p. Ph.D. thesis.

Bortel, Michael Foster. 1977. Five *Betula papyrifera* community types in the Whatcom Lowland, northern Puget Trough, Washington. Bellingham, WA: Western Washington University. 54 p. M.S. thesis.

Boss, Theodore R. 1981. Intertidal salt marshes of Oregon. SG 63. Corvallis, OR: Oregon State University Extension Service, Marine Advisory Program, Land Grant/Sea Grant Cooperative. 8 p.

Boss, Theodore R. 1982. Vegetation ecology and net primary productivity of selected freshwater wetlands in Oregon. Corvallis, OR: Oregon State University. 236 p. Ph.D. thesis.

Boule, Marc E. 1981. Tidal wetlands of the Puget Sound region, Washington. Wetlands. 1: 47-60.

Breckon, Gary J.; Barbour, Michael G. 1974. Review of North American Pacific Coast beach vegetation. Madrono. 22: 333-360.

Brockway, Dale G.; Topik, Christopher; Hemstrom, Miles A.; Emmingham, William H. 1983. Plant association and management guide for the Pacific Silver Fir Zone, Gifford Pinchot National Forest. R6-Ecol-130a-1983. Portland, OR: U.S. Department of Agriculture, Forest Service, Pacific Northwest Region. 122 p.

Brubaker, Linda B. 1980. Spatial patterns of tree growth anomalies in the Pacific Northwest. Ecology. 61(4): 798-807.

Burg, Mary E.; Tripp, Donald R.; Rosenberg, Eric S. 1980. Plant associations and primary productivity of the Nisqually salt marsh on southern Puget Sound, Washington. Northwest Science. 54(3): 222-236.

Burke, Constance J. 1979. Historic fires in the central western Cascades, Oregon. Corvallis, OR: Oregon State University. 130 p. M.S. thesis.

Burkhardt, J. Wayne; Tisdale, E.W. 1976. Causes of juniper invasion in southwestern Idaho. Ecology. 57: 472-484.

Burns, Russell M., tech. compiler. 1983. Silvicultural systems for the major forest types of the United States. Agric. Handb. 445. Washington, DC: U.S. Department of Agriculture, Forest Service. 191 p.

Campbell, Alsie Gilbert; Franklin, Jerry F. 1979. Riparian vegetation in Oregon's western Cascade mountains: composition, biomass, and autumn phenology. Bull. 14. Seattle: Coniferous Forest Biome, Ecosystem Analysis Studies, University of Washington. 90 p.

Canaday, B.B.; Fonda, R.W. 1974. The influence of subalpine snowbanks on vegetation pattern, production, and phenology. Bulletin of the Torrey Botanical Club. 101(6): 340-350.

Chabot, B.F.; Mooney, H.A., eds. 1985. Physiological ecology of North American plant communities. New York: Chapman and Hall. 351 p.

Cholewa, Anita F.; Johnson, Frederic D. 1983. Secondary succession in the *Pseudotsuga menziesii/Physocarpus malvaceus* Association. Northwest Science. 57(4): 273-282.

Christy, E. Jennifer. 1982. Population dynamics of understory *Tsuga heterophylla*, western hemlock, in the Cascade Mountains, Oregon. Pullman, WA: Washington State University. 112 p. Ph.D. thesis.

Christy, E. Jennifer. 1986. Effect of root competition and shading growth of suppressed western hemlock (*Tsuga heterophylla*). Vegetatio. 65: 21-28.

Christy, E. Jennifer; Mack, Richard N. 1984. Variation in demography of juvenile *Tsuga heterophylla* across the substratum mosaic. Journal of Ecology. 72: 75-91.

Christy, John A. 1979. Report on a preliminary survey of sphagnum-containing wetlands of the Oregon coast. Salem, OR: Oregon Natural Area Preserves Advisory Committee to the State Land Board. 92 p.

Christy, John A. 1980. Additions to the moss flora of Oregon. The Bryologist. 83(3): 355-358.

Christy, John A.; Lyford, John H.; Wagner, David H. 1982. Checklist of Oregon mosses. The Bryologist. 85(1): 22-36.

Cline, J.F.; Uresk, D.W.; Rickard, W.H. 1977. Plants and soil of a sagebrush community on the Hanford Reservation. Northwest Science. 51(1): 60-70.

Cole, David. 1977. Ecosystem dynamics in the coniferous forest of the Willamette Valley, Oregon, U.S.A. Journal of Biogeography. 4: 181-192.

Cole, David N. 1982. Vegetation of two drainages in Eagle Cap Wilderness, Wallowa Mountains, Oregon. Res. Pap. INT-288. Ogden, UT: U.S. Department of Agriculture, Forest Service, Intermountain Forest and Range Experiment Station. 42 p.

Conard, Susan G.; Jaramillo, Annabelle E.; Cromack, Kermit, Jr.; Rose, Sharon, comps. 1985. The role of the genus *Ceanothus* in western forest ecosystems. Gen. Tech. Rep. PNW 182. Portland, OR: U.S. Department of Agriculture, Forest Service, Pacific Northwest Forest and Range Experiment Station. 72 p.

Cook, Stanton A. 1982. Stand development in the presence of a pathogen, *Phellinus weirii*. In: Means, Joseph E., ed. Forest succession and stand development research in the Northwest: Proceedings of the symposium (part of the Northwest Scientific Association annual meetings); 1981 March 26; Corvallis, OR. Corvallis, OR: Forest Research Laboratory, Oregon State University: 159-163.

Copeland, William N. 1979. Harney Lake Research Natural Area. Suppl. 9. Portland, OR: U.S. Department of Agriculture, Forest Service, Pacific Northwest Forest and Range Experiment Station. 21 p. Supplement to: Federal Research Natural Areas in Oregon and Washington: a guidebook for scientists and educators.

Copeland, William N.; Greene, Sarah E. 1982. Stinking Lake Research Natural Area. Suppl. 12. Portland, OR: U.S. Department of Agriculture, Forest Service, Pacific Northwest Forest and Range Experiment Station. 21 p. Supplement to: Federal Research Natural Areas in Oregon and Washington: a guidebook for scientists and educators.

Cornelius, Lynn. 1982. Checklist of vascular plants of Sister Rocks Research Natural Area. Adm. Rep. PNW-2. Portland, OR: U.S. Department of Agriculture, Forest Service, Pacific Northwest Forest and Range Experiment Station. 8 p.

Cornelius, Lynn C.; Schuller, Reid. 1982. Checklist of the vascular plants of Cedar Flats Research Natural Area. Adm. Rep. PNW-5. Portland, OR: U.S. Department of Agriculture, Forest Service, Pacific Northwest Forest and Range Experiment Station. 14 p.

Coville, Frederick V. 1898. Forest growth and sheep grazing in the Cascade Mountains of Oregon. Bull. 15. Washington, DC: U.S. Department of Agriculture. 54 p.

Cox, George W.; Allen, Douglas W. 1987. Sorted stone nets and circles of the Columbia Plateau: a hypothesis. Northwest Science. 61(3): 179-185.

Critchfield, W.B. 1985. The late Quaternary history of lodgepole and jack pines. Canadian Journal of Forest Research. 15: 749-772.

Curtis, Alan. 1986a. Camas Swale Research Natural Area. Suppl. 21. Portland, OR: U.S. Department of Agriculture, Forest Service, Pacific Northwest Research Station. 18 p. Supplement to: Federal Research Natural Areas in Oregon and Washington: a guidebook for scientists and educators.

Curtis, Alan. 1986b. Fox Hollow Research Natural Area. Suppl. 22. Portland, OR: U.S. Department of Agriculture, Forest Service, Pacific Northwest Research Station. 19 p. Supplement to: Federal Research Natural Areas in Oregon and Washington: a guidebook for scientists and educators.

Curtis, Alan. 1986c. Mohawk Research Natural Area. Suppl. 23. Portland, OR: U.S. Department of Agriculture, Forest Service, Pacific Northwest Research Station. 17 p. Supplement to: Federal Research Natural Areas in Oregon and Washington: a guidebook for scientists and educators.

Curtis, Alan. 1986d. Upper Elk Meadows Research Natural Area. Suppl. 18. Portland, OR: U.S. Department of Agriculture, Forest Service, Pacific Northwest Research Station. 19 p. Supplement to: Federal Research Natural Areas in Oregon and Washington: a guidebook for scientists and educators.

Cushman, Martha J. 1981. The influence of recurrent snow avalanches on vegetation patterns in the Washington Cascades. Seattle: University of Washington. 175 p. Ph.D. dissertation.

Dahl, B.E.; Tisdale, E.W. 1975. Environmental factors related to medusahead distribution. Journal of Range Management. 28(6): 463-468.

Dahlgreen, Matthew Craig. 1984. Observations on the ecology of *Vaccinium membranaceum* Dougl. on the southeast slope of the Washington Cascades. Seattle: University of Washington. 108 p. M.S. thesis.

Dale, Virginia H.; Hemstrom, Miles. 1984. CLIMACS: a computer model of forest stand development for western Oregon and Washington. Res. Pap. PNW 327. Portland, OR: U.S. Department of Agriculture, Forest Service, Pacific Northwest Forest and Range Experiment Station. 60 p.

Dale, Virginia H.; Hemstrom, Miles; Franklin, Jerry. 1986. Modeling the long-term effects of disturbances on forest succession, Olympic Peninsula, Washington. Canadian Journal of Forest Research. 16: 56-67.

Daubenmire, R. 1975. Floristic plant geography of eastern Washington and northern Idaho. Journal of Biogeography. 2: 1-18.

Daubenmire, R. 1976. Derivation of the flora of the Pacific Northwest. In: Andrews, Rollin D., III [and others], eds. Proceedings of the symposium on terrestrial and aquatic ecological studies of the Northwest; 1976 March 26-27. Cheney, WA: Eastern Washington State College Press: 159-171.

Daubenmire, R. 1977. A bibliography on vegetation of the State of Washington: supplement 1. Northwest Science. 52(2): 111-113.

Daubenmire, R. 1978. Plant geography with special reference to North America. New York: Academic Press. 338 p.

Daubenmire, R. 1982. The distribution of *Artemisia rigida* in Washington: a challenge to ecology and geology. Northwest Science. 56(3): 162-164.

Daubenmire, R. 1984. Snowdrift vegetation in the shrub-steppe of Washington. Phytocoenologia. 11(4): 449-454.

Daubenmire, Rexford. 1985. The western limits of the range of the American bison. Ecology. 66(2): 622-624.

Dealy, J. Edward. 1971. Habitat characteristics of the Silver Lake Mule Deer Range. Res. Pap. PNW-125. Portland, OR: U.S. Department of Agriculture, Forest Service, Pacific Northwest Forest and Range Experiment Station. 99 p.

Dealy, J. Edward. 1978. Autecology of curlleaf mountain-mahogany (*Cercocarpus ledifolius*). In: Proceedings of the 1st International Rangeland Congress. [Location of publisher unknown]: [Publisher unknown]: 398-400.

Dealy, J. Edward; Geist, J. Michael. 1978. Conflicting vegetational indicators on some central Oregon scablands. Journal of Range Management. 31(1): 56-59.

Dealy, J. Edward; Geist, J. Michael; Driscoll, Richard S. 1978a. Communities of western juniper in the intermountain Northwest. In: Proceedings of the western juniper ecology and management workshop; 1977 January; Bend, OR. Gen. Tech. Rep. PNW-74. Portland, OR: U.S. Department of Agriculture, Forest Service, Pacific Northwest Forest and Range Experiment Station: 11-29.

Dealy, J. Edward; Geist, J. Michael; Driscoll, Richard S. 1978b. Western juniper communities on rangelands of the Pacific Northwest. In: Proceedings of the 1st International Rangeland Congress. [Location of publisher unknown]: [Publisher unknown]: 201-204.

Dealy, J. Edward; Leckenby, Donavin A.; Concannon, Diane M. 1981. Wildlife habitats in managed rangelands—the Great Basin of southeastern Oregon: plant communities and their importance to wildlife. Gen. Tech. Rep. PNW-120. Portland, OR: U.S. Department of Agriculture, Forest Service, Pacific Northwest Forest and Range Experiment Station. 66 p.

DeBell, Dean S.; Franklin, Jerry F. 1987. Old-growth Douglas-fir and western hemlock: a 36-year record of growth and mortality. Western Journal of Applied Forestry. 2(4): 111-114.

del Moral, R. 1979a. High elevation species clusters in the Enchantment Lakes basin, Washington. Madrono. 26: 164-172.

del Moral, R. 1981. Life returns to Mount St. Helens. Natural History. 90(5): 36-49.

del Moral, R. 1983a. Initial recovery of subalpine vegetation on Mount St. Helens, Washington. American Midland Naturalist. 109: 72-80.

del Moral, R. 1985. Competitive effects on the structure of subalpine meadow communities. Canadian Journal of Botany. 63: 1444-1452.

del Moral, R.; Clampitt, C.A. 1985. Growth of native plant species on recent substrates from Mount St. Helens. American Midland Naturalist. 114: 374-383.

del Moral, R.; Clampitt, C.A.; Wood, D.M. 1985. Does interference cause niche differentiation? Evidence from subalpine plant communities. American Journal of Botany. 72: 1891-1901.

del Moral, R.; Watson, A.F.; Fleming, R.S. 1976. Vegetation structure in the Alpine Lakes region of Washington State: classification of vegetation on granitic rocks. Syesis. 9: 291-316.

del Moral, Roger. 1972. Diversity patterns in forest vegetation of the Wenatchee Mountains, Washington. Bulletin of the Torrey Botanical Club. 99: 57-64.

del Moral, Roger. 1976. Wilderness in ecological research: an example from the alpine lakes. In: Andrews, Rollin D., III [and others], eds. Proceedings of the symposium on terrestrial and aquatic ecological studies of the Northwest; 1976 March 26-27. Cheney, WA: Eastern Washington State College Press: 173-194.

del Moral, Roger. 1974. Species patterns in the upper North Fork Teanaway River drainage, Wenatchee Mountains, Washington. Syesis. 7: 13-30.

del Moral, Roger. 1979b. High elevation vegetation of the Enchantment Lakes basin, Washington. Canadian Journal of Botany. 57(10): 1111-1130.

del Moral, Roger. 1982. Control of vegetation on contrasting substrates: herb patterns on serpentine and sandstone. American Journal of Botany. 69(2): 227-238.

del Moral, Roger. 1983b. Competition as a control mechanism in subalpine meadows. American Journal of Botany. 70(2): 232-245.

del Moral, Roger. 1983c. Vegetation ordination of subalpine meadows using adaptive strategies. Canadian Journal of Botany. 61(12): 3117-3127.

del Moral, Roger. 1984. The impact of the Olympic marmot on subalpine vegetation structure. American Journal of Botany. 71(9): 1228-1236.

del Moral, Roger; Deardorff, David C. 1976. Vegetation of the Mima Mounds, Washington State. Ecology. 57(3): 520-530.

del Moral, Roger; Denton, Melinda F. 1977. Analysis and classification of vegetation based on family composition. Vegetatio. 34(3): 155-165.

del Moral, Roger; Fleming, Richard S. 1980. Structure of coniferous forest communities in western Washington: diversity and ecotope properties. Vegetatio. 41(3): 143-154.

del Moral, Roger; Long, James N. 1977. Classification of montane forest community types in the Cedar River drainage of western Washington, U.S.A. Canadian Journal of Forest Research. 7: 217-225.

del Moral, Roger; Standley, Lisa A. 1979. Pollination of angiosperms in contrasting coniferous forests. American Journal of Botany. 66(1): 26-35.

del Moral, Roger; Watson, Alan F. 1978. Gradient structure of forest vegetation in the central Washington Cascades. Vegetatio. 38(1): 29-48.

Dennis, La Rae June. 1959. A taxonomic study of the vascular flora of Ashland Peak, Jackson County, Oregon. Corvallis, OR: Oregon State University. 114 p. M.S. thesis.

Denton, Melinda F.; del Moral, Roger. 1976. Comparison of multivariate analyses using taxonomic data of *Oxalis*. Canadian Journal of Botany. 54(14): 1637-1646.

Dickman, Alan. 1978. Reduced fire frequency changes species composition of a ponderosa pine stand. Journal of Forestry. January: 24-25.

Disraeli, D.J.; Fonda, R.W. 1979. Gradient analysis of the vegetation in a brackish marsh in Bellingham Bay, Washington. Canadian Journal of Botany. 57(5): 465-475.

Doescher, Paul S.; Eddleman, Lee E.; Vaitkus, Milda R. 1987. Evaluation of soil nutrients, pH, and organic matter in rangelands dominated by western juniper. Northwest Science. 61(2): 97-102.

Douglas, George W.; Ballard, T.M. 1971. Effects of fire on alpine plant communities in the North Cascades, Washington. Ecology. 52(6): 1058-1064.

Douglas, George W.; Bliss, L.C. 1977. Alpine and high subalpine plant communities in the North Cascades Range, Washington and British Columbia. Ecological Monographs. 47(2): 113-150.

Dunwiddie, Peter W. 1987. Macrofossil and pollen representation of coniferous trees in modern sediments from Washington. Ecology. 68(1): 1-11.

Dyrness, C.T.; Franklin, Jerry F.; Maser, Chris. 1973. Wheeler Creek Research Natural Area. Suppl. 1. Portland, OR: U.S. Department of Agriculture, Forest Service, Pacific Northwest Forest and Range Experiment Station. 16 p. Supplement to: Federal Research Natural Areas in Oregon and Washington: a guidebook for scientists and educators.

Dyrness, C.T.; Franklin, Jerry F.; Maser, Chris [and others]. 1975. Research natural area needs in the Pacific Northwest. Gen. Tech. Rep. PNW-38. Portland, OR: U.S. Department of Agriculture, Forest Service, Pacific Northwest Forest and Range Experiment Station. 231 p.

Dyrness, C.T.; Franklin, Jerry F.; Moir, W.H. 1976. A preliminary classification of forest communities in the central portion of the western Cascades of Oregon. Bull. 4. Seattle: Coniferous Forest Biome, Ecosystem Analysis Studies, University of Washington. 123 p.

Edgerton, Paul J. 1983. Response of the bitterbrush understory of a central Oregon lodgepole pine forest to logging disturbance. In: Proceedings—research and management of bitterbrush and cliffrose in western North America; 1982 April 13-15; Salt Lake City, UT. Gen. Tech. Rep. INT-152. Ogden, UT: U.S. Department of Agriculture, Forest Service, Intermountain Forest and Range Experiment Station: 99-106.

Edmonds, R.L., ed. 1982. Analysis of coniferous ecosystems in the western United States. U.S. IBP Synthesis Ser. 14. Stroudsburg, PA: Hutchinson Ross Publishing Co. 419 p.

Edmonds, Robert L. 1980. Litter decomposition and nutrient release in Douglas-fir, red alder, western hemlock, and Pacific silver fir ecosystems in western Washington. Canadian Journal of Forest Research. 10(3): 327-337.

Egler, F.E. 1934. Communities and successional trends in the vegetation of the Coos Bay sand dunes, Oregon. Minneapolis, MN: University of Minnesota. 39 p. M.S. thesis.

Eilers, H. Peter. 1976. The ecological biogeography of an Oregon coastal salt marsh. In: Yearbook of the Association of Pacific Coast Geographers. Corvallis, OR: Oregon State University Press; 19-32. Vol. 38.

Eilers, H. Peter. 1979. Production ecology in an Oregon coastal salt marsh. Estuarine and Coastal Marine Science. 8: 399-410.

Eilers, H. Peter; Taylor, Alan; Sanville, William. 1983. Vegetative delineation of coastal salt marsh boundaries. Environmental Management. 7(5): 443-452.

Eilers, Hio Peter, III. 1974. Plants, plant communities, net production and tide levels: the ecological biogeography of the Nehalem salt marshes, Tillamook County, Oregon. Corvallis, OR: Oregon State University. 368 p. Ph.D. thesis.

Emmingham, William H.; Waring, Richard H. 1977. An index of photosynthesis for comparing forest sites in western Oregon. Canadian Journal of Forest Research. 7(1): 165-174.

Evans, Raymond David. 1986. The relationship between snowmelt and subalpine meadow community pattern, Excelsior Ridge, western North Cascades. Bellingham, WA: Western Washington University. 44 p. M.S. thesis.

Ewing, Kern. 1982a. A comparison of treeless areas in the forested foothills of the Cascade Mountains of western Washington. Northwest Science. 56(3): 180-189.

Ewing, Kern. 1982b. Plant response to environmental variation in the Skagit Marsh. Seattle: University of Washington. 203 p. Ph.D. thesis.

Ewing, Kern. 1983. Environmental controls in Pacific Northwest intertidal marsh plant communities. Canadian Journal of Botany. 61(4): 1105-1116.

Eyre, F.H., ed. 1980. Forest cover types of the United States and Canada. Washington, DC: Society of American Foresters. 148 p.

Fahnestock, G.; Agee, J.K. 1983. Biomass consumption and smoke production from prehistoric and modern forest fires in western Washington. Journal of Forestry. 81: 653-657.

Finney, M.A. 1986. Effects of low intensity fire on the successional development of seral lodgepole pine forests in the North Cascades. Seattle: University of Washington. 140 p. M.S. thesis.

Fonda, R.W. 1974. Forest succession in relation to river terrace development in Olympic National Park, Washington. Ecology. 55(5): 927-942.

Fonda, R.W. 1976a. Ecology of alpine timberline in Olympic National Park. Proceedings of the Conference on Scientific Research in National Parks. 1: 209-212.

Fonda, R.W. 1976b. Fire-resilient forests of Douglas-fir in Olympic National Park: an hypothesis. Proceedings of the Conference on Scientific Research in National Parks. 1: 1239-1242.

Fonda, R.W.; Bernardi, J.A. 1976. Vegetation of Sucia Island in Puget Sound, Washington. Bulletin of the Torrey Botanical Club. 103(3): 99-109.

Frank, Douglas A.; del Moral, Roger. 1986. Thirty-five years of secondary succession in *Festuca viridula-Lupinus latifolius* dominated meadow at Sunrise, Mount Rainier National Park, Washington. Canadian Journal of Botany. 64: 1232-1236.

Franklin, J.F. 1976. Scientific reserves in the Pacific Northwest and their significance for ecological research. In: Proceedings of the symposium on terrestrial and aquatic ecological studies of the Northwest; 1976 March 26-27. Cheney, WA: Eastern Washington State College Press: 195-208.

Franklin, J.F.; MacMahon, J.A.; Swanson, F.J.; Sedell, J.R. 1985. Ecosystem responses to the eruption of Mount St. Helens. National Geographic Research. Spring: 198-216.

Franklin, Jerry F. 1982. Ecosystem studies in the Hoh River drainage, Olympic National Park. In: Ecological research in national parks of the Pacific Northwest: Proceedings of the 2d conference on scientific research in the national parks; 1979 November; San Francisco, CA. p. 1-8. Available from: National Park Service Cooperative Park Studies Unit, Forestry Sciences Laboratory, Corvallis, OR.

Franklin, Jerry F. 1983. Ecology of noble fir. In: Oliver, Chadwick Dearing; Kenady, Reid M., eds. The biology and management of true fir in the Pacific Northwest: Proceedings of the symposium; 1981 February 24-26; Seattle. Seattle: University of Washington, College of Forest Resources; Portland, OR: U.S. Department of Agriculture, Forest Service, Pacific Northwest Forest and Range Experiment Station: 59-69.

Franklin, Jerry F. [In press]. Pacific Northwest forests. In: North American terrestrial vegetation. New York: Cambridge University Press: 103-130.

Franklin, Jerry F.; Cromack, Kermit, Jr.; Denison, William [and others]. 1981. Ecological characteristics of old-growth Douglas-fir forests. Gen. Tech. Rep. PNW-118. Portland, OR: U.S. Department of Agriculture, Forest Service, Pacific Northwest Forest and Range Experiment Station. 417 p.

Franklin, Jerry F.; DeBell, Dean S. 1987. Thirty-six years of tree population change in an old-growth *Pseudotsuga-Tsuga* forest. Unpublished manuscript; submitted to: Canadian Journal of Forest Research; on file at: Forestry Sciences Laboratory, Corvallis, OR.

Franklin, Jerry F.; Forman, Richard T.T. 1987. Creating landscape patterns by forest cutting: ecological consequences and principles. Landscape Ecology. 1(1): 5-18.

Franklin, Jerry F.; Frenzen, Peter M.; Swanson, Frederick J. [In press]. Re-creation of ecosystems at Mount St. Helens: contrasts in artificial and natural approaches. In: Rehabilitating damaged ecosystems. Vol. 2. Boca Raton, FL: CRC (Chemical Rubber Company) Press.

Franklin, Jerry F.; Hemstrom, Miles A. 1981. Aspects of succession in coniferous forests of the Pacific Northwest. In: West, Darrell C.; Shugart, Herman H.; Botkin, Daniel B., eds. Forest succession: concepts and application. New York: Springer-Verlag: 212-229.

Franklin, Jerry F.; Moir, William H.; Hemstrom, Miles A.; Greene, Sarah E.; Smith, Bradley G. 1987a. Forest communities of Mount Rainier National Park. Sci. Monogr. Unpublished manuscript; submitted to: U.S. Department of the Interior, National Park Service, Atlanta, GA; on file at: Forestry Sciences Laboratory, Corvallis, OR.

Franklin, Jerry F.; Shugart, H.H.; Harmon, Mark E. 1987b. Tree death as an ecological process: the causes, consequences, and variability of tree mortality. BioScience. 37(8): 550-556.

Franklin, Jerry F.; Spies, Thomas A. 1986. The ecology of old-growth Douglas-fir forests. Oregon Birds. Eugene, OR: University of Oregon Press; 12(2): 79-90.

Franklin, Jerry F.; Swanson, Frederick J.; Sedell, J.R. 1982. Relationships within the valley floor ecosystems in western Olympic National Park: a summary. In: Ecological research in national parks of the Pacific Northwest: Proceedings of the 2d conference on scientific research in the national parks; 1979 November; San Francisco, CA. p. 43-45. Available from: National Park Service Cooperative Park Studies Unit, Forestry Sciences Laboratory, Corvallis, OR.

Franklin, Jerry F.; Waring, Richard W. 1980. Distinctive features of the northwestern coniferous forest: development, structure, and function. In: Forests: fresh perspectives from ecosystem analysis: Proceedings, 40th annual biological colloquium; 1979 April 27-28; Corvallis, OR. Corvallis, OR: Oregon State University Press; 59-85.

Franklin, Jerry F.; Wiberg, Curt. 1979. Goat Marsh Research Natural Area. Suppl. 10. Portland, OR: U.S. Department of Agriculture, Forest Service, Pacific Northwest Forest and Range Experiment Station. 19 p. Supplement to: Federal Research Natural Areas in Oregon and Washington: a guidebook for scientists and educators.

Frenkel, Robert E. 1974. An isolated occurrence of Alaska-cedar (*Chamaecyparis nootkatensis* (D. Don) Spach) in the Aldrich Mountains, central Oregon. Northwest Science. 48: 29-37.

Frenkel, Robert E. 1980. Natural area inventory and assessment: Blacklock Point, Oregon. In: Yearbook of the Association of Pacific Coast Geographers. Corvallis, OR: Oregon State University Press; 119-129. Vol 42.

Frenkel, Robert E. 1985. Vegetation. In: Kimerling, A. Jon; Jackson, Philip L., eds. Atlas of the Pacific Northwest. 7th ed. Corvallis, OR: Oregon State University Press: 58-66.

Frenkel, Robert E. 1987a. Introduction and spread of cordgrass (*Spartina*) into the Pacific Northwest. Northwest Environmental Journal. 3: 152-154.

Frenkel, Robert E. 1987b. Valley to the coast: a biogeographic transect: field guide prepared for the 83rd annual meeting of the Association of American Geographers, Portland, Oregon, April 21-25, 1987. Corvallis, OR: Department of Geography, Oregon State University. 74 p.

Frenkel, Robert E.; Boss, Theodore; Schuller, S. Reid. 1978. Transition-zone vegetation between intertidal marsh and upland in Oregon and Washington. 320 p. Unpublished report; prepared for: U.S. Environmental Protection Agency, Corvallis, OR.

Frenkel, Robert E.; Eilers, H. Peter; Jefferson, Carol A. 1981. Oregon coastal salt marsh upper limits and tidal datums. Estuaries. 4(3): 198-205.

Frenkel, Robert E.; Hanson, David; Kolar, Scott. 1976. Vegetation. In: Loy, W.G., ed. Atlas of Oregon. Eugene, OR: University of Oregon: 144-148.

Frenkel, Robert E.; Heinitz, Eric F. 1987. Composition and structure of Oregon ash (*Fraxinus latifolia*) forest in William L. Finley National Wildlife Refuge, Oregon. Northwest Science. 61(4): 203-212.

Frenkel, Robert E.; Kiilsgaard, Christen W. 1984. Vegetation classification and map of the central Siskiyou Mountains, Oregon. Final report for NASA-Ames, Univ. Joint Res. Interchange SEA-1600/T-4885. Moffett Field, CA: NASA-Ames. 101 p.

Frenkel, Robert E.; Moir, William H.; Christy, John A. 1986. Vegetation of Torrey Lake Mire, central Cascade Range, Oregon. Madrono. 33(1): 24-39.

Frenkel, Robert E.; Wickramaratne, S. Nimal; Heinitz, Eric F. 1984. Vegetation and land cover change in the Willamette River Greenway in Benton and Linn Counties, Oregon: 1972-1981. In: Yearbook of the Association of Pacific Coast Geographers. Corvallis, OR: Oregon State University Press; 63-77. Vol. 46.

Frenzen, P.M.; Franklin, J.F. 1985. Establishment of conifers from seed on tephra deposited by the 1980 eruptions of Mount St. Helens, Washington. American Midland Naturalist. 114: 84-97.

Frenzen, Peter M.; Franklin, Jerry F. 1987. Tree mortality and plant succession following a mudflow in an old-growth forest, Cedar Flats, Washington. Unpublished manuscript; submitted to: American Midland Naturalist; on file at: Forestry Sciences Laboratory, Corvallis, OR.

Frenzen, Peter M.; Krasny, Marianne E.; Rigney, Lisa P. [In press]. Thirty-three years of plant succession on the Kautz Creek mudflow, Mount Rainier National Park, Washington. Canadian Journal of Botany.

Fujimori, Takao; Jawanabe, Saburo; Saito, Hideki [and others]. 1976. Biomass and primary production in forests of three major vegetation zones of the northwestern United States. Journal of the Japanese Forestry Society. 58(10): 360-373.

Gallagher, John L.; Kibby, Harold V. 1981. The streamside effect in a *Carex lyngbyei* estuarine marsh: the possible role of recoverable underground reserves. Estuarine, Coastal and Shelf Science. 12: 451-460.

Gara, R.I.; Agee, J.K.; Littke, W.R.; Geiszler, D.R. 1986. Fire wounds and beetle scars: distinguishing between the two can help reconstruct past disturbances. Journal of Forestry. 84: 47-50.

Gara, R.I.; Littke, W.R.; Agee, J.K. [and others]. 1985. Influence of fires, fungi, and mountain pine beetles on development of a lodgepole pine forest in south-central Oregon. In: Baumgartner, D.M. [and others], eds. Lodgepole pine: the species and its management: Proceedings of a symposium; 1984 May 8-10 and 14-16; Spokane, WA, and Vancouver, BC. Pullman, WA: Washington State University: 153-162.

Gholz, H.L. 1980. Structure and productivity of *Juniperus occidentalis* in central Oregon. American Midland Naturalist. 103(2): 251-261.

Gholz, Henry L. 1982. Environmental limits on aboveground net primary production, leaf area, and biomass in vegetation zones of the Pacific Northwest. Ecology. 63(2): 469-481.

Gholz, Henry L.; Hawk, Glenn M.; Campbell, Alsie; Cromack, Kermit, Jr. 1985. Early vegetation recovery and element cycles on a clear-cut watershed in western Oregon. Canadian Journal of Forest Research. 15: 400-409.

Gholz, Henry Lewis. 1979. Limits on above ground net primary production, leaf area, and biomass in vegetational zones of the Pacific Northwest. Corvallis, OR: Oregon State University. 61 p. Ph.D. thesis.

Gilleland, Cevin Lee. 1980. Vegetation ecology of solifluction areas in the North Cascades, Washington. Bellingham, WA: Western Washington University. 56 p. M.S. thesis.

Graham, Robin Lee Lambert. 1981. Biomass dynamics of dead Douglas-fir and western hemlock boles in mid-elevation forests of the Cascade Range. Corvallis, OR: Oregon State University. 152 p. Ph.D. thesis.

Graham, Robin Lee Lambert. 1982. The biomass, coverage, and decay rates of dead boles in terrace forests, South Fork Hoh River, Olympic National Park. In: Ecological research in national parks of the Pacific Northwest: Proceedings of the 2d conference on scientific research in the national parks; 1979 November; San Francisco, CA. p. 15-21. Available from: National Park Service Cooperative Park Studies Unit, Forestry Sciences Laboratory, Corvallis, OR.

Greene, Sarah E. 1982a. Indian Creek Research Natural Area. Suppl. 14. Portland, OR: U.S. Department of Agriculture, Forest Service, Pacific Northwest Forest and Range Experiment Station. 15 p. Supplement to: Federal Research Natural Areas in Oregon and Washington: a guidebook for scientists and educators.

Greene, Sarah E. 1982b. Neskowin Crest Research Natural Area. Suppl. 13. Portland, OR: U.S. Department of Agriculture, Forest Service, Pacific Northwest Forest and Range Experiment Station. 17 p. Supplement to: Federal Research Natural Areas in Oregon and Washington: a guidebook for scientists and educators.

Greene, Sarah E. 1985. New rangeland research areas in Oregon. Rangelands. 7(4): 165-166.

Greene, Sarah E.; Blinn, Tawny; Franklin, Jerry F. 1986. Research natural areas in Oregon and Washington: past and current research and related literature. Gen. Tech. Rep. PNW-197. Portland, OR: U.S. Department of Agriculture, Forest Service, Pacific Northwest Research Station. 115 p.

Greene, Sarah E.; Copeland, Bill. 1984. Poker Jim Ridge Research Natural Area. Suppl. 16. Portland, OR: U.S. Department of Agriculture, Forest Service, Pacific Northwest Forest and Range Experiment Station. 19 p. Supplement to: Federal Research Natural Areas in Oregon and Washington: a guidebook for scientists and educators.

Greene, Sarah E.; Franklin, Jerry F. 1987. Middle Santiam Research Natural Area. Suppl. 24. Portland, OR: U.S. Department of Agriculture, Forest Service, Pacific Northwest Research Station. 22 p. Supplement to: Federal Research Natural Areas in Oregon and Washington: a guidebook for scientists and educators.

Greene, Sarah E.; Frenkel, Robert E. 1986. Steamboat Mountain Research Natural Area. Suppl. 20. Portland, OR: U.S. Department of Agriculture, Forest Service, Pacific Northwest Forest and Range Experiment Station. 19 p. Supplement to: Federal Research Natural Areas in Oregon and Washington: a guidebook for scientists and educators.

Greene, Sarah E.; Klopsch, Mark. 1985. Soil and air temperatures for different habitats in Mount Rainier National Park. Res. Pap. PNW-342. Portland, OR: U.S. Department of Agriculture, Forest Service, Pacific Northwest Forest and Range Experiment Station. 50 p.

Greene, Sarah E.; Lesher, Robin; Wasem, Robert. 1984. Silver Lake Research Natural Area. Suppl. 15. Portland, OR: U.S. Department of Agriculture, Forest Service, Pacific Northwest Forest and Range Experiment Station. 15 p. Supplement to: Federal Research Natural Areas in Oregon and Washington: a guidebook for scientists and educators.

Grier, C.C.; Logan, R.S. 1977. Old-growth *Pseudotsuga menziesii* communities of a western Oregon watershed: biomass distribution and production budgets. Ecological Monographs. 47: 373-400.

Grier, Charles C. 1978. A *Tsuga heterophylla-Picea sitchensis* ecosystem of coastal Oregon: decomposition and nutrient balances of fallen logs. Canadian Journal of Forest Research. 8(2): 198-206.

Grier, Charles C.; Running, Steven. 1977. Leaf area of mature northwestern coniferous forests: relation to site water balance. Ecology. 58(4): 893-899.

Grier, Charles C.; Vogt, Kristiina A.; Keyes, Michael R.; Edmonds, Robert L. 1981. Biomass distribution and above- and below-ground production in young and mature *Abies amabilis* zone ecosystems of the Washington Cascades. Canadian Journal of Forest Research. 11(1): 155-167.

Gruell, George E. 1985. Fire on the early western landscape: an annotated record of wildland fires 1776-1900. Northwest Science. 59(2): 97-107.

Hall, F.C. 1978a. Western juniper in association with other tree species. In: Martin, R.E.; Dealy, J.E.; Caraher, D.L., tech. eds. Proceedings, western juniper ecology and management workshop. Gen. Tech. Rep. PNW-74. Portland, OR: U.S. Department of Agriculture, Forest Service, Pacific Northwest Forest and Range Experiment Station: 31-36.

Hall, F.C. 1980a. Western forest types and avian management practices. In: DeGraff, R.M., tech. ed. Proceedings, management of western forests and grasslands for nongame birds; Salt Lake City, UT. Gen. Tech. Rep. INT-86. Ogden, UT: U.S. Department of Agriculture, Forest Service, Intermountain Forest and Range Experiment Station: 27-37.

Hall, F.C. [In press]. Pacific Northwest ecoclass codes for plant associations. R6-Ecol-T.P.-289-87. Portland, OR: U.S. Department of Agriculture, Forest Service, Pacific Northwest Region.

Hall, Frederick C. 1973. Plant communities of the Blue Mountains in eastern Oregon and southeastern Washington. R6 Area Guide 3-1. Portland, OR: U.S. Department of Agriculture, Forest Service, Pacific Northwest Region. 62 p.

Hall, Frederick C. 1976a. Classification, designation, identification equal confusion. R6 Regional Guide 4. Portland, OR: U.S. Department of Agriculture, Forest Service, Pacific Northwest Region. 11 p.

Hall, Frederick C. 1976b. Range trend sampling by photographs. R6 Regional Guide 2-1. Portland, OR: U.S. Department of Agriculture, Forest Service, Pacific Northwest Region. 50 p.

Hall, Frederick C. 1978b. Applicability of rangeland management concepts to forest-range in the Pacific Northwest. In: Hyder, Donald N., ed. Proceedings of the 1st International Rangeland Congress; 1978 August 14-18; Denver, CO. Denver, CO: Society for Range Management: 496-499.

Hall, Frederick C. 1978c. Pacific Northwest ecoclass vegetation identification: concept and codes. R6 Regional Guide 1-3. Portland, OR: U.S. Department of Agriculture, Forest Service, Pacific Northwest Region. 60 p.

Hall, Frederick C. 1979a. Codes for Pacific Northwest ecoclass vegetation classification. R6-Ecol-79-002. Portland, OR: U.S. Department of Agriculture, Forest Service, Pacific Northwest Region. 62 p.

Hall, Frederick C. 1979b. Ecology of natural underburning in the Blue Mountains of Oregon. R6-Ecol-79-001. Portland, OR: U.S. Department of Agriculture, Forest Service, Pacific Northwest Region. 11 p.

Hall, Frederick C. 1980b. Applications of a classification system based on plant community types (associations) with special reference to wildlife, range, and timber management. In: Proceedings, land-use allocation; processes, people, politics, professionals: SAF national convention; Spokane, WA. Washington, DC: Society of American Foresters: 163-169.

Hall, Frederick C. 1984. Ecoclass coding system for Pacific Northwest plant associations. R6-Ecol-173-1984. Portland, OR: U.S. Department of Agriculture, Forest Service, Pacific Northwest Region. 83 p.

Halpern, Charles; Harmon, M.E. 1983. Early plant succession on the Muddy River Mudflow, Mount St. Helens, Washington. American Midland Naturalist. 110: 97-106.

Halpern, Charles B. 1987. Twenty-one years of secondary succession in *Pseudotsuga* forests of the western Cascade Range, Oregon. Corvallis, OR: Oregon State University. 239 p. Ph.D. thesis.

Halverson, N.M.; Lecher, R.D.; McClure, R.H., Jr. 1986. Major indicator shrubs and herbs on national forests of western Oregon and southwestern Washington. R6-TM-229-1986. Portland, OR: U.S. Department of Agriculture, Forest Service, Pacific Northwest Region. 196 p. Looseleaf; limited distribution.

Halverson, N.M.; Topik, C.; VanVickle, R. 1986. Plant association and management guide for the western hemlock zone, Mt. Hood National Forest. R6-Ecol-232A-1986. Portland, OR: U.S. Department of Agriculture, Forest Service, Pacific Northwest Region. 111 p.

Halverson, Nancy M.; Emmingham, William E. 1982. Reforestation in the Cascades Pacific silver fir zone: a survey of sites and management experiences on the Gifford Pinchot, Mt. Hood, and Willamette National Forests. R6-Ecol-091-1982. Portland, OR: U.S. Department of Agriculture, Forest Service, Pacific Northwest Region. 36 p.

Hamet Ahti, Leena; Ahti, Teuvo. 1969. The homologies of the Fennoscandian Mountain and coastal birch forests in Eurasia and North America. Vegetatio. 19: 208-219.

Harcombe, P.A. 1986. Stand development in a 130-year-old spruce hemlock forest based on age structure and 50 years of mortality data. Forest Ecology and Management. 14: 41-58.

Harmon, M.E.; Franklin, J.F.; Swanson, F.J. [and others]. 1986. Ecology of coarse woody debris in temperate ecosystems. In: Advances in ecological research. Orlando, FL: Academic Press: 133-302. Vol. 15.

Harmon, Mark E. 1986. Logs as sites of tree regeneration in *Picea sitchensis-Tsuga heterophylla* forests of coastal Washington and Oregon. Corvallis, OR: Oregon State University. 183 p. Ph.D. dissertation.

Harmon, Mark E. 1987. Retention of needles and seeds on logs in *Picea sitchensis-Tsuga heterophylla* forests of coastal Oregon and Washington. Unpublished manuscript; submitted to: Canadian Journal of Botany; on file at: Forestry Sciences Laboratory, Corvallis, OR.

Harmon, Mark E. [In press]. The influence of litter accumulations and canopy openness on *Picea sitchensis* (Bong.) Carr. and *Tsuga heterophylla* (Raf.) Sarg. seedlings growing on logs. Canadian Journal of Forest Research.

Harmon, Mark E.; Franklin, Jerry F. 1987. Tree seedlings on logs in *Picea sitchensis-Tsuga heterophylla* forests of Washington and Oregon. Unpublished manuscript in editorial review; on file at: Forestry Sciences Laboratory, Corvallis, OR.

Harniss, Roy O.; Harvey, Stephen J.; Murray, Robert B. 1981. A computerized bibliography of selected sagebrush species (genus *Artemisia*) in western North America. Gen. Tech. Rep. INT-102. Ogden, UT: U.S. Department of Agriculture, Forest Service, Intermountain Forest and Range Experiment Station. 107 p.

Harniss, Roy O.; Murray, Robert B. 1973. 30 years of vegetal change following burning of sagebrush-grass range. Journal of Range Management. 26(5): 322-325.

Harris, Larry D. 1984. The fragmented forest: island biogeography theory and the preservation of biotic diversity. Chicago: University of Chicago Press. 211 p.

Hawk, G.M. 1973. Forest vegetation and soils of terraces and floodplains along the McKenzie River, Oregon. Corvallis, OR: Oregon State University. 188 p. M.S. thesis.

Hawk, G.M.; Franklin, J.F.; McKee, W.A.; Brown, R.B. 1978. H.J. Andrews Experimental Forest reference stand system: establishment and use history. Bull. 12. Seattle: Coniferous Forest Biome, Ecosystem Analysis Studies, University of Washington. 79 p.

Hawk, G.M.; Zobel, D.B. 1974. Forest succession on alluvial landforms of the McKenzie River Valley, Oregon. Northwest Science. 48: 245-265.

Hawk, Glenn M. 1974. Little Sink Research Natural Area. Suppl. 4. Portland, OR: U.S. Department of Agriculture, Forest Service, Pacific Northwest Forest and Range Experiment Station. 14 p. Supplement to: Federal Research Natural Areas in Oregon and Washington: a guidebook for scientists and educators.

Hawk, Glenn M. 1979. Vegetation mapping and community description of a small western Cascade watershed. Northwest Science. 53(3): 200-212.

Heady, Harold F.; Bartolome, James. 1977. The Vale rangeland rehabilitation program: the desert repaired in southeastern Oregon. Resour. Bull. PNW-70. Portland, OR: U.S. Department of Agriculture, Forest Service, Pacific Northwest Forest and Range Experiment Station. 139 p.

Heikkinen, Olavi. 1984a. Climatic changes during recent centuries as indicated by dendrochronological studies, Mount Baker, Washington, U.S.A. In: Morner, N.-A.; Karlen, W., eds. Climatic changes on a yearly to millenial basis. [Place of publication unknown]: D. Reidel Publishing Co.; 353-361.

Heikkinen, Olavi. 1984b. Dendrochronological evidence of variations of Coleman Glacier, Mount Baker, Washington, U.S.A. Arctic and Alpine Research. 16(1): 53-64.

Heikkinen, Olavi. 1984c. Forest expansion in the subalpine zone during the past hundred years, Mount Baker, Washington, U.S.A. Erdkunde. 38: 194-202.

Heikkinen, Olavi. 1985. Relationships between tree growth and climate in the subalpine Cascade Range of Washington, U.S.A. Annales Botanici Fennici. 22: 1-14.

Hemstrom, M.A.; Logan, S.E.; Pavlat, W. 1987. Plant association and management guides for the Willamette National Forest. R6-Ecol-257-1986. Portland, OR: U.S. Department of Agriculture, Forest Service, Pacific Northwest Region. 312 p.

Hemstrom, Miles; Adams, Virginia Dale. 1982. Modeling long-term forest succession in the Pacific Northwest. In: Means, Joseph E., ed. Forest succession and stand development research in the Northwest: Proceedings of the symposium (part of the Northwest Scientific Association annual meetings); 1981 March 26; Corvallis, OR. Corvallis, OR: Forest Research Laboratory, Oregon State University: 14-23.

Hemstrom, Miles A. 1982. Fire in the forests of Mount Rainier National Park. In: Ecological research in national parks of the Pacific Northwest: Proceedings of the 2d conference on scientific research in the national parks; 1979 November; San Francisco, CA. p. 121-126. Available from: National Park Service Cooperative Park Studies Unit, Forestry Sciences Laboratory, Corvallis, OR.

Hemstrom, Miles A.; Emmingham, W.E.; Halverson, N.M.; Logan, S.E.; Topik, C. 1982. Plant association and management guide for the Pacific silver fir zone, Mt. Hood and Willamette National Forests. R6-Ecol-100-1982a. Portland, OR: U.S. Department of Agriculture, Forest Service, Pacific Northwest Region. 104 p.

Hemstrom, Miles A.; Franklin, Jerry F. 1982. Fire and other disturbances of the forests in Mount Rainier National Park. Journal of Quaternary Research. 18: 32-51.

Hemstrom, Miles A.; Logan, Sheila E. 1986. Plant association and management guide, Siuslaw National Forest. R6-Ecol-220-1986a. Portland, OR: U.S. Department of Agriculture, Forest Service, Pacific Northwest Region. 121 p.

Hemstrom, Miles Arthur. 1979. A recent disturbance history of forest ecosystems at Mount Rainier National Park. Corvallis, OR: Oregon State University. 67 p. Ph.D. dissertation.

Henderson, J.A. 1974. Composition, distribution and succession of subalpine meadows in Mount Rainier National Park. Corvallis, OR: Oregon State University. 123 p. Ph.D. dissertation.

Henderson, J.A.; Smith, B.; Mauk, R. 1978. Plant communities of the Hoh and Dosewallips drainages, Olympia National Park. Progress report. Logan, UT: Utah State University, Department of Forestry and Outdoor Recreation; National Park Service Contract CX-9000-7-0063. 141 p.

Henderson, Jan A. 1978. Plant succession on the *Alnus rubra/Rubus spectabilis* habitat type in western Oregon. Northwest Science. 52(3): 156-167.

Henderson, Jan A. 1982. Succession on two habitat types in western Washington. In: Means, Joseph E., ed. Forest succession and stand development research in the Northwest: Proceedings of the symposium (part of the Northwest Scientific Association annual meetings); 1981 March 26; Corvallis, OR. Corvallis, OR: Forest Research Laboratory, Oregon State University: 80-86.

Henderson, Jan A.; Peter, David. 1981a. Preliminary plant associations and habitat types of the Green and Cedar River drainages, North Bend District, Mt. Baker-Snoqualmie National Forest. Portland, OR: U.S. Department of Agriculture, Forest Service, Pacific Northwest Region. 55 p.

Henderson, Jan A.; Peter, David. 1981b. Preliminary plant associations and habitat types of the Quinault Ranger District, Olympic National Forest. Portland, OR: U.S. Department of Agriculture, Forest Service, Pacific Northwest Region. 96 p.

Henderson, Jan A.; Peter, David. 1981c. Preliminary plant associations and habitat types of the Shelton Ranger District, Olympic National Forest. Portland, OR: U.S. Department of Agriculture, Forest Service, Pacific Northwest Region. 113 p.

Henderson, Jan A.; Peter, David. 1981d. Preliminary plant associations and habitat types of the White River District, Mt. Baker Snoqualmie National Forest. Portland, OR: U.S. Department of Agriculture, Forest Service, Pacific Northwest Region. 114 p.

Henderson, Jan A.; Peter, David. 1982a. Preliminary plant associations and habitat types of the Snoqualmie and adjacent Skykomish River drainages, Mt. Baker-Snoqualmie National Forest. Portland, OR: U.S. Department of Agriculture, Forest Service, Pacific Northwest Region. 148 p.

Henderson, Jan A.; Peter, David. 1982b. Preliminary plant associations and habitat types of the Soleduck Ranger District, Olympic National Forest. Portland, OR: U.S. Department of Agriculture, Forest Service, Pacific Northwest Region. 141 p.

Henderson, Jan A.; Peter, David. 1983a. Preliminary plant associations and habitat types of the Hoodsport and Quilcene Ranger Districts, Olympic National Forest. Portland, OR: U.S. Department of Agriculture, Forest Service, Pacific Northwest Region. 145 p.

Henderson, Jan A.; Peter, David. 1983b. Preliminary plant associations and habitat types of the Northern Skykomish Ranger District, Mt. Baker-Snoqualmie National Forest. Portland, OR: U.S. Department of Agriculture, Forest Service, Pacific Northwest Region. 128 p.

Henderson, Jan A.; Peter, David. 1984. Preliminary plant associations and habitat types of the Darrington Ranger District, Mt. Baker-Snoqualmie National Forest. Portland, OR: U.S. Department of Agriculture, Forest Service, Pacific Northwest Region. 129 p.

Henderson, Jan A.; Peter, David. 1985. Preliminary plant associations and habitat types of the Mt. Baker Ranger District, Mt. Baker-Snoqualmie National Forest. Portland, OR: U.S. Department of Agriculture, Forest Service, Pacific Northwest Region. 134 p.

Henderson, Jan A.; Peter, David H.; Lesher, R.D. 1986. Preliminary plant associations of the Olympic National Forest. Portland, OR: U.S. Department of Agriculture, Forest Service, Pacific Northwest Region. 136 p.

Heusser, C.J. 1973. Modern pollen spectra from Mt. Rainier, Washington. Northwest Science. 47: 1-8.

Heusser, C.J.; Heusser, L.E. 1980. Sequence of pumiceous tephra layers and the late Quaternary environmental record near Mount St. Helens. Science. 210: 1007-1009.

Hickman, James C. 1976. Non-forest vegetation of the central western Cascade Mountains of Oregon. Northwest Science. 50(3): 145-155.

Hillaby, F.B.; Barrett, D.T. 1976. Vegetation communities of a Fraser River salt marsh. Tech. Rep. Ser. PAC/T-76-14. Vancouver, BC: Environment Canada, Habitat Protection Directorate, Fisheries and Marine Service, Department of the Environment, Pacific Region. 20 p.

Hinds, W.T. 1975. Energy and carbon balances in cheatgrass: an essay in autecology. Ecological Monographs. 45: 367-388.

Hironaka, M. 1979. Basic synecological relationships of the Columbia River sagebrush type. In: The sagebrush ecosystem: a symposium: Proceedings; 1978 April. Logan, UT: Utah State University, College of Natural Resources: 27-32.

Hoffnagle, John R. 1980. Estimates of vascular plant primary production in a west coast saltmarsh-estuary ecosystem. Northwest Science. 54(1): 68-79.

Hopkins, W.E.; Rawlings, R.C. 1985. Major indicator shrubs and herbs on National Forests of eastern Oregon. R6-TM-190-1985. Portland, OR: U.S. Department of Agriculture, Forest Service, Pacific Northwest Region. 188 p. Looseleaf; limited distribution.

Hopkins, William E. 1979a. Plant associations of the Fremont National Forest. R6-Ecol-79-004. Portland, OR: U.S. Department of Agriculture, Forest Service, Pacific Northwest Region. 106 p.

Hopkins, William E. 1979b. Plant associations of the south Chiloquin and Klamath Ranger Districts, Winema National Forest. R6-Ecol-79-005. Portland, OR: U.S. Department of Agriculture, Forest Service, Pacific Northwest Region. 96 p.

Hopkins, William E.; Kovalchik, Bernard L. 1983. Plant associations of the Crooked River National Grassland. R6-Ecol-133-1983. Portland, OR: U.S. Department of Agriculture, Forest Service, Pacific Northwest Region. 98 p.

Howe, Kent Donald. 1978. Distribution and abundance of terrestrial and arboreal lichens in old-growth coniferous forests of the western Cascades of Oregon. Eugene, OR: University of Oregon. [Pages unknown]. M.S. thesis.

Hunn, E.S.; Norton, H.H. 1983. Impact of Mt. St. Helens ashfall on fruit yield of mountain huckleberry, *Vaccinium membranaceum*, important native American food. Economic Botany. 38: 121-127.

Imper, D.K.; Zobel, D.B. 1983. Soils and foliar analysis of *Chamaecyparis lawsoniana* and *Thuja plicata* in southwestern Oregon. Canadian Journal of Forest Research. 13: 1219-1227.

Imper, David Kimberly. 1981. The relation of soil characteristics to growth and distribution of *Chamaecyparis lawsoniana* and *Thuja plicata* in southwestern Oregon. Corvallis, OR: Oregon State University. 100 p. M.S. thesis.

Jackson, Louise E. 1982. Comparison of phenological patterns in prairie and subalpine meadow communities. Northwest Science. 56(4): 316-328.

Jackson, M.T.; Faller, Adolph. 1973. Structural analysis and dynamics of the plant communities of Wizard Island, Crater Lake National Park. Ecological Monographs. 43: 441-461.

Jefferson, C.A. 1975. Plant communities and succession in Oregon coastal salt marshes. Corvallis, OR: Oregon State University. 192 p. Ph.D. dissertation.

Jenkins, Kurt J.; Starkey, Edward E. 1982. Home range and habitat use by nonmigratory elk (*Cervus elaphus roosevelti*) in Olympic National Park. In: Ecological research in national parks of the Pacific Northwest: Proceedings of the 2d conference on scientific research in the national parks; 1979 November; San Francisco, CA. p. 69-76. Available from: National Park Service Cooperative Park Studies Unit, Forestry Sciences Laboratory, Corvallis, OR.

Jenkins, Kurt J.; Starkey, Edward E. 1984. Habitat use by Roosevelt elk in unmanaged forests of the Hoh Valley, Washington. The Journal of Wildlife Management. 48(2): 642-646.

Johannessen, Carl L. 1964. Marshes prograding in Oregon: aerial photographs. Science. 146(3651): 1575-1578.

Johnson, Charles G.; Simon, Steven A. 1987. Plant associations of the Wallowa-Snake Province, Wallowa-Whitman National Forest. R6-ECOL-TP-255A-86. Baker, OR: U.S. Department of Agriculture, Forest Service, Pacific Northwest Region, Wallowa Whitman National Forest. 400 p. plus appendices.

Johnson, Janet L.; Franklin, Jerry F.; Krebill, Richard G., coords. 1984. Research natural areas: baseline monitoring and management: Proceedings of the symposium; 1984 March 21; Missoula, MT. Gen. Tech. Rep. INT-173. Ogden, UT: U.S. Department of Agriculture, Forest Service, Intermountain Forest and Range Experiment Station. 84 p.

Johnson, Mark E. 1979. Morphology, genesis, and classification of soils forming in recent age tephra deposits from Mt. St. Helens volcano. Corvallis, OR: Oregon State University. 99 p. M.S. thesis.

Joyal, E. 1984. Ecology and reproduction in *Collomia macrocalyx* Brand (Polemoniaceae). Corvallis, OR: Oregon State University. 97 p. M.S. thesis.

Juday, G.P. 1977. The location, composition, and structure of old-growth forests of the Oregon Coast Range. Corvallis, OR: Oregon State University. 206 p. Ph.D. dissertation.

Keller, S.A.C., ed. 1982. Mount St. Helens: one year later. Cheney, WA: Eastern Washington University Press. 243 p.

Keller, S.A.C., ed. 1987. Mount St. Helens: five years later. Cheney, WA: Eastern Washington University Press. 441 p.

Kemp, Lois; Schuller, S. Reid. 1982. Checklist of the vascular plants of Thornton T. Munger Research Natural Area. Adm. Rep. PNW-4. Portland, OR: U.S. Department of Agriculture, Forest Service, Pacific Northwest Forest and Range Experiment Station. 16 p.

Kertis, J. 1986. Vegetation dynamics and disturbance history of Oak Patch Natural Area Preserve, Mason County, Washington. Seattle: University of Washington. 93 p. M.S. thesis.

Keyes, Michael R.; Grier, Charles C. 1981. Above- and below-ground net production in 40-year-old Douglas-fir stands on law and high productivity sites. Canadian Journal of Forest Research. 11(3): 599-605.

Kindschy, Robert R.; Maser, Chris. 1978. Jordan Crater Research Natural Area. Suppl. 7. Portland, OR: U.S. Department of Agriculture, Forest Service, Pacific Northwest Forest and Range Experiment Station. 18 p. Supplement to: Federal Research Natural Areas in Oregon and Washington: a guidebook for scientists and educators.

Klopsch, Mark W. 1985. Structure of mature Douglas-fir stands in a western Oregon watershed and implications for interpretation of disturbance history and succession. Corvallis, OR: Oregon State University. 52 p. M.S. thesis.

Kovalchik, Bernard L. 1987. Riparian zone associations of the Deschutes, Ochoco, Fremont, and Winema National Forests. R6-Ecol-TP-279-87. Portland, OR: U.S. Department of Agriculture, Forest Service, Pacific Northwest Region. 171 p.

Kratz, Andrew Michael. 1975. Vegetation analysis of the coastal *Picea sitchensis* forest zone in Olympic National Park, Washington. Bellingham, WA: Western Washington University. 41 p. M.S. thesis.

Kruckeberg, A.R. 1977. Manzanita (*Arctostaphylos*) hybrids in the Pacific Northwest: effects of human and natural disturbance. Systematic Botany. 2(4): 233-250.

Kruckeberg, A.R. 1980. Golden chinquapin (*Chrysolepis chrysophylla*) in Washington State: a species at the northern limit of its range. Northwest Science. 54: 9-16.

Kruckeberg, A.R. 1985. California serpentines: flora, vegetation, geology, soils, and management problems. Publ. in Bot. 78. Berkeley, CA: University of California. 180 p.

Langham, G. 1970. Ecology of colonizing plant species on unstable and stabilized sand dunes on the Hanford Reservation. Pullman, WA: Washington State University. [Pages unknown]. Ph.D. dissertation.

Larson, Bruce C. 1982. Development of even-aged and uneven-aged mixed conifer stands in eastern Washington. In: Means, Joseph E., ed. Forest succession and stand development research in the Northwest: Proceedings of the symposium (part of the Northwest Scientific Association annual meetings); 1981 March 26; Corvallis, OR. Corvallis, OR: Forest Research Laboratory, Oregon State University: 113-118.

Larson, Bruce C.; Oliver, Chadwick Dearing. 1982. Forest dynamics and fuelwood supply of the Stehekin Valley, Washington. In: Ecological research in national parks of the Pacific Northwest: Proceedings of the 2d conference on scientific research in the national parks; 1979 November; San Francisco, CA. p. 127-134. Available from: National Park Service Cooperative Park Studies Unit, Forestry Sciences Laboratory, Corvallis, OR.

Lebednik, Gretchen K.; del Moral, Roger. 1976. Vegetation surrounding Kings Lake Bog, Washington. Madrono. 23(7): 386-400.

Lesher, Robin. 1984. Botanical reconnaissance of Silver Lake Research Natural Area, North Cascades National Park, Washington. Res. Note PNW-410. Portland, OR: U.S. Department of Agriculture, Forest Service, Pacific Northwest Forest and Range Experiment Station. 27 p.

Liverman, Marc C. 1981. Multivariate analysis of a tidal marsh ecosystem at Netarts Spit, Tillamook County, Oregon. Corvallis, OR: Oregon State University. 88 p. M.S. thesis.

Loneragan, William A.; del Moral, Roger. 1984. The influence of microrelief on community structure of subalpine meadows. Bulletin of the Torrey Botanical Club. 111(2): 209-216.

Long, James N. 1977. Trends in plant species diversity associated with development in a series of *Pseudotsuga menziesii/Gaultheria shallon* stands. Northwest Science. 51(2): 119-130.

Long, James N.; Schreiner, Edward G.; Manuwal, Naomi J. 1979. The role of actively moving sand dunes in the maintenance of an azonal, juniper-dominated community. Northwest Science. 53(3): 170-179.

Lueck, D. 1980. Ecology of *Pinus albicaulis* on Bachelor Butte, Oregon. Corvallis, OR: Oregon State University. 90 p. M.S. thesis.

Luken, J.O.; Fonda, R.W. 1983. Nitrogen accumulation in a chronosequence of red alder communities along the Hoh River, Olympic National Park, Washington. Canadian Journal of Forest Research. 13: 1228-1237.

Luken, James O. 1979. Biomass and nitrogen accretion in red alder communities along the Hoh River, Olympic National Park. Bellingham, WA: Western Washington University. 56 p. M.S. thesis.

MacDonald, Keith B. 1977. Plant and animal communities of Pacific North American salt marshes. In: Chapman, V.J., ed. Wet coastal ecosystems. Amsterdam, Great Britain: Elsevier Scientific Publishing Company; 167-191.

Mack, R.N. 1981a. Initial effects of ashfall from Mount St. Helens on vegetation in eastern Washington and adjacent Idaho. Science. 213: 537-539.

Mack, R.N. 1984. Invaders at home on the range. Natural History. 93: 40-47.

Mack, R.N. 1986. Alien plant invasion into the intermountain west: a case history. In: Mooney, H.A.; Drake, J., eds. Ecology and biological invasion of North America and Hawaii. New York: Springer-Verlag: 192-213.

Mack, R.N. 1987. Effects of Mount St. Helens ashfall in steppe communities of eastern Washington. In: Bilderback, D.E.; Leviton, A.E., eds. The biological effects of the Mount St. Helens and other volcanic eruptions. Los Angeles: University of California Press: 262-281.

Mack, R.N.; Pyke, D.A. 1983. The demography of *Bromus tectorum* L.: variation in time and space. Journal of Ecology. 71: 69-93.

Mack, R.N.; Rutter, N.W.; Bryant, V.M.; Valastro, S. 1978a. Late Quaternary pollen record from Big Meadow, Pend Oreille Co., Washington. Ecology. 59: 956-966.

Mack, R.N.; Rutter, N.W.; Bryant, V.M.; Valastro, S. 1978b. Postglacial vegetation history of Waits Lake, Stevens Co., Washington. Botanical Gazette. 139: 499-506.

Mack, R.N.; Rutter, N.W.; Valastro, S. 1978c. Late Quaternary pollen record from the San Poil River valley, Washington. Canadian Journal of Botany. 56: 1642-1650.

Mack, R.N.; Rutter, N.W.; Valastro, S. 1979. Holocene vegetation history of the Okanogan Valley, Washington. Quaternary Research. 12: 212-225.

Mack, R.N.; Thompson, J.N. 1982. Evolution in steppe with few large hooved mammals. American Naturalist. 119: 757-773.

Mack, Richard N. 1981b. Invasion of *Bromus tectorum* L. into western North America: an ecological chronicle. Agro-Ecosystems. 7: 145-165.

Mack, Richard N.; Bryant, Vaughn M., Jr.; Fryxell, Roald. 1976. Pollen sequence from the Columbia Basin, Washington: reappraisal of postglacial vegetation. American Midland Naturalist. 95(2): 390-397.

Mack, Richard N.; Pyke, David A. 1984. The demography of *Bromus tectorum:* the role of microclimate, grazing and disease. Journal of Ecology. 72(3): 731-748.

Magee, T.K. 1985. Invasion by trees into a grass bald on Mary's Peak, Oregon Coast Range. Corvallis, OR: Oregon State University. 165 p. M.S. thesis.

Mairs, John W. 1979. Plant communities and snow cover in the subalpine grassland, Steens Mountain, Oregon. In: Yearbook of the Association of Pacific Coast Geographers. Corvallis, OR: Oregon State University Press: 65-79. Vol. 41.

Mairs, John William. 1977. Plant communities of the Steens Mountain subalpine grassland and their relationship to certain environmental elements. Corvallis, OR: Oregon State University. 195 p. Ph.D. thesis.

Marquis, R.J. 1978. An investigation into the ecology and distribution of *Kalmiopsis leachiana* (Hend.) Rehder. Corvallis, OR: Oregon State University. 271 p. M.S. thesis.

Maser, C.; Trappe, J.M. 1984. The seen and unseen world of the fallen tree. Gen. Tech. Rep. PNW 164. Portland, OR: U.S. Department of Agriculture, Forest Service, Pacific Northwest Forest and Range Experiment Station. 56 p.

Maser, Chris; Mate, Bruce R.; Franklin, Jerry F.; Dyrness, C.T. 1981. Natural history of Oregon coast mammals. Gen. Tech. Rep. PNW-133. Portland, OR: U.S. Department of Agriculture, Forest Service, Pacific Northwest Forest and Range Experiment Station. 496 p.

Maser, Chris; Strickler, Gerald S. 1978. The sage vole, *Lagurus curtatus*, as an inhabitant of subalpine sheep fescue, *Festuca ovina*, communities of Steens Mountain—an observation and interpretation. Northwest Science. 52(3): 276-284.

Massman, W.J. 1982. Foliage distribution in old-growth coniferous tree canopies. Canadian Journal of Forest Research. 12(1): 10-17.

Mastrogiuseppe, J.D.; Gill, S.J. 1983. Steppe by step: understanding Priest Rapids plants. Douglasia Occasional Papers. 1: 1-68.

Matson, Pamela A.; Boone, Richard D. 1984. Natural disturbance and nitrogen mineralization: wave form dieback of mountain hemlock in the Oregon Cascades. Ecology. 65(5): 1511-1516.

Matson, Pamela A.; Waring, Richard H. 1984. Effects of nutrient and light limitation on mountain hemlock: susceptibility to laminated root rot. Ecology. 65(5): 1517-1524.

Mayfield, Molly Morton; Kjelmyr, Janet. 1984. Boardman Research Natural Area. Suppl. 17. Portland, OR: U.S. Department of Agriculture, Forest Service, Pacific Northwest Forest and Range Experiment Station. 19 p. Supplement to: Federal Research Natural Areas in Oregon and Washington: a guidebook for scientists and educators.

McArthur, E. Durant; Plummer, A. Perry. 1978. Biogeography and management of native western shrubs: a case study, section Tridentatae of *Artemisia*. Great Basin Naturalist Memoirs. 2: 229-243.

McCauley, Kevin J.; Cook, S.A. 1980. *Phellinus weirii* infestation of two mountain hemlock forests in the Oregon Cascades. Forest Science. 26(1): 23-29.

McCorquodale, Scott M. 1987. Fall-winter habitat use by elk in the shrub-steppe of Washington. Northwest Science. 61(3): 171-173.

McKee, Arthur; Bierlmaier, Frederick. 1987. H.J. Andrews Experimental Forest, Oregon. In: Greenland, David, ed. The climates of the long-term ecological research sites. Occas. Pap. 44. Boulder, CO: Institute of Arctic and Alpine Research, University of Colorado; 11-17.

McKee, Arthur; Knutson, Donald. 1987. A disjunct ponderosa pine stand in southeastern Oregon. Great Basin Naturalist. 47(1): 163-167.

McKee, Arthur; LaRoi, George; Franklin, Jerry F. 1982. Structure, composition, and reproductive behavior of terrace forests, South Fork Hoh River, Olympic National Park. In: Ecological research in national parks of the Pacific Northwest: Proceedings of the 2d conference on scientific research in the national parks; 1979 November; San Francisco, CA. p. 22-29. Available from: National Park Service Cooperative Park Studies Unit, Forestry Sciences Laboratory, Corvallis, OR.

McKee, Arthur; Stonedahl, Gary M.; Franklin, Jerry F.; Swanson, Frederick J., comps. 1987. Research publications of the H.J. Andrews Experimental Forest, Cascade Range, Oregon, 1948 to 1986. Gen. Tech. Rep. PNW-GTR-201. Portland, OR: U.S. Department of Agriculture, Forest Service, Pacific Northwest Forest and Range Experiment Station. 74 p.

McNeil, R.C. 1976. Vegetation and fire history of a ponderosa pine-white fir forest in Crater Lake National Park. Corvallis, OR: Oregon State University. 171 p. M.S. thesis.

McNeil, Robert C.; Zobel, Donald B. 1980. Vegetation and fire history of ponderosa pine-white fir forest in Crater Lake National Park. Northwest Science. 54(1): 30-46.

Means, Joseph E. 1982a. Developmental history of dry coniferous forests in the western Oregon Cascades. In: Means, Joseph E., ed. Forest succession and stand development research in the Northwest: Proceedings of the symposium (part of the Northwest Scientific Association annual meetings); 1981 March 26; Corvallis, OR. Corvallis, OR: Forest Research Laboratory, Oregon State University: 142-158.

Means, Joseph E., ed. 1982b. Forest succession and stand development research in the Northwest: Proceedings of the symposium (part of the Northwest Scientific Association annual meetings); 1981 March 26; Corvallis, OR. Corvallis, OR: Forest Research Laboratory, Oregon State University. 170 p.

Means, Joseph Earl. 1980. Dry coniferous forests in the western Oregon Cascades. Corvallis, OR: Oregon State University. 268 p. Ph.D. thesis.

Mehringer, P.J., Jr.; Blinman, E.; Petersen, K.L. 1977. Pollen influx and volcanic ash. Science. 198: 257-261.

Miles, D.W.R.; Swanson, F.J. 1986. Vegetation composition on recent landslides in the Cascade Mountains of western Oregon. Canadian Journal of Forest Research. 16: 739-744.

Milko, Robert J.; Bell, M.A.M. 1986. Subalpine meadow vegetation of south central Vancouver Island. Canadian Journal of Botany. 64: 815-821.

Miller, M.M.; Miller, J.W. 1976. Succession after wildfire in the North Cascades National Park Complex. Tall Timbers Fire Ecology Conference. 15: 71-83.

Milliren, Patricia Ann. 1983. Vegetation ecology of an early snowmelt meadow at timberline in Olympic National Park, Washington. Bellingham, WA: Western Washington University. 58 p. M.S. thesis.

Minore, D. 1979. Comparative autecological characteristics of northwestern tree species—a literature review. Gen. Tech. Rep. PNW-87. Portland, OR: U.S. Department of Agriculture, Forest Service, Pacific Northwest Forest and Range Experiment Station. 72 p.

Minore, Don. 1983. Western redcedar—a literature review. Gen. Tech. Rep. PNW-150. Portland, OR: U.S. Department of Agriculture, Forest Service, Pacific Northwest Forest and Range Experiment Station. 70 p.

Mitchell, Diane Lynne. 1981. Salt marsh reestablishment following dike breaching in the Salmon River estuary, Oregon. Corvallis, OR: Oregon State University. 171 p. Ph.D. thesis.

Mitchell, Rod. 1979. A checklist of the vascular plants in Abbott Creek Research Natural Area, Oregon. Res. Note PNW-341. Portland, OR: U.S. Department of Agriculture, Forest Service, Pacific Northwest Forest and Range Experiment Station. 18 p.

Mitchell, Rod; Moir, Will. 1976. Vegetation of the Abbott Creek Research Natural Area, Oregon. Northwest Science. 50(1): 42-58.

Mitchell, Roderic James. 1972. An analysis of vegetation of Abbott Creek Research Natural Area, Oregon. Corvallis, OR: Oregon State University. 131 p. Ph.D. thesis.

Moir, William H.; Franklin, Jerry F.; Maser, Chris. 1973a. Lost Forest Research Natural Area. Suppl. 3. Portland, OR: U.S. Department of Agriculture, Forest Service, Pacific Northwest Forest and Range Experiment Station. 17 p. Supplement to: Federal Research Natural Areas in Oregon and Washington: a guidebook for scientists and educators.

Moir, William H.; Maser, Chris; Franklin, Jerry F. 1973b. Bagby Research Natural Area. Suppl. 2. Portland, OR: U.S. Department of Agriculture, Forest Service, Pacific Northwest Forest and Range Experiment Station. 12 p. Supplement to: Federal Research Natural Areas in Oregon and Washington: a guidebook for scientists and educators.

Moran, Morley Stanton. 1967. Reconstruction of vegetation on rangelands in the Klamath Basin of Oregon. 42 p. Unpublished research paper; on file at: Department of Geography, Oregon State University, Corvallis, OR.

Morris, William G. 1934. Lightning storms and fires on the National Forests of Oregon and Washington. Portland, OR: U.S. Department of Agriculture, Forest Service, Pacific Northwest Forest and Range Experiment Station. 27 p.

Morrison, Elizabeth. 1973. The Blackwater Island Research Natural Area, a description of the vegetation and environment. Portland, OR: Portland State University. 45 p. B.S. thesis.

Morrow, Robert J. 1985. Age structure and spatial pattern of old-growth ponderosa pine in Pringle Falls Experimental Forest, central Oregon. Corvallis, OR: Oregon State University. 80 p. M.S. thesis.

Nadkarni, N.M. 1984. Biomass and mineral capital of epiphytes in an *Acer macrophyllum* community of a temperate moist coniferous forest, Olympic Peninsula, Washington State. Canadian Journal of Botany. 62(11): 2223-2228.

Newman, K.W. 1974. The relation of time and the water table to plant distribution on deflation plains along the central Oregon coast. Corvallis, OR: Oregon State University. 59 p. M.S. thesis.

Old-Growth Definition Task Group. 1986. Interim definitions for old-growth Douglas-fir and mixed-conifer forests in the Pacific Northwest and California. Res. Note PNW-447. Portland, OR: U.S. Department of Agriculture, Forest Service, Pacific Northwest Research Station. 7 p.

Oliver, C.D.; Adams, A.B.; Zasoski, R.J. 1985. Disturbance patterns and forest development in a recently deglaciated valley in the northwestern Cascade Range of Washington, U.S.A. Canadian Journal of Forest Research. 15(1): 221-232.

Oliver, Chadwick Dearing; Kenady, Reid M. 1982. True fir: proceedings of the biology and management of true fir in the Pacific Northwest symposium. Contrib. 45. Seattle: University of Washington; Portland, OR: U.S. Department of Agriculture, Forest Service, Pacific Northwest Forest and Range Experiment Station. 344 p.

Olsen, Sigund; Olsen, Ingrith Deyrup. 1983. Observations on the biology of *Boschniakia hookeri* (Orobanchaceae). Nordic Journal of Botany. 1(5): 585-594.

Olson, Thomas E.; Knopf, Fritz L. 1986. Naturalization of Russian-olive in the western United States. Western Journal of Applied Forestry. 1(3): 65-69.

Omernik, James M.; Gallant, Alisa L. 1986. Ecoregions of the Pacific Northwest. EPA/600/3-86/033. Corvallis, OR: U.S. Environmental Protection Agency, Environmental Research Laboratory. 39 p. plus map.

Oregon State Land Board. 1981. Oregon natural heritage plan. Salem, OR. 141 p.

Oregon State University. [n.d.] Proceedings: western juniper management short course, October 15-16, 1984, Bend, Oregon. Corvallis, OR: Oregon State University Extension Service and Department of Rangeland Resources. 98 p.

Ossinger, Mary C. 1983. The *Pseudotsuga-Tsuga/Rhododendron* community in the northeast Olympic Mountains. Bellingham, WA: Western Washington University. 50 p. M.S. thesis.

Owens, T.E. 1982. Postburn regrowth of shrubs related to canopy mortality. Northwest Science. 56(1): 34-40.

Padgett, Wayne G. 1982. Ecology of riparian plant communities in southern Malheur National Forest. Corvallis, OR: Oregon State University. 143 p. M.S. thesis.

Pike, Lawrence H. 1973. Lichens and bryophytes of a Willamette Valley oak forest. Northwest Science. 47(3): 149-158.

Pike, Lawrence H.; Rydell, Robert A.; Denison, William C. 1977. A 400-year-old Douglas-fir tree and its epiphytes: biomass, surface area, and their distributions. Canadian Journal of Forest Research. 7(4): 680-699.

Price, Larry W. 1978. Mountains of the Pacific Northwest, U.S.A.: a study in contrasts. Arctic and Alpine Research. 10(2): 465-478.

Proctor, Charles M. [and others] [names of other authors unknown]. 1980. An ecological characterization of the Pacific Northwest coastal region. Report prepared for: National Coastal Ecosystems Team, Office of Biological Sciences, Fish and Wildlife Service, U.S. Department of the Interior, Washington, DC; FWS/OBS-79/11, 12, 13, 14, 15; contract no. 14-16-0009-77-019. 5 vol.

Puritch, George S. 1973. Effect of water stress on photosynthesis, respiration, and transpiration of four *Abies* species. Canadian Journal of Forest Research. 3: 293-298.

Pyke, David A. 1986. Demographic responses of *Bromus tectorum* and seedlings of *Agropyron spicatum* to grazing by small mammals: occurrence and severity of grazing. Journal of Ecology. 74: 739-754.

Quaye, Eric Charles. 1982. The structure and dynamics of old-growth Sitka spruce (*Picea sitchensis*) forest of the Oregon Coast Range. Corvallis, OR: Oregon State University. 109 p. Ph.D. dissertation.

Ramborger, Terry Douglas. 1980. The influence of two tree species on the prairie soils of Pierce County, Washington. Seattle: University of Washington. 161 p. M.S. thesis.

Rayburn, W.R.; Mack, R.N.; Metting, B. 1982. Conspicuous algal colonization of the ash from Mount St. Helens. Journal of Phycology. 18: 537-543.

Reid, Elbert H.; Strickler, Gerald S.; Hall, Wade B. 1980. Green fescue grassland: 40 years of secondary succession. Res. Pap. PNW-274. Portland, OR: U.S. Department of Agriculture, Forest Service, Pacific Northwest Forest and Range Experiment Station. 39 p.

Rhoades, Frederick M. 1981. Biomass of epiphytic lichens and bryophytes on *Abies lasiocarpa* on a Mt. Baker lava flow, Washington. The Bryologist. 84(1): 39-47.

Rickard, W.H. 1964. Demise of sagebrush through soil changes. BioScience. 14: 43-44.

Rickard, W.H. 1965. The influence of greasewood on soil-moisture penetration and soil chemistry. Northwest Science. 39: 36-42.

Rickard, W.H. 1967. Seasonal soil moisture patterns in adjacent greasewood and sagebrush stands. Ecology. 48: 1034-1038.

Rickard, W.H. 1985. Biomass and shoot production in an undisturbed sagebrush-bunchgrass community. Northwest Science. 59(2): 126-133.

Rickard, W.H.; Cline, J.F. 1980. Cheatgrass communities: effect of plowing on species composition and productivity. Northwest Science. 54(3): 216-221.

Rickard, W.H.; Cushing, C.E. 1982. Recovery of streamside woody vegetation after exclusion of livestock grazing. Journal of Range Management. 35: 360-361.

Rickard, W.H.; Keough, R.F. 1968. Soil-plant relationships of two steppe desert shrubs. Plant and Soil. 29: 205-212.

Rickard, W.H.; Sauer, R.H. 1982a. Primary production and canopy cover in bitterbrush-cheatgrass communities. Northwest Science. 56(3): 250-256.

Rickard, W.H.; Sauer, R.H. 1982b. Self-revegetation of disturbed ground in the deserts of Nevada and Washington. Northwest Science. 56(1): 41-47.

Rickard, W.H.; Uresk, D.W.; Cline, J.F. 1976. Productivity response to precipitation by native and alien plant communities. In: Andrews, Rollin D., III [and others], eds. Proceedings of the symposium on terrestrial and aquatic ecological studies of the Northwest; 1976 March 26-27. Cheney, WA: Eastern Washington State College Press: 1-7.

Riegel, Gregg; Smith, Brad; Franklin, Jerry; Atzet, Tom; Wheeler, David. 1985. The Oregon White Oak Series of southwest Oregon. FIR Report. Corvallis, OR: Oregon State University, Extension Service, Forestry Intensified Research; 7(1): 5-7.

Ripley, James Douglas. 1983. Description of the plant communities and succession of the Oregon coast grasslands. Corvallis, OR: Oregon State University. 234 p. Ph.D. thesis.

Robichaux, R.H.; Taylor, D.W. 1977. Vegetation-analysis techniques applied to late Tertiary fossil floras from the western United States. Journal of Ecology. 65: 643-660.

Rotenberry, J.T.; Hinds, W.T.; Thorp, J.M. 1975. Microclimatic patterns on the Arid Lands Ecology Reserve. Northwest Science. 50(2): 120-130.

Ryan, Bruce. 1985. Lichens of Chowder Ridge, Mt. Baker, Washington. Northwest Science. 59: 279-294.

Sanville, William D.; Eilers, H. Peter; Boss, Theodore R.; Pfleeger, Thomas G. 1986. Environmental gradients in Northwest freshwater wetlands. Environmental Management. 10(1): 125-134.

Sauer, R.H.; Rickard, W.H. 1979. Vegetation of steep slopes in the shrub-steppe region of southcentral Washington. Northwest Science. 53(1): 5-11.

Sauer, Ronald H.; Uresk, Daniel W. 1976. Phenology of steppe plants in wet and dry years. Northwest Science. 50(3): 133-139.

Schreiner, Edward George. 1982. The role of exotic species in plant succession following human disturbance in an alpine area of Olympic National Park, Washington. Seattle: University of Washington. 132 p. Ph.D. dissertation.

Schuller, Reid; Evans, Shelley. 1986. Botanical reconnaissance of Meeks Table Research Natural Area, Washington. Res. Note PNW-451. Portland, OR: U.S. Department of Agriculture, Forest Service, Pacific Northwest Research Station. 22 p.

Schuller, S. Reid. 1978. Vegetation ecology of selected mountain hemlock (*Tsuga mertensiana*) communities along the eastern high Cascades, Oregon. Corvallis, OR: Oregon State University. 79 p. M.S. thesis.

Schuller, S. Reid; Cornelius, Lynn C. 1982. Checklist of the vascular plants of Goat Marsh Research Natural Area. Adm. Rep. PNW-3. Portland, OR: U.S. Department of Agriculture, Forest Service, Pacific Northwest Forest and Range Experiment Station. 18 p.

Schuller, S. Reid; Frenkel, Robert E. 1981. Checklist of vascular plants of Steamboat Mountain Research Natural Area. Res. Note PNW-375. Portland, OR: U.S. Department of Agriculture, Forest Service, Pacific Northwest Forest and Range Experiment Station. 46 p.

Scott, W. Frank. 1980. Geology of the Columbia Basin. Journal of Forestry. September: 537-541.

Seliskar, D.M.; Gallagher, J.L. 1983. The ecology of tidal marshes of the Pacific Northwest coast: community profile. FWS/OBS-82/32. Washington, DC: U.S. Department of the Interior, Fish and Wildlife Service. 65 p.

Seliskar, Denise M. 1983. Root and rhizome distribution as an indicator of upper salt marsh wetland limits. Hydrobiologia. 107: 231-236.

Seliskar, Denise M. 1985a. Effect of reciprocal transplanting between extremes of plant zones on morphometric plasticity of five plant species in an Oregon salt marsh. Canadian Journal of Botany. 63: 2254-2262.

Seliskar, Denise M. 1985b. Morphometric variations of five tidal marsh halophytes along environmental gradients. American Journal of Botany. 72(9): 1340-1352.

Seliskar, Denise Martha. 1980. Morphological and anatomical responses of selected coastal salt marsh plants to soil moisture. Corvallis, OR: Oregon State University. 186 p. M.S. thesis.

Seyer, Susan Cornelia. 1979. Vegetative ecology of a montane mire, Crater Lake National Park, Oregon. Corvallis, OR: Oregon State University. 187 p. M.S. thesis.

Shaw, David. 1982. Pollination ecology of an alpine fellfield—Chowder Ridge, Washington. Bellingham, WA: Western Washington University. 60 p. M.S. thesis.

Shaw, David; Taylor, Ronald J. 1986. Pollination ecology of an alpine fellfield, Mt. Baker area, Washington. Northwest Science. 60(1): 21-31.

Shelly, Stephen. 1985. Biosystematic studies of *Phacelia capitata* (Hydrophyllaceae), a species endemic to serpentine soils in southwestern Oregon. Corvallis, OR: Oregon State University. 120 p. M.S. thesis.

Shinn, Dean A. 1980. Historical perspectives on range burning in the inland Pacific Northwest. Journal of Range Management. 33(6): 415-423.

Shinn, Dean Allison. 1977. Man and the land: an ecological history of fire and grazing on eastern Oregon rangelands. Corvallis, OR: Oregon State University. 92 p. M.A. thesis.

Shipley, S.; Sarna-Wojcicki, A.M. 1983. Distribution, thickness and mass of late Pleistocene and Holocene tephra from major volcanoes in the northwestern United States: a preliminary assessment of hazards from volcanic ejecta to nuclear reactors in the Pacific Northwest. Misc. Field Stud. Map MF-1435. Washington, DC: U.S. Geological Survey.

Smith, Brad; Atzet, Tom; Wheeler, David; Franklin, Jerry. 1984. The Jeffrey Pine Series of the Siskiyou Region of southwest Oregon. In: Adaptive FIR annual report: October 1, 1983-September 30, 1984. Corvallis, OR: Oregon State University, Forest Research Laboratory, Forestry Intensified Research: 45-46.

Smith, Brad; Atzet, Tom; Wheeler, David; Franklin, Jerry. 1985. The Western Hemlock Series of the Siskiyou Region. FIR Report. Corvallis, OR: Oregon State University Extension Service, Forestry Intensified Research; 7(3): 8-10.

Smith, Bradley G.; Henderson, Jan. 1986. Baseline vegetation survey of the Hoh and Dosewallips drainages, Olympic National Park, Washington. 350 p. Unpublished manuscript; on file at: Olympic National Park, Port Angeles, WA.

Smith, Paul Winston. 1985. Plant association within the interior valleys of the Umpqua River basin, Oregon. Journal of Range Management. 38(6): 526-530.

Snow, B.D. 1984. Plant communities of the grassy balds of Mary's Peak, Oregon. Corvallis, OR: Oregon State University. 105 p. M.S. thesis.

Sollins, P.; Grier, C.; McCorison, F.M.; Cromack, K.; Fogel, R.; Fredriksen, R.L. 1980. The internal element cycles of an old-growth Douglas-fir ecosystem in western Oregon. Ecological Monographs. 50: 261-285.

Sollins, Phillip. 1982. Input and decay of coarse woody debris in coniferous stands in western Oregon and Washington. Canadian Journal of Forest Research. 12(1): 18-28.

Sowell, J.B. 1985. A predictive model relating North American plant formations and climate. Vegetatio. 60: 103-111.

Speaker, R.W.; Luchessa, Karen J.; Franklin, Jerry F. 1987. The effect of riparian shrub communities on leaf retention in a coastal Oregon stream. Unpublished manuscript; submitted to: American Midland Naturalist; on file at: Forestry Sciences Laboratory, Corvallis, OR.

Starkey, Edward E.; Franklin, Jerry F.; Matthews, Jean W. 1982. Ecological research in national parks of the Pacific Northwest: compiled from proceedings of the 2d conference on scientific research in the national parks; 1979 November; San Francisco, CA. Corvallis, OR: Forest Research Laboratory, Oregon State University. 142 p.

Stevens, Leslie H. 1975. The distribution and ecology of epiphytic lichens on ponderosa pine in eastern Washington. Cheney, WA: Eastern Washington University. [Pages unknown]. M.S. thesis.

Stevenson, Patrick W. 1983. Physiological response of subalpine fir to summer drought at timberline in the Olympic Mountains. Bellingham, WA: Western Washington University. 35 p. M.S. thesis.

Stewart, Glenn H. 1985. Forest structure and regeneration in the *Tsuga heterophylla-Abies amabilis* transition zone, central western Cascades, Oregon. Corvallis, OR: Oregon State University. 148 p. Ph.D. dissertation.

Stewart, Glenn H. 1986a. Forest development in canopy openings in old-growth *Pseudotsuga* forests of the western Cascade Range, Oregon. Canadian Journal of Forest Research. 16(3): 558-568.

Stewart, Glenn H. 1986b. Population dynamics of a montane conifer forest, western Cascade Range, Oregon, USA. Ecology. 67(2): 534-544.

St. John, H.; Douglas, G. 1986. Census of the flora of Glacier Peak. In: Plant life of the North Cascades. Douglasia Occasional Papers 2: 79-97.

Strickland, Richard, ed. 1986. Wetland functions, rehabilitation, and creation in the Pacific Northwest: the state of our understanding: Proceedings of a conference; 1986 April 30-May 2; Port Townsend, WA. Publ. 8614. Olympia, WA: Washington State Department of Ecology. 184 p.

Strickler, Gerald S.; Hall, Wade B. 1980. The Standley allotment: a history of range recovery. Res. Pap. PNW-278. Portland, OR: U.S. Department of Agriculture, Forest Service, Pacific Northwest Forest and Range Experiment Station. 35 p.

Stuart, J.D.; Geiszler, D.R.; Gara, R.I.; Agee, J.K. 1983. Mountain pine beetle scarring of lodgepole pine in south-central Oregon. Forest Ecology and Management. 5: 207-214.

Stuart, John David. 1984. Hazard rating of lodgepole pine stands to mountain pine beetle outbreaks in southcentral Oregon. Canadian Journal of Forest Research. 14: 666-671.

Stuth, Jerry W.; Winward, A.H. 1976. Logging impacts on bitterbrush in the lodgepole pine-pumice region of central Oregon. Journal of Rangeland Management. 29(6): 453-456.

Swanson, Frederick J.; Lienkaemper, George W. 1982. Interactions among fluvial processes, forest vegetation, and aquatic ecosystems, South Fork Hoh River, Olympic National Park. In: Ecological research in national parks of the Pacific Northwest: Proceedings of the 2d conference on scientific research in the national parks; 1979 November; San Francisco, CA. p. 30-34. Available from: National Park Service Cooperative Park Studies Unit, Forestry Sciences Laboratory, Corvallis, OR.

Swedburg, Kenneth C. 1973. A transition coniferous forest in the Cascade mountains of northern Oregon. American Midland Naturalist. 89(1): 1-25.

Talbott Roche, Cindy Jo; Busacca, Alan J. 1987. Soil-vegetation relationships in a subalpine grassland in northeastern Washington. Northwest Science. 61(3): 139-147.

Tappeiner, John C., II; McDonald, Philip M. 1984. Development of tanoak understories in conifer stands. Canadian Journal of Forest Research. 14: 271-277.

Tappeiner, John C., II; McDonald, Philip M.; Hughes, Thomas F. 1986. Survival of tanoak (*Lithocarpus densiflorus*) and Pacific madrone (*Arbutus menziesii*) seedlings in forests of southwestern Oregon. New Forests. 1: 43-55.

Taylor, Alan Henry. 1980. Plant communities and elevation in the diked portion of Joe Ney Slough: a baseline assessment of a marsh restoration project in Coos Bay, Oregon. Corvallis, OR: Oregon State University. 118 p. M.S. thesis.

Taylor, R.J. 1986. Floristics of the Stehekin River Riparian Zone. In: Plant life of the North Cascades. Douglasia Occasional Papers. 2: 64-78.

Taylor, Ronald J.; Boss, Theodore R. 1975. Biosystematics of *Quercus garryana* in relation to its distribution in the State of Washington. Northwest Science. 49(2): 49-57.

Taylor, Ronald J.; Douglas, George W. 1978. Plant ecology and natural history of Chowder Ridge, Mt. Baker: a potential alpine research natural area in the western North Cascades. Northwest Science. 52(1): 35-50.

Taylor, Ronald J.; Naas, Dorothy B.; Naas, Ralph W.; Douglas, George W. 1978. Contributions to the flora of Washington. III. Northwest Science. 52(3): 220-225.

Thomas, Duncan W. 1984. The vascular flora of the Columbia River Estuary. The Wasmann Journal of Biology. 42(1-2): 92-106.

Thompson, Ralph L. 1979. Vegetation and flora of Myrtle Island Research Natural Area, Douglas County, Oregon. Bulletin of the Ecological Society of America. 26(2): 93. Abstract.

Thornburgh, Dale A. 1982. Succession in the mixed evergreen forests of northwestern California. In: Means, Joseph E., ed. Forest succession and stand development research in the Northwest: Proceedings of the symposium (part of the Northwest Scientific Association annual meetings); 1981 March 26; Corvallis, OR. Corvallis, OR: Forest Research Laboratory, Oregon State University: 87-91.

Tiedemann, Arthur R.; Driver, Charles H. 1983. Snow eriogonum; a native halfshrub to revegetate winter game ranges. Reclamation and Revegetation Research. 2: 31-39.

Tiedemann, Arthur R.; Gjertson, Joseph O.; McColley, Phillip D. 1977. Thompson Clover Research Natural Area. Suppl. 5. Portland, OR: U.S. Department of Agriculture, Forest Service, Pacific Northwest Forest and Range Experiment Station. 16 p. Supplement to: Federal Research Natural Areas in Oregon and Washington: a guidebook for scientists and educators.

Tiedemann, Arthur R.; Klock, Glen O. 1977. Meeks Table Research Natural Area reference sampling and habitat classification. Res. Pap. PNW-223. Portland, OR: U.S. Department of Agriculture, Forest Service, Pacific Northwest Forest and Range Experiment Station. 19 p.

Tisdale, E.W. 1979. A preliminary classification of Snake River canyon grasslands in Idaho. Note 32. Moscow, ID: University of Idaho, Forest, Wildlife and Range Experiment Station. 8 p.

Tisdale, E.W.; Hironaka, M. 1981. The sagebrush-grass region: a review of the ecological literature. Bull. 33. Moscow, ID: University of Idaho, Forest, Wildlife and Range Experiment Station. 31 p.

Topik, C.; Halverson, N.M.; Brockway, D.G. 1986. Plant association and management guide for the western hemlock zone, Gifford Pinchot National Forest. R6-Ecol-230A-1986. Portland, OR: U.S. Department of Agriculture, Forest Service, Pacific Northwest Region. 133 p.

Topik, Christopher. 1982. Forest floor accumulation and decomposition in the western Cascades of Oregon. Eugene, OR: University of Oregon. 172 p. Ph.D. thesis.

Towle, J.C. 1982. Changing geography of Willamette Valley woodlands. Oregon Historical Quarterly. 83(1): 66-87.

Turner, David P. 1985. Successional relationships and a comparison of biological characteristics among six northwestern conifers. Bulletin of the Torrey Botanical Club. 112(4): 421-428.

Turner, David P.; Franz, Eldon H. 1985a. The influence of western hemlock and western redcedar on microbial numbers, nitrogen mineralization, and nitrification. Plant and Soil. 88: 259-267.

Turner, David P.; Franz, Eldon H. 1985b. Size class structure and tree dispersion patterns in old-growth cedar-hemlock forests of the northern Rocky Mountains (USA). Oecologia. 68: 52-56.

Turner, David P.; Franz, Eldon H. 1986. The influence of canopy dominants on understory vegetation patterns in an old-growth cedar-hemlock forest. American Midland Naturalist. 116(2): 387-393.

Ugolini, F.C.; Schlichte, A.K. 1973. The effect of Holocene environmental changes on selected western Washington soils. Soil Science. 116(3): 218-227.

Utah State University. 1979. The sagebrush ecosystem: a symposium, April, 1978. Logan, UT: Utah State University, College of Natural Resources. 251 p.

Vale, Thomas R. 1981. Tree invasion of montane meadows in Oregon. American Midland Naturalist. 105(1): 61-69.

Van Vuren, Dirk. 1987. Bison west of the Rocky Mountains: an alternative explanation. Northwest Science. 61(2): 65-69.

Veirs, Stephen D. 1982. Coast redwood forest: stand dynamics, successional status, and the role of fire. In: Means, Joseph E., ed. Forest succession and stand development research in the Northwest: Proceedings of the symposium (part of the Northwest Scientific Association annual meetings); 1981 March 26; Corvallis, OR. Corvallis, OR: Forest Research Laboratory, Oregon State University: 119-141.

Vogt, Kristiina A.; Edmonds, Robert L.; Grier, Charles C. 1981a. Biomass and nutrient concentrations of sporocarps produced by mycorrhizal and decomposer fungi in *Abies amabilis* stands. Oecologia. 50: 170-175.

Vogt, Kristiina A.; Edmonds, Robert L.; Grier, Charles C. 1981b. Dynamics of ectomycorrhizae in *Abies amabilis* stands: the role of *Cenococcum graniforme*. Holarctic Ecology. 4: 167-173.

Vogt, Kristiina A.; Edmonds, Robert L.; Grier, Charles C. 1981c. Seasonal changes in biomass and vertical distribution of mycorrhizal and fibrous-textured conifer fine roots in 23- and 180-year-old subalpine *Abies amabilis* stands. Canadian Journal of Forest Research. 11(2): 223-229.

Volland, L.A. 1974. Relation of pocket gophers to plant communities in the pine region of central Oregon. In: Proceedings, wildlife and forest management in the Pacific Northwest. Corvallis, OR: Forest Research Laboratory, Oregon State University: 149-166.

Volland, L.A. 1976. Plant communities of the central Oregon pumice zone. R6 Area Guide 4-2. Portland, OR: U.S. Department of Agriculture, Forest Service, Pacific Northwest Region. 113 p.

Volland, L.A. 1978. Trends in standing crop and species composition of a rested Kentucky bluegrass meadow over an 11-year period. In: Proceedings, 1st international rangeland congress. Denver, CO: Society for Range Management: 526-529.

Volland, L.A. 1985a. Guidelines for forage resource evaluation within central Oregon pumice zone. R6-Ecol-177-1985. Portland, OR: U.S. Department of Agriculture, Forest Service, Pacific Northwest Region. 216 p.

Volland, L.A. [In press]. Ecological classification of lodgepole pine in the USA. In: The species and its management: Proceedings, the lodgepole pine symposium; May 8-10; Spokane, WA; May 14-16; Vancouver, BC. Pullman, WA: Washington State University, Cooperative Extensive Service.

Volland, Leonard A. 1985b. Plant associations of the central Oregon pumice zone. R6-Ecol-104-1985. Portland, OR: U.S. Department of Agriculture, Forest Service, Pacific Northwest Region. 138 p.

Volland, Leonard A.; Dell, John D. 1981. Fire effects on Pacific Northwest forest and range plants. R6-Rm-067-1981. Portland, OR: U.S. Department of Agriculture, Forest Service, Pacific Northwest Region. 13 p.

Waggoner, Gary S. 1980. Vegetation map of Stetattle Creek watershed, North Cascades National Park. Denver, CO: National Park Service, Denver Service Center. 1 p. plus map.

Wagstaff, Steve; Taylor, Ronald J. 1980. Botanical reconnaissance in the Stetattle Creek Research Natural Area, North Cascades National Park, Washington. Bellingham, WA: Western Washington University. 21 p.

Waring, R.H.; Emmingham, W.H.; Gholz, H.L.; Grier, C.C. 1978. Variation in maximum leaf areas of coniferous forests in Oregon and its ecological significance. Forest Science. 24(1): 131-140.

Waring, R.H.; Emmingham, W.H.; Running, S.W. 1975. Environmental limits of an endemic spruce, *Picea breweriana*. Canadian Journal of Botany. 53(15): 1599-1613.

Waring, R.H.; Franklin, J. F. 1979. Evergreen coniferous forests of the Pacific Northwest. Science. 204(4400): 1380-1386.

Waring, Richard H., ed. 1980. Forests: fresh perspectives from ecosystem analysis: proceedings of the 40th annual biology colloquium. Corvallis, OR: Oregon State University Press. 199 p.

Waring, Richard H. 1982. Land of the giant conifers. Natural History. 91(10): 54-63.

Washington State Department of Natural Resources. 1987. Final State of Washington natural heritage plan. Olympia, WA. 108 p. plus appendices.

Wellner, Charles A. 1986. Salmo Research Natural Area. Suppl. 19. Portland, OR: U.S. Department of Agriculture, Forest Service, Pacific Northwest Forest and Range Experiment Station. 15 p. Supplement to: Federal Research Natural Areas in Oregon and Washington: a guidebook for scientists and educators.

West, Neil E.; Rea, Kenneth H.; Harniss, Roy O. 1979. Plant demographic studies in sagebrush-grass communities of southeastern Idaho. Ecology. 60(2): 376-388.

Wheeler, David; Atzet, Tom; Smith, Brad; Franklin, Jerry. 1986a. The White Fir Series of the Siskiyou Mountain Province. FIR Report. Corvallis, OR: Oregon State University, Extension Service, Forestry Intensified Research; 8(2): 4-6.

Wheeler, David; Atzet, Tom; Smith, Brad; Franklin, Jerry. 1986b. The White Fir Series of the Siskiyou Mountain Province: [part 2]. FIR Report. Corvallis, OR: Oregon State University, Extension Service, Forestry Intensified Research; 8(3): 6-9.

Wiberg, Curt; Greene, Sarah. 1981. Blackwater Island Research Natural Area. Suppl. 11. Portland, OR: U.S. Department of Agriculture, Forest Service, Pacific Northwest Forest and Range Experiment Station. 20 p. Supplement to: Federal Research Natural Areas in Oregon and Washington: a guidebook for scientists and educators.

Wiberg, Curt; McKee, Arthur. 1978. Boston Glacier Research Natural Area. Suppl. 6. Portland, OR: U.S. Department of Agriculture, Forest Service, Pacific Northwest Forest and Range Experiment Station. 14 p. Supplement to: Federal Research Natural Areas in Oregon and Washington: a guidebook for scientists and educators.

Wickramaratne, Siri Nimal. 1983. Vegetation changes in the Willamette River Greenway, Benton and Linn Counties, Oregon: 1972-1981. Corvallis: Oregon State University. 118 p. M.S. thesis.

Williams, C. 1980. An approach and philosophy to habitat type classification using the U.S. Forest Service Region Six computer programs. In: Sondheim, M.W.; Daykin, P.M., tech. eds. Proceedings of the ecological data processing and interpretation workshop; Victoria, BC. Victoria, BC: British Columbia Ministry of Environments, Assessment and Planning Division, Terrestrial Studies Branch: 161-167.

Williams, Clinton K.; Lillybridge, Terry R. 1983. Forested plant associations of the Okanogan National Forest. R6-Ecol-132-1983. Portland, OR: U.S. Department of Agriculture, Forest Service, Pacific Northwest Region. 100 p.

Williams, Clinton Karl. 1978. Vegetation classification for the Badger Allotment, Mt. Hood National Forest. Corvallis, OR: Oregon State University. 141 p. Ph.D. thesis.

Wilson, M.L. 1982a. Identification and mapping of habitat types in the vicinity of Mount Tolman in eastern Washington. Corvallis, OR: Oregon State University. 103 p. M.S. thesis.

Wilson, Mark Virgil. 1982b. Microhabitat influences on species distributions and community dynamics in the conifer woodland of the Siskiyou Mountains, Oregon. Ithaca, NY: Cornell University. 148 p. Ph.D. thesis.

Winward, A.H. 1980. Taxonomy and ecology of sagebrush in Oregon. Bull. 642. Corvallis, OR: Agricultural Experiment Station, Oregon State University. 15 p.

Winward, A.H.; Tisdale, E.W. 1977. Taxonomy of the *Artemisia tridentata* complex in Idaho. Bull. 19. Moscow, ID: University of Idaho, College of Forestry, Wildlife and Range Sciences. 15 p.

Wolfe, J.A. 1978. A paleobotanical interpretation of Tertiary climates in the northern hemisphere. American Scientist. 66: 694-703.

Wolfe, J.A. 1979. Temperature parameters of humid to mesic forests of eastern Asia and relation to forests of other regions of the northern hemisphere and Australia. Prof. Pap. 1106. Washington, DC: U.S. Geological Survey. 37 p.

Wood, David M.; del Moral, Roger. 1987. Mechanisms of early primary succession in subalpine habitats on Mount St. Helens. Ecology. 68(4): 780-790.

Yamaguchi, D.K. 1986. The development of old-growth Douglas-fir forests northeast of Mount St. Helens, Washington, following an A.D. 1480 eruption. Seattle: University of Washington. 100 p. Ph.D. dissertation.

Young, J.A.; Evans, R.A.; Eckert, R.E., Jr. 1981. Environmental quality and the use of herbicides on *Artemisia*/grasslands of the U.S. intermountain area. Agriculture and Environment. 6: 53-61.

Young, James A.; Evans, Raymond A. 1981. Demography and fire history of a western juniper stand. Journal of Range Management. 34(6): 501-506.

Zamaro, Benjamin A. 1982. Understory development in forest succession: an example from the inland Northwest. In: Means, Joseph E., ed. Forest succession and stand development research in the Northwest: Proceedings of the symposium (part of the Northwest Scientific Association annual meetings); 1981 March 26; Corvallis, OR. Corvallis, OR: Forest Research Laboratory, Oregon State University: 63-69.

Zeigler, R.S. 1978. The vegetation dynamics of *Pinus contorta* forest, Crater Lake National Park. Corvallis, OR: Oregon State University. 182 p. M.S. thesis.

Zobel, D.B.; Antos, J.A. 1986. Survival of prolonged burial by subalpine forest and understory plants. American Midland Naturalist. 115: 282-287.

Zobel, Donald B. 1986. Port-Orford-cedar: a forgotten species. Journal of Forest History. 30(1): 29-36.

Zobel, Donald B.; Antos, Joseph A. 1985. Response of conifer shoot elongation to tephra from Mount St. Helens. Forest Ecology and Management. 12: 38-91.

Zobel, Donald B.; McKee, Arthur; Hawk, Glenn M.; Dyrness, C.T. 1976. Relationships of environment to composition structure and diversity of forest communities of the central western Cascades of Oregon. Ecological Monographs. 46(2): 135-156.

Zobel, Donald B.; Roth, Lewis F.; Hawk, Glenn M. 1985. Ecology, pathology, and management of Port-Orford-cedar. Gen. Tech. Rep. PNW-184. Portland, OR: U.S. Department of Agriculture, Forest Service, Pacific Northwest Research Station. 161 p.

Zobel, Donald B.; Wasem, C. Robert. 1979. Pyramid Lake Research Natural Area. Suppl. 8. Portland, OR: U.S. Department of Agriculture, Forest Service, Pacific Northwest Forest and Range Experiment Station. 17 p. Supplement to: Federal Research Natural Areas in Oregon and Washington: a guidebook for scientists and educators.